ENERGY IN AGROECOSYSTEMS

A TOOL FOR ASSESSING SUSTAINABILITY

Advances in Agroecology
Series Editor: Clive A. Edwards

ENERGY IN AGROECOSYSTEMS

A TOOL FOR ASSESSING SUSTAINABILITY

Gloria I. Guzmán Casado
Universidad Pablo de Olavide, Seville, Spain

Manuel González de Molina
Universidad Pablo de Olavide, Seville, Spain

CRC Press
Taylor & Francis Group
Boca Raton London New York

CRC Press is an imprint of the
Taylor & Francis Group, an **informa** business

CRC Press
Taylor & Francis Group
6000 Broken Sound Parkway NW, Suite 300
Boca Raton, FL 33487-2742

First issued in paperback 2019

ISBN-13: 978-1-4987-7476-5 (hbk)
ISBN-13: 978-0-367-43604-9 (pbk)

Library of Congress Cataloging-in-Publication Data
Names: Guzmán Casado, Gloria I., editor. | González de Molina Navarro, Manuel, editor.   Title: Energy in agroecosystems : a tool for assessing sustainability / edited by Gloria Isabel Guzmán Casado and Manuel González de Molina.   Description: New York : Taylor & Francis, 2017. | Series: Advances in agroecology   Identifiers: LCCN 2016027442 (print) | LCCN 2016041545 (ebook) | ISBN 9781498774765 (hardback) | ISBN 9781315317465 (E-book)   Subjects: LCSH: Agriculture and energy. | Sustainable agriculture. | Agricultural ecology.   Classification: LCC S494.5.E5 E536 2016 (print) | LCC S494.5.E5 (ebook) | DDC 577.5/5--dc23   LC record available at https://lccn.loc.gov/2016027442

Visit the Taylor & Francis Web site at
http://www.taylorandfrancis.com

and the CRC Press Web site at
http://www.crcpress.com

Contents

Gloria I. Guzmán and Marta Astier

**Gloria I. Guzmán, Eduardo Aguilera, Leticia Paludo Vargas, and
Romina Iodice**

Foreword

Agroecology is deeply enriched by interaction and communication between disciplines and different systems of knowledge. This interactive enrichment is called transdisciplinarity. I remember sitting in a room full of graduate students from Latin America and Spain listening to Manuel González de Molina (Manolo), a historian, and Eduardo Sevilla Guzmán, a rural sociologist, lead a discussion on agroecology. This took place at La Rábida, a center of the Universidad Internacional de Andalucía near Huelva in southern Spain. The year was 1996, and I had been invited to present my ecologically based focus on agroecology in a master's program entitled "Agroecology: A Sustainable Approach for Ecological Agriculture." The students in the course came primarily from backgrounds in agronomy, with a few from sociology and anthropology, and at the time agroecology was a newly emerging focus for graduate study.

It was amazing to me as an agroecologist to share concepts and understanding with González de Molina who approached agriculture from the point of view of a historian with a focus on agroecosystem change over time. I was also impressed by Guzmán's views as a Marxist who was deeply moved by the imbalance of power (political, economic, and social) that had come about in modern day industrial agriculture, and how agroecology and its holistic view offered an alternative agrarian vision. Among the students listening to our transdisciplinary exchange was Gloria Guzmán, an agronomist who has become one of the leading agroecologists in Spain and beyond, building and directing programs that are as much about agroecological farming practices as they are about social change in food systems.

The common ground that I found with Manuel González de Molina and Gloria Guzmán, as well as Eduardo Sevilla Guzmán, is built on the understanding that agroecosystems are much more than systems that produce food. They are ecosystems with a "purpose," and that purpose is socially constructed and changes over time. For that purpose, however, to be sustainable, important indicators of sustainability must be achieved and maintained. The natural resource base on which agriculture depends must be maintained, providing ecological sustainability. The environmental services that all ecosystems provide for our planet (such as biodiversity conservation, soil and water protection, carbon sequestration, etc.) must all be maintained. The economic viability, affordability, and access for all is also a high priority. But perhaps most importantly, the social sustainability of the food system must become a primary focus of food system change, with what we now call food justice, food security, and food sovereignty being the key goals.

As an ecologist, the concept of metabolism was part of my understanding of energy. Plants are capable of capturing solar energy through photosynthesis, converting this energy into simple sugars, and then through various metabolic pathways, transforming these sugars into biomass or primary productivity. Some of this energy-containing productivity then becomes the energy source for organisms higher up the food chain, using a part of this biomass energy for their own growth and development, but ultimately releasing most of it as the by-products of respiration. Needless to say, the first time I heard González de Molina mention the concept of social metabolism, I was not sure what he meant.

The biomass produced in agroecosystems comes from the same pathways as in natural ecosystems, but with one big difference. We have learned to increase the movement of energy stored in biomass from net primary productivity *out* of the ecosystem as harvestable products by subsidizing the capture and flow with human-derived energy, often termed cultural energy. In indigenous, traditional, and most smallholder farming systems around the world, this "cultural" energy comes from what can be called renewable biological cultural energy such as human labor, animal labor, and animal manure. But modern agriculture uses energy derived from nonbiological sources dependent primarily on fossil fuels, which can be termed nonrenewable industrial cultural energy. As a result, it is no wonder that modern food systems are responsible for as much as 30% of global greenhouse gas emissions, contributing significantly to global climate change.

The other problem with the export of so much biomass from modern farming systems is that very little, if any, biomass is left over to return to the ecosystem to perform important functions such as maintaining soil organic matter, preventing soil erosion, stimulating soil biological activity, sequestering carbon, and other important ecosystem services.

The challenge then is to work toward agroecosystems that perform the dual roles of providing sustenance for humans, as well as performing the ecosystem services needed for a healthy planet. González de Molina and Guzmán meet this challenge in this book.

When we apply agroecosystem analysis for energy use, efficiency, and ultimately, for sustainability, social metabolism takes on an important role for agroecologists concerned with agrarian sustainability. Since agroecosystems are dependent on human management, the quantity and quality of ecosystem services depend on how they are managed. Sustainability means that an agroecosystem should be able to provide an optimal level of biomass production over time without deteriorating the basis of its fundamental functional elements, all the while maintaining an optimal provision of ecosystem services. Therefore, the objective of agroecological energy analysis and their indicators, which González de Molina and Guzmán call the energy return on investment or EROIs, is to determine whether a given agroecosystem is capable of simultaneously maintaining its biomass production and ecosystem services or whether it degrades them, requiring increasing amounts of external energy to compensate for their loss. What they propose in this book, therefore, is a different approach to the question of energy efficiency in agroecosystems, one which is "complementary" to traditional methods, but also one which aims to bring agroecological and social metabolism perspectives to energy analysis.

The energy indicators (EROIs) can also be more than a mere indicator of energy efficiency. If designed appropriately, EROIs can, in effect, become a measurement of metabolic efficiency, that is to say, of the exchange of energy between agrarian systems and the environment, in order to establish whether this metabolic exchange is sustainable over time. This book considers EROIs that go beyond the social benefits offered by increasing investment of energy in agriculture. This requires us to recognize that not only is it necessary to invest energy in the production of biomass useful to society or to farmers, but also in maintaining the agroecosystem so that it can continue to produce biomass under the best possible conditions. The key lies in considering not only the energy cost of the production of socially useful biomass, but also the maintenance cost of the ecosystem services provided by an agroecosystem. This cost does not end with the reuse of seeds or the production of animal feed (which corresponds only to the supply services provided by agroecosystems), but also extends to the maintenance of the remaining ecosystem services. It is therefore necessary to adopt a broader, agroecological focus.

The interaction of different disciplines, knowledge systems, social perspectives, and field experiences that come together in this book has been occurring for several decades. Out of this interaction, many of us, and especially myself, have learned how to engage in a type of transdisciplinarity that rarely cuts across normal discipline boundaries and ways of thinking. González de Molina and Guzmán have done an admirable service for agroecology by preparing this book. In it we have a powerful tool for promoting deep food system transformation.

Steve Gliessman
Santa Cruz, California

Editors

Gloria I. Guzmán Casado, PhD, earned a PhD in agronomy from the University of Córdoba (Andalusia, Spain) in 2002. Currently, she is an associate professor of Agroecology and Environmental History at Pablo de Olavide University, Seville (Spain). From 2002 to 2009, she was the director of the Research and Training Center for Organic Farming and Rural Development of Andalusia (depending on Regional Government). From 2011, she has been a coordinator of the master's program on organic farming at the Universidad Internacional de Andalucía. Dr. Guzmán's research revolves around organic farming from an agroecological approach, traditional peasant knowledge and local crop varieties, energy analysis in agroecosystems, agriculture and climate change, and agroecological transition.

Manuel González de Molina, PhD, earned a PhD in history. Currently, he is a full professor of Environmental History and the head of the Agroecosystems History Laboratory linked to the Pablo de Olavide University (Seville, Spain), which is devoted to promoting an improved understanding of the function of agroecosystems from a historical perspective. He is also a principal investigator of several national and international research projects. Dr. González de Molina has published many articles, chapters, and books over the years. He has studied the socioecological transitions of Spanish agriculture that has been occurring since the eighteenth century and developed new interpretive and methodological instruments to evaluate agrarian sustainability from a metabolic and agroecological point of view. Dr. González de Molina has also carried out a reconstruction of Spanish agricultural soil fertilization systems (González de Molina and Garrabou, eds., 2010) and has published a methodological manual for the analysis of nutrient balances from a historical perspective (González de Molina et al., 2010, 2015). He has contributed to developing the theory and methodology of the social metabolism approach summarized in the book (Springer, 2014) along with Víctor M. Toledo, an agroecologist from Mexico.

Contributors

Juan Infante Amate
Department of Geography, History and
Philosophy
Universidad Pablo de Olavide
Seville, Spain

Marta Astier
Research Center on Environmental
Geography
Universidad Naciona Autónoma de
Mexico
Michoacán, Mexico

David Soto Fernández
Department of Geography, History and
Philosophy
Universidad Pablo de Olavide
Seville, Spain

Eduardo Aguilera Fernández
Department of Geography, History and
Philosophy
Universidad Pablo de Olavide
Seville, Spain

Romina Iodice
Universidad Nacional de Luján
Buenos Aires, Argentina

Wilson Picado
History School
Universidad Nacional de Costa Rica
Heredia, Costa Rica

Leticia Paludo Vargas
Universidade Federal de Santa Maria
(UFSM)
Camobi, Santa Maria, Brazil

Introduction

Industrial civilization has radically transformed the role of agrarian activities in the social metabolism. The production of biomass no longer provides the bulk of energy required by society to function. Biomass supplied between 95% and 100% of the energy consumed by preindustrial societies; today, on the other hand, it only provides between 10% and 30% (Krausmann et al., 2008a). In fact, agriculture has been excluded from the energy metabolism of industrial societies. From providing a surplus of energy that was essential for society, it has now become a demander of energy. Without inputs of external energy, a large proportion of global agriculture simply could not function (Gliessman, 1998). This explains the increase in land productivity, allowing the agrarian system to feed a global population that has increased sixfold since the start of the nineteenth century. According to Smil (2001, p. 256), cultivated land area in the world grew by a third during the twentieth century, while productivity multiplied fourfold, and annual production grew sixfold during this period. This is principally due to the fact that the amount of energy used in agriculture has multiplied by eight.

The scientific community and international organizations question whether agrarian production will be sufficient to feed the global population predicted for the year 2050: over 9000 million inhabitants. The gloomy outlook for fossil energies, responsible for the huge increase in agricultural productivity, casts reasonable doubt on its capacity to do so. In a world where land will undoubtedly once again play a crucial role, we must understand the efficiency of energy use in agriculture. Furthermore, given that the endosomatic metabolism of humans and the production of raw materials that are difficult to produce synthetically can only be satisfied through the production of biomass, the sustainable use of energy in agriculture has become a fundamental question (Tello et al., 2015, p. 9).

The most widely used tool in energy analysis has been the energy input/output balance. In recent decades, two different indicators with a similar foundation have begun to be used to measure efficiency. The net energy balance (NEB) is the result of deducting from the energy produced by a system the energy invested to produce it (Pérez-Soba et al., 2015), generating a net amount of energy available for possible use. The energy return on investment (EROI) has been more successful, which is the result of dividing the energy obtained by the energy invested in its production (Hall, 2011). It has become the most widely used instrument for measuring the efficiency of energy usage in all kinds of productive activities (Gupta and Hall, 2011, p. 28; Pervanchon et al., 2002, p. 150), especially the conversion of oil and other primary energy sources into fuel and other energy products (Cleveland et al. 1984; Hall et al. 1986; Cleveland, 1992; Hall et al. 2008, 2009; Mulder and Hagens, 2008; Giampietro et al., 2010; Hall, 2011, pp. 2–3). It emerged as a consequence of the growing scarcity of fossil fuels, particularly oil, and rising oil prices, owing among other things to the increasing investment required for extraction (Murphy and Hall, 2010). It provides a numerical indicator, which can be quickly and easily used for comparison with other similar energy processes, in both space and time (Murphy et al., 2011a, p. 8).

In doing so, it also provides information about decision making in this vital aspect of the operation of productive activities.

The application of energy analysis to agricultural systems also has a long tradition. When applied to agriculture, it measures the amount of energy obtained in the form of biomass per unit of energy invested. Since the 1970s, numerous studies have been examining the use of energy in agriculture, highlighting the growing inefficiency of industrial agrarian systems and input-intensive crop management (Leach, 1976; Pimentel and Pimentel, 1979; Pimentel et al., 1983, 1990, 2005; Fluck and Baird, 1980; Dovring, 1985; Schahczenski, 1984; Jones, 1989). From the first calculations of efficiency based on simple balances of inputs and outputs or the net efficiency of agrarian systems, this area of study now also widely uses EROIs (Schramski et al., 2013; Markussen and Østergård, 2013, Martinez-Alier, 2011, Moore, 2010). These have been used in different ways: synchronically, to compare agrarian systems or crops within the same period of time; and also diachronically, to compare a single agrarian system over time. Furthermore, they have been widely used to compare different forms of crop management, as well as an agrarian system as a whole. They are also very frequently used to compare organic and conventional production (for a review, see Smith et al., 2015).

There have been a large number of studies conducted, but the methodologies used have varied greatly, making comparison extremely difficult if not impossible (Murphy et al., 2011a). Furthermore, energy analyses are often based on efficiency indicators where the calculations are not transparent, using different criteria that are not made explicit: if embodied energy is used or not, whether human labor or animal labor is calculated, etc. There are no unified criteria in this regard (Murphy et al., 2011a; Pérez-Soba et al., 2015). In addition, the indicators are not designed for an agroecological context and, therefore, they measure efficiency without taking into account basic aspects of the way agroecosystems function.

The EROI is *one* indicator of efficiency in the use of energy or in the generation of net energy. Depending on the system boundaries we choose, it could even yield contradictory results. Efficiency in the use of energy cannot be reduced to a single number or one single analysis criterion, as highlighted by Giampietro et al. (2010), and this is particularly true in the case of agriculture. This is a paradigmatic case of the need for multicriteria analysis. The EROI is economic in origin, based on the same valuation criteria as monetary investments, that is to say, on cost–benefit analysis (Hall et al. 2009, p. 26), setting aside other potentialities of this indicator. Indeed, the perspective commonly adopted by analysts has been that of society in general (macroeconomic perspective), assessing the returns on energy investments made to obtain a certain amount of useful biomass; or the perspective of the farmer (microeconomic perspective), also evaluating the return on the total investment made in the agroecosystem. This is undoubtedly a useful and necessary perspective, but an insufficient one. Energy efficiency should be based on different perspectives to aid coherent decision making regarding energy use. One of the most relevant is the perspective of the agroecosystem itself, trying to measure both its efficient management and the state of health of its different components. This way of measuring efficiency falls squarely within the realm of agroecology. Nevertheless, agroecology has not yet developed a theoretical and methodological proposal for energy analyses specifically adapted to agroecosystems. This book focuses on this important issue for environmental scientists and for agroecologists concerned with agrarian sustainability.

Sustainability means that an agroecosystem should provide an optimal level of biomass production over time without deteriorating the basis of its funds elements while maintaining an optimal provision of ecosystem services. Thus, the objective of agroecological energy analyses and their indicators (EROIs) is to ascertain whether a given agroecosystem is capable of maintaining its biomass production and ecosystem services or whether it degrades them, requiring increasing amounts of external energy in order to compensate for the loss only partially.

Therefore, energy indicators (EROIs) can also be more than a mere indicator of energy efficiency. If designed appropriately, EROIs can, in effect, become a measurement of metabolic efficiency, that is to say, of the exchange of energy between agrarian systems and the environment, in order to establish whether this metabolic exchange is sustainable over time. This book considers EROIs that go beyond the social benefits offered by increasing investment of energy in agriculture. As we shall see, the key lies in considering not only the energy cost of the production of socially useful biomass, but also the maintenance cost of the ecosystem services provided by an agroecosystem: this cost does not end with the reuse of seeds or the production of animal feed (which corresponds only to the supply services provided by agroecosystems), but also extends to the maintenance of the remaining ecosystem services.

As shown in the next chapters, the starting point of this agroecological approach is the laws of thermodynamics as a central criterion for energy analysis within agroecosystems. Indeed, from a thermodynamic perspective, we consider agroecosystems as complex adaptive systems that dissipate energy to compensate for the law of entropy. To do so, they exchange flows of energy and materials with their environment. Since agroecosystems are dependent on human management, the quantity and quality of ecosystem services depend on how they are managed. An adequate provision of services will depend on the health of the agroecosystem. Conversely, the degradation of the fundamental elements of an agroecosystem can lead to the reduction of ecosystem service supply and sustainability. Low entropy systems can be obtained by recirculating energy within the agroecosystem, since the agroecosystems in which internal recirculation processes have been reduced require large amounts of external energy, and are thereby converted into highly entropic agroecosystems. All of these situations can be measured using suitable indicators from an agroecological point of view.

The purpose of this book, therefore, is twofold. On the one hand, it aims to provide an agroecological perspective on the usual energy analysis, which has focused perhaps too much on the social utility of the energy reaped in relation to the energy invested. On the other hand, it endeavors to develop a theoretical and methodological proposal adapted to the interests of Agroecology, given that energy analysis is usually conducted within this field using conventional tools, barely taking into account a rigorous development of the throughput of energy within agroecosystems and their implications for the calculation of efficiency. The intention is not to contrast indicators of agroecological efficiency with the usual indicators, which highlight the perspective of the farmer or society. We propose examining both types in conjunction so that farmers and policy makers will be better able to make the right decisions.

This book, therefore, dedicates a great deal of attention to the theoretical grounding and methodological development of agroecological indicators (EROIs). The intention is to analyze and understand the structure, functioning and dynamic of agroecosystems at different scales and over different periods of time. Using these indicators, we can study the transition from traditional organic agriculture to industrial agriculture. They are also particularly interesting when it comes to understanding and planning, as much as we can, the transition from industrialized agriculture to modern organic agriculture. In this respect, the EROIs proposed in this book could be a very useful instrument in designing the sustainable agroecosystems of the future on different scales, at the scale of an individual farm or holding (see, for example, Chapters 7 through 9), as well as the scale of a community (see Chapter 5) and even at more aggregated scales (Chapter 6). The final chapter offers a few lessons learned from the case studies set out in this book, which point in this direction. Finally, as stated before, agroecological indicators have been designed to ascertain the state of the fund elements of agroecosystems and, therefore, provide a very useful tool to evaluate the sustainability of agroecosystem management. This can be seen in the different agrarian systems studied here: olive groves, coffee plantations, avocado groves, livestock farming, and so on. It is precisely for this reason that certification bodies and their supervisory organizations could use them as a tool to ascertain the real state and condition of certified agroecosystems.

The book is divided into two parts, with a chapter on conclusions and two appendices. In the first part, we develop the theoretical basis of the approach and the methodological procedures to calculate several indicators of energy efficiency and sustainability of energy use in agroecosystems. Chapter 1 is devoted to exploring energy flows between nature and society, where agricultural activity plays a major role, producing not only biomass but also ecosystem services. We use the social metabolism approach to analyze the energy flows through society and the fund elements that dissipate them for generating goods and services for society. In this chapter, we make a first attempt to apply this metabolic approach to the energy throughput in agriculture as "agrarian metabolism" and discuss the methodological tools proposed to calculate it. We consider the functioning of agroecosystems from a thermodynamic point of view, the most appropriate way of understanding the energy flows that run and operate agroecosystems, understood here as dissipative structures that create order or negentropy as biomass and ecosystem services. Finally, we have tried to highlight synthetically the major milestones of metabolic change in agriculture up to its complete industrialization. By doing so, we aim to provide a broad framework in which to contextualize the main changes in both the quality and the organization of energy flows that run agroecosystems.

Chapter 2 focuses on energy flows through agroecosystems and their relationship with the whole energy metabolism of society. In this chapter, we propose various indicators of energy efficiency, depending on the perspective adopted. The most common perspective is that of society or the economic point of view. In other words, the returns obtained in terms of socialized biomass for the investment made by society in agrarian production. Here, we develop our proposal jointly with other colleagues working on the project "Sustainable Farm System: Long-Term Socio-Ecological Metabolism in Western Agriculture," published recently as a Working Paper (Tello et al., 2015) and as a paper in the journal *Ecological Economics* (Tello et al., 2016).

However, in this chapter we endeavor to go beyond this approach. In line with our agroecological perspective, we are particularly interested in the energy functioning of agroecosystems and, within these, the analysis of energy flows that allows them to function as dissipative structures. In this respect, EROIs can also be more than a mere indicator of energy efficiency; they could be an indicator of whether the exchange of energy between an agrarian system and the environment is sustainable over time. An adequate provision of services depends on the health of the agroecosystem, that is, on the sustainability of its fund elements. These funds are maintained by means of adequate biomass flows. So, the quantity and quality of fund elements and the rate at which they provide services depend on how they are managed. And, conversely, the degradation of the fund elements of an agroecosystem can lead to the reduction of its supply of ecosystem services. Agroecological EROIs try to measure whether a given agroecosystem is capable of maintaining its ecosystem services or whether it degrades them, requiring increasing amounts of external energy in order only partially to compensate for the loss. What we propose in this chapter is a different way of approaching the question of energy efficiency in agroecosystems, which is "complementary" to traditional methods and which aims to bring an agroecological perspective to energy analysis.

Chapters 3 and 4 contain a methodological guide to assessing indicators (EROIs) of energy use efficiency in agroecosystems both from an economic and agroecological point of view. Chapter 3 explains how to calculate actual net primary productivity (NPP_{act}) and its components, that is to say, the input side of the EROI calculation. NPP_{act} is the sum total of vegetable or plant biomass that will be returned to society in the form of food, energy, or industrial products; plus the biomass that is reused within the agroecosystems (for animal consumption, as seeds for the next harvest, etc.), as well as the unharvested biomass that is available for the maintenance of trophic chains and, in general, for the reproduction of the agroecosystem's fund elements; and finally, the biomass that accumulates annually in the aerial structure (trunk and crown) and in the roots of perennial species. In accordance with our agroecological approach, and moving away from the usual perspective, when calculating NPP_{act} we have taken into account not only aerial biomass but also root biomass and biomass accumulated in agroecosystems when their rate of extraction is lower than their rate of production. All these categories must be taken into account to ensure a correct measurement of NPP_{act}. Chapter 3 shows the way in which different types of biomass are turned into gross energy on the basis of the values compiled in Appendix I. Finally, Chapter 3 provides an exercise in calculating NPP_{act} and its components by way of an example, taking Santa Fe as a case study, which will be the subject of Chapter 5.

Chapter 4 is dedicated to calculating the other side of the balance, in other words, the inputs required for agrarian production. In this chapter, we offer a detailed description of all energy inputs that have been used over time in agrarian production. We also show the way to calculate embodied energy. Thus, the embodied energy of a given input refers to the sum of the higher heating value (gross energy) of the input plus the energy requirements for the production and delivery of the input. For this calculation, we have taken into account the changes in inputs (fertilizers, machinery, pesticides, etc.) embodied energy over time as a consequence of the efficiency gains of manufacturing processes. We offer a complete and coherent dataset based on a wide literature review, which has been complemented with our own estimations, including all direct and indirect energy linked to the main agricultural inputs with the maximum possible level of disaggregation (Appendix II). At the end of this chapter, as in the previous chapter, we apply this methodology to a case study of Spanish agriculture, which is developed more broadly in Chapter 6.

Thus, thanks also to the two appendices included at the end, this book offers rigorous tools for a quick calculation of energy efficiency in agroecosystems, providing more consistent indicators to assess their sustainability. Appendix I includes a complete and updated compilation of gross energy converters of different crops, pastures, and woods, as well as animal biomass, to be used anywhere, avoiding a long and tedious consultation of secondary sources of information. The second appendix collects the embodied energy of the main inputs used in agricultural production. When compiling Appendices I and II, we consulted a huge number of monographic studies and journal articles.

The second part contains case studies of both Europe and the Americas using the methodology presented previously. The studies selected were chosen because they are representative of the most commonly used scales of analysis (crop, local, or national), the two main agricultural sectors (agriculture and livestock), and organic versus conventional agroecosystem management. Almost all the cases studied have a historic dimension, with the intention of comparing the management and organization of traditional and modern agroecosystems, evaluating whether the major agrarian transformations experienced in the past two and a half centuries have entailed gains or losses of efficiency in the use of energy. Only by using truly historical analysis is it possible to compare such different approaches to farming management, with very different crop management intensities, operating within the same agroecosystem. This long-term dimension of analysis allows us to ascertain the state of the fund elements, in other words, the dissipation structures that remain over time and which enable the sustainable provision of ecosystem services.

The first of them, set out in Chapter 5, applies our proposal to a case study on a local scale. It is an agroecosystem that is representative of Mediterranean agroclimatic conditions at four key moments in the past two and a half centuries: one of balanced organic agriculture (1752), at the start of the use of chemical inputs (1904), in the middle of the industrialization process (1934), and a time of agricultural production decoupled from its territory (1997). Chapter 6 conducts a similar exercise but on a much more aggregated scale. It studies the evolution of efficiency in the use of energy in Spanish agriculture between 1900 and the present day. We compiled economic and agroecological EROIs for three key moments: at the start of the study, in 1900, when Spanish agriculture was essentially organic; in 1960, when the process of industrialization was already underway; and finally, in 2008, which illustrates the current situation of an essentially industrialized agriculture that is representative of the intensive management approach used in European agroecosystems. This change of scale allows us to observe differences in efficiency seen on a local and national scale, with different and even contradictory values for the same moments in history.

Chapter 7 is dedicated to studying one crop, coffee, which is representative of tropical climates, very different to the Mediterranean climate that is characteristic of Santa Fe and also of Spain. The intention has been to ascertain the functioning of the energy indicators proposed within a very different environmental context and to see whether the tendencies observed are the same or not. In this chapter, we compare traditional and modern management approaches to this crop and the ways in which coffee plantations were organized before and after industrialization. The case chosen is that of coffee agroecosystems in Costa Rica between 1935 and 2010, since this is one of the most important coffee growing countries with a long tradition in this area of agriculture.

Chapter 8 offers a comparative study of two woody crops, olives and avocados, representative of two very different soil types and climates. Energy efficiency is compared in the management of dry-farmed and irrigated olive groves in Sierra de Mágina (Jaén, Spain) with the cultivation of avocados in the Cupatitzio River Basin (Michoacán, Mexico). In both cases, comparisons are made between organic and conventionally managed holdings in order to draw consistent conclusions about the differences in energy efficiency between the two types of management in very different environmental contexts, making the conclusions reached more robust.

Chapter 9 applies our theoretical and methodological approach to livestock farming, an area that is completely different from plant production, which is the main orientation in the case studies tackled up until now. Obviously, the peculiarities of animal production and its well-known inefficiency in converting plant biomass into animal biomass pose a very relevant challenge when it comes to confirming the validity and utility of the proposed indicators. To this end, we chose a group of extensive farms dedicated to ruminant livestock breeding in Argentina, Brazil, and Spain, which present major differences in terms of the structure of their respective agroecosystems and also in the level of input intensification.

Finally, Chapter 10 compiles the main conclusions and highlights the utility of these tools. It compares the results obtained from each of the case studies, with a view to describing common or diverging patterns and explaining their significance, not only in terms of energy use efficiency, but also from the perspective of agrarian sustainability. This has been made possible by developing a unified theoretical proposal that goes beyond measuring efficiency in the use of energy and turns some of the indicators proposed into sustainability energy indicators. This is also thanks to the effort we have made in unifying conversion coefficients (set out in the two appendices to this book) and the biomass classification categories, in other words, standardizing the values and establishing the same indicators, thereby making all the case studies fully comparable. The results are convergent, displaying a fundamental coincidence in the trends manifested in energy efficiency over time and also in the levels of efficiency that characterize different agrarian orientations and management styles: low levels of efficiency in industrial management, high levels of efficiency in traditional management, and intermediate levels in organic management. Furthermore, it has also allowed us to assess the extent to which the transformations that have taken place in agroecosystems during the socioecological transition processes have allowed the fund elements of agroecosystems to be conserved, or vice versa. Using these indicators, it is also possible to identify in detail specific practices (those that provide, for example, greater density of low entropy internal loops), which are key in the functioning of agroecosystems, maintaining fund elements and at the same time increasing the returns for society. Therefore, these indicators enable us to make recommendations regarding management and to design public policies to guide the socioecological transition in which we are currently immersed.

We would like to thank all of our colleagues from the History of Agro-Ecosystems Lab at Pablo de Olavide University for their collaboration in putting together this book. Without their enthusiastic help and assistance, this book would not have come to light: in particular David Soto, Antonio Herrera, Inmaculada Villa, Inmaculada Zamora, Antonio Cid, Guiomar Carranza, and Roberto García-Ruiz. For their dedication and commitment to this book, we would especially like to thank Eduardo Aguilera and Juan Infante, who contributed not only through the corresponding chapters, but who also provided tremendous support in the writing of the entire book. We should also like to thank Stephen Gliessman and Ernesto Méndez, Marta Astier, and Omar Masera for their editorial work and their support in getting this book published, and for reviewing parts of the manuscript. Finally, we must express our gratitude for the financial and scientific support provided by the members of the international project Sustainable Farm Systems: Long-Term Socio-Ecological Metabolism in Western Agriculture, financed by the Social Sciences and Humanities Research Council (SSHRC) of Canada: Geoff Cunfer, Enric Tello, Fridolin Krausmann, and all the other members as well.

PART I

Concept and Methods

GLORIA I. GUZMÁN
MANUEL GONZÁLEZ DE MOLINA

CHAPTER 1

The Energetic Metabolism
of Human Societies

CONTENTS

1.1 INTRODUCTION

Drawing an analogy with the biological concept of metabolism, different scientific disciplines have developed the concept of social metabolism, which aims to structure relationships between society and nature. All human beings draw from nature sufficient quantities of oxygen, water, and biomass per time unit to survive as an organism, and they excrete heat, water, carbon dioxide, and mineralized and organic substances back into nature. Similarly, individuals connected through social relations organize themselves to guarantee their subsistence and reproduction, also drawing energy from nature through meta-individual structures or artifacts, and excreting all manner of waste (González de Molina and Toledo, 2014). Hence, social metabolism alludes to the exchange of energy, materials, and information that every society engages in with its physical environment to produce and reproduce its material conditions of existence.

The idea of using the concept of metabolism in a socioecological approach to social reality has gained ground over the past decade, owing to its growing importance as a theoretical and methodological tool. This concept has been used recurrently since the nineteenth century, but remained in a latent state until the late 1960s, when a handful of economists "reinvented" it (Ayres and Simonis, 1994). In recent years, the number of studies using this concept has increased substantially, applying it principally as a tool to evaluate sustainability by studying flows of energy and materials between societies and their environments in the past and present day. Today, there are methodological proposals available that offer methods, indices, and sources of statistical information (Giampietro et al., 2012) to calculate in detail the flows of energy and materials on a national scale, even managing to quantify the energy and/or material metabolism of certain countries and their changes over time, providing a historical analysis, or commercial relations among countries measured in terms of physical or energy magnitudes.

However, very few studies have attempted to apply this tool to agriculture and agroecosystems. Agriculture, as a human activity intended principally to meet the food requirements of the population, is a particularly suitable activity on account of its peculiarities. Some authors have argued the need to study not only the input and output flows of energy that allow society to function, but also the circulation and destination of those flows within it. This need is even more evident in the case of agriculture, since agroecosystems are physical and biological entities that exhibit peculiarities that other economic activities simply do not possess. In any case, the concept of social metabolism is an ideal way of studying the use of energy within agroecosystems and of measuring its efficiency.

Furthermore, social metabolism provides agroecology with a powerful tool for analysis and a theoretical support capable of grounding the hybrid nature—among culture, communication, and the material world—of any agroecosystem, whose dynamics are explained by the interaction of rural societies with their environment. Transferring this approach to the field of agriculture implies considering the *"agrarian metabolism"* as the part of the social metabolism that specializes in the generation of biomass and environmental services for human consumption. This metabolic approach also allows agroecosystems to be integrated at different scales with other landscape units with which they also exchange biophysical flows, and with other social units (information flows), without which it would be impossible to explain their dynamics and organization. In this book, however, we will focus purely on the energy aspects of this exchange.

This chapter looks at the thermodynamic foundations of the metabolic approach to explore its possible application to agriculture. It also discusses the specific place occupied by agrarian activity within the metabolic relationship between society and environment. It describes the components of this relationship, distinguishing between fund elements and the energy flows that nourish them. It pays particularly close attention to agroecosystems, as the center of the socioecological relationship. Finally, we examine the recent history of agriculture from a metabolic perspective to contextualize the major changes that have occurred in the energy functionality of biomass and its changes in efficiency.

1.2 A THERMODYNAMIC APPROACH TO HUMAN SOCIETIES

In his book, *What Is Life?* Schrödinger (1944 [1984]), stated that living organisms are neither exempt from nor in opposition to thermodynamic laws, but rather they retain or increase their complexity by exporting the entropy they generate. Human societies are also self-maintained (autopoietic) systems, having emergent forms of stable organization in space and time, but within a process of dynamic configuration, since adaptive complex systems are nonlinear, dynamic systems capable of learning and transforming themselves through cumulative experience. The existence, configuration, maintenance, and reproduction of societies require a continual supply of energy and materials, along with the *dissipation* of part of that energy. Entropy is also the key element in the functioning of societies: by exchanging information, energy, and matter with their environment, societies are also subject to the laws of thermodynamics. So, we assume that entropy is common to all natural processes, be these human or of any other nature. This grounds our understanding of the material structure, functioning, and dynamics of human societies as based on a thermodynamic understanding as biological systems, which they also are. The laws of nature operate on and affect human beings and the devices they build. So the principle of entropy applies to social practice, and therefore social systems are subject to the laws of thermodynamics, which is perhaps the most relevant physical law when it comes to explaining human evolution over time.

Although human societies share the same evolutionary precepts as physical and biological systems, they represent an *innovation* that differentiates them and makes their dynamic specific, adding complexity and connectivity to the whole evolutionary process. Social systems cannot be explained by a simple application of the laws of physics, even though human acts are subject to them. The reason for this is that although evolution is a unified process, human society is an evolutionary innovation emerged from the reflective (self-referring) capacity possessed by human beings, which is more developed than in any other species. The most direct consequence of this human mental feature is the capacity—not exclusive among higher-order animals, but rare—for building tools and, therefore, for using energy *outside* the organism, that is, the use of exosomatic energy. To build and use tools, information and knowledge need to be generated and transmitted, that is, the generation of culture is required. Culture involves a symbolic dimension containing, besides knowledge, beliefs, rules and regulations, technologies, and so on. Accordingly, evolutionary innovation encompasses human capability regarding the exosomatic use of information, energy, and materials, also giving rise to a new type of complex system: the *reflexive complex system* (Martínez-Alier et al., 1998, p. 282) or *self-reflexive system* and *self-aware system* (Kay et al., 1999; Ramos-Martin, 2003). This feature will be instrumental because it gives social systems a unique *neopoietic* capacity absent from other systems or species, and that confers an essential, creative dimension to human individual and—more so—collective actions.

In analogy to living organisms, culture is the transmission of information by nongenetic means, a metaphor that became popular in the academic world. It has been said that cultural evolution is an extension of biological information *by other means* (Sahlins and Service, 1960; Margalef, 1980), and a parallel has been drawn between the diffusion of genes and of culture. Culture can then be seen as an innovative manifestation of the adaptive complexity of social systems; it is the name of a new genus of complexity provided by the environment for perpetuating and reorganizing a particular kind of dissipative system: social systems (Tyrtania, 2008, p. 51). Culture is but an emergent property of human societies. Its performative or neopoietic character, and its creative nature (Maturana and Varela, 1980; Rosen, 1985, 2000; Pattee, 1995; Giampietro et al., 2006) enable the configuration of new and more complex dissipative structures at even larger scales by means of technology (Adams, 1988).

Although biological systems have a limited capacity for processing energy—mainly endosomatically—due to the availability in the environment and genetic load, human societies exhibit a less constrained dissipative capacity that is only limited by the environment. Human beings can thus dissipate energy by means of artifacts or tools, that is, through knowledge and technology, and can do it faster and with greater mobility than any other species. Societies adapt to the environment by changing their structures and frontiers by means of association, integration, or conquest of other societies, something biological organisms cannot do. In other words, different from the biological systems with well-defined boundaries, human societies can organize and reorganize, thus acquiring the capability of avoiding or overcoming local limitations from the environment. That explains why some societies maintain exosomatic consumption levels that are beyond the provision means of their local environments without entering into a steady state. What is specifically human is the exosomatic consumption of energy. Since no genetic load regulates such exosomatic consumption, it becomes codified by culture, which involves a faster but less predictable evolutionary rate.

From that perspective, the theoretical key is the consideration that human societies, according to the evolutionary innovations they represent, build structures—in the sense of Prigogine—that dissipate heat (entropy) to the environment, obtaining free energy from it. These structures are not only biological but also technological, thanks to the species' capacity for building tools and mechanical, electronic, and digital artifacts. As we have seen, while biological metabolism is genetically determined, technological metabolism is culturally determined and, therefore, subject to purely social constraints in addition to environmental constraints. Hence, the metabolism of a society will be the sum of the biological and the technological metabolisms built by society itself over time, enabling the individual metabolisms of its members.

From a thermodynamic point of view, all human societies share with other physical and biological systems the need for controlled, efficient processing of energy extracted from the surroundings. Such is the proposal of Prigogine (1983) regarding nonequilibrium systems (thermodynamics of irreversible processes), which is one of the basic concepts of our agroecological approach to energy in agricultural systems: generation of *order out of chaos*. Since the natural trend of societies—as any physical and biological system—is toward a state of maximum entropy, social systems depend on building dissipative structures for balancing this trend and keeping away from maximum entropy. Dissipative structures transfer entropy to the outside environment and thus gain internal order or negentropy.

As Prigogine (1947, 1955, and 1962) said, all complex adaptive systems are kept away from thermodynamic equilibrium by means of *controlled dissipation*, which entails transferring part of their entropy to the environment. The structures of an open system are maintained thanks to the transfer by the system of a part of the energy being dissipated by its conversion processes; hence the name *dissipative systems* (Glansdorff and Prigogine, 1971, p. 288). Such transference is made by means of building dissipative structures using the flows of energy, materials, and information for performing work and dissipating heat, consequently increasing their inner organization. Order emerges from temporal patterns (systems) within a universe that, as a whole, moves slowly toward thermodynamic dissipation (Swanson et al., 1997, p. 47). Prigogine described this configuration of dissipative structures as a process of self-organization of the system.

Human societies give priority to performing two basic tasks: on the one hand, producing goods and services and distributing them among its individual members, and on the other hand, reproducing the conditions that make production possible to gain stability over time. In thermodynamic terms, this implies building dissipative structures and exchanging with the environment energy, materials, and information so that these structures may function. An important number of social relations are geared toward organizing and maintaining this exchange of energy, materials, and information. In fact, the interaction between the components of a system is no more than the exchange of energy, materials, and information. For analytical purposes, let us distinguish between two types of exchanges: a purely physical exchange of energy and materials, and a second type of exchange that, despite its physical costs, is more ideal or *immaterial*, the exchange of information. Adams (1975) considered human societies as a conglomerate of human and nonhuman forms, forms in equilibrium and out of equilibrium, living forms, which we may say constitute the equipment of a society, its material or symbolic fund. Therefore, from a biophysical perspective, human societies can be viewed as dissipative structures, or more precisely, as being made up of dissipative structures exchanging energy, materials, and information with their environment. It is this exchange that gives rise to the metabolic relationship (González de Molina and Toledo, 2014).

The exchange of energy, materials, and information that governs the metabolic dynamic is an asymmetrical and always a unidirectional exchange in which some structures become more ordered while others become more disordered, elevating the level of local entropy. Open systems such as human societies have managed to create order through their assurance of an uninterrupted flow of energy from their environment, transferring the resulting entropy back to their surroundings. This behavior, as indicated by Prigogine (1947, 1955, 1962), grounds the theoretical and methodological proposal of social metabolism. From a thermodynamic perspective, the functioning and physical dynamic of societies can be understood on the basis of this metabolic simile: any change in the total entropy of a system is the sum of external entropy production and internal entropy production owing to the irreversibility of the processes that occur within.

$$\Delta S_t = S_{in} + S_{out}$$

where ΔS_t is the increase in total entropy, S_{in} is the internal entropy, and S_{out} is the external entropy.

To put it another way, the generation of order within a society is achieved at the expense of increasing the total entropy of the system through the consumption of energy, materials, and information by its dissipative structures or fund elements. This level of order will remain constant or will increase if sufficient quantities of energy and materials or information are added to the system, creating new dissipative structures. This will in turn increase total entropy and, paradoxically, will reduce order or make it even more costly. Complex adaptive systems have resolved this dilemma by capturing from their surroundings the flows of energy, materials, and information required to maintain and increase their level of negentropy, transferring the entropy generated to their surroundings. In other words, the total entropy of the system tends to increase, reducing at the same time the internal entropy, if external entropy increases. To put it another way

$$\nabla S_{in} = \Delta S_{out}$$

Accordingly, we could say that the level of negentropy maintained by a society—in other words, the distance it remains from thermodynamic equilibrium—is the product of the sum of levels of internal and external entropy. Entropy is reduced by extracting energy and materials from one's own environment (domestic extraction [DE]) or by importing from another environment. The greater the flow of energy and materials extracted from its own territory or imported from others (or both at the same time), the more complex order a society will create, increasing its metabolic profile.

Consequently, a society's level of entropy is always a function of the relationship between internal and external entropy and, therefore, it is a function of the natural asymmetrical relationship established between a society and its environment, or between one society and another. This is not to say that this relationship is proportional or that an increase in one will always give rise to an increase in the other. To understand this, there is a useful distinction between "high-entropy" and "low-entropy" dissipative structures. A society that requires low amounts of energy and materials to maintain its fund elements reduces its internal entropy, generating in turn low entropy in its environment; in other words, low levels of domestic extraction and/or imports. In this case, the society would produce low total entropy. In contrast, another society might need large amounts of energy and materials from its environment and, if these are not sufficient, it might need to import energy and materials on a large scale to reduce its internal entropy. In this case, such a society would generate a much higher level of total entropy. This asymmetrical relationship between society and the environment also translates into differentials of complexity between the environment and system, whereby the system is always much less complex than the environment. This forms the basis of the strategy of "biomimicry" (Benyus, 1997) developed intentionally by humans and other high-order species in the extraction of information from the environment, and unintentionally by other living organisms. In fact, biomimicry is perhaps the most determining basic principle on which agroecology is based.

1.3 THE SOCIAL METABOLISM OF SOCIETIES

Originally, the concept of metabolism was widely used in biochemistry and biology especially when referring to cells and organisms. However, the same term has also been extrapolated to the study of human society since at least the nineteenth century (for a history of the use of this concept, see Fischer-Kowalski, 1998; Fisher-Kowalski and Hüttler, 1999). The concept attempts to grasp the set of interrelated processes by which human societies, independent of their situation in space (social formation) or time (historical moment), *appropriate, circulate, transform, consume, and excrete* materials and/or energies derived from the natural world (Figure 1.1). By performing these activities, human beings consummate two acts. First, they "socialize" segments or parts of nature, and second, they "naturalize" society by producing and reproducing linkages with nature. Likewise, a dynamic of *reciprocal determination* becomes established between society and nature during this general metabolic process, principally due to the fact that as human beings organize themselves into societies, they transform nature, and as nature becomes progressively modified by human beings, its altered condition comes to configure societies.

The material relationships that human beings establish with nature are dual: they are both individual and social. At the individual level, human beings extract amounts of oxygen, water, and biomass per unit of time from nature, and at the same time, they excrete or release heat, carbon dioxide, water, and various mineralized

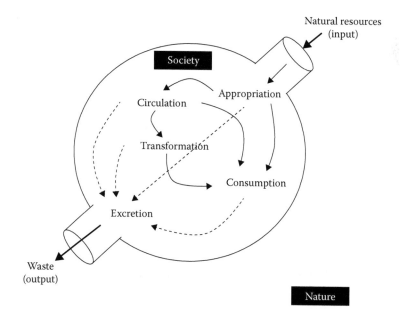

Figure 1.1 General diagram showing the metabolic processes and the relationship between society and nature. (From González de Molina, M. and Toledo, V.M., *Social Metabolisms: A Theory on Socio-Ecological Transformations*, Springer, New York, 2014, p. 62.)

or organic substances. At the social level, an ensemble of human beings (connected by some type of relationship nexus and organized to guarantee survival and reproduction) also extracts materials and energy from nature to create artifacts and meta-individual structures (buildings, tools, machines, factories, roads). Similarly, these human originated artifacts release, during their fabrication and/or function, a certain amount of waste and contaminants.

These two levels of metabolism (individual and social) correspond to what Lotka (1956), Georgescu-Roegen (1975), and later Margalef (1980, 1993) called endosomatic and exosomatic energy, a distinction that has an axiomatic value for the foundations of Ecological Economics (Martinez-Alier and Roca-Jusmet, 2000). These two flows also represent the biometabolic and sociometabolic energy flows, respectively, which together make up the general process of metabolism between nature and society. "The flow of endosomatic metabolism is fairly constant in time, and especially when considered per capita, and is directly related to population size. On the other hand, the exosomatic metabolism is highly variable and depends on the amount of technological capital present in society and its usually heterogeneous distribution across the various compartments distinguished within the society" (Giampietro et al., 2012, p. 187). A detailed discussion of these concepts and their application can also be found in Giampietro (2004).

The history of humanity is, therefore, simply the history of the expansion of sociometabolism beyond the addition of the biometabolisms of all its members. In other terms, human societies throughout time have been obliged to increase exosomatic energy over endosomatic energy; hence, the exo:endo ratio can be used as an indicator of the material complexity of societies (Giampietro, 2004). During the early societal stages, endosomatic energy was practically the only type of energy extracted from nature—with only a minimal amount of energy transformed into instruments for domestic use, clothing, and housing materials—but in modern industrial societies, exosomatic energy is 30 or 40 times larger than the overall endosomatic energy used by the individuals that make up these societies (Naredo, 1999, 2000). On the global scale, the extraction of mineral resources (fossil fuels, metallic and nonmetallic ores) measured in tons, doubles the extraction of biomass (photosynthetic products) obtained through agriculture, livestock, fishing, gathering, and extraction (Krausmann et al., 2009).

According to Fisher-Kowalski (1997), social metabolism describes the particular way in which societies establish and maintain their material and energy input from and output to nature and the way in which they organize the exchange of matter and energy with their natural environment. Social metabolism has been used as a set of methodological tools useful for analyzing the biophysical behavior of economies (Layke et al., 2000; Haberl, 2001; Weisz, 2007), but it is also being used as a theory to explain socioecological change (Fisher-Kowalsky and Haberl, 1997, 2007; Sieferle, 2001, 2011; González de Molina and Toledo, 2011, 2014). To all intents and purposes, social metabolism provides a new perspective for analyzing relations between society and nature from its material bases, mainly through the study of flows of energy and materials (Fisher-Kowalski and Haberl, 1997; Fisher-Kowalski et al., 2014; Giampietro and Mayumi, 2000; Giampietro et al., 2012).

Karl Marx was the first social scientist to apply this concept (Schmidt, 1971) when he utilized the German term *stoffweschel* (literally, "the interchange of substances") in his monumental analysis of capitalist society. Marx, in turn, obtained the concept from the seminal works of Moleschött, a German naturalist who wrote the first treatises on ecology. However, the concept remained virtually dormant for decades until the 1960s, when Wolman (1965) applied it to the biophysical analysis of cities, as did Boulding (1966), and the economists Ayres and Kneese (1969) applied it to industrial countries. But it was Marina Fisher-Kowalski who formally relaunched the concept in a chapter of her *Handbook of Environmental Sociology* published in 1997 (Redclift and Woodgate, 1997), presenting it as a stellar concept useful for analyzing flows of materials. The same author also wrote accounts of the historical trajectory of the concept (Fisher-Kowalski, 1998; Fisher-Kowalski and Hûttler, 1999). By that time, other concepts had appeared such as industrial metabolism, societal metabolism, socioeconomic metabolism, urban metabolism, and more recently, agrarian or rural metabolism, and hydraulic metabolism. These terms correspond to the study of fractions or dimensions of the general metabolic process (see Chapters 5 through 8). In the context of the emergence and development of new hybrid disciplines that predicate and practice interdisciplinarity, the concept of social metabolism and its equivalents was placed predominantly—but not exclusively—as a tool and method of ecological economics and industrial economics.

Metabolism can be measured in terms of the mass transferred from nature to a social system per unit time, and also as the energy flow integrated by a society per unit time. Similarly, it is possible to calculate the amount of materials and energy released as waste by societies. Given the former argument, nature and societies could be said to be connected in two ways: by the *inputs* that society obtains from nature, and by the *outputs* that society returns to nature. This dual relationship also defines the two basic functions or services that nature offers human societies: a *resource function*, which is determined by the regenerative capacity of ecosystems, and a *waste-processing function*, provided by the absorptive capacity of ecosystems. These two connections with nature (input and output) allow each society to develop social (or socioeconomic) processes, in which the materials and energy originally obtained from nature (natural capital) is converted into materials and energy that is socially utilizable (man-made capital). This, in turn, is ultimately transformed into waste or unusable by-products and released again into nature.

According to Nicholas Georgescu-Roegen, two fundamental elements must be distinguished when representing social–natural metabolic processes: funds and flows. Flows involve the energy and materials consumed or dissipated by the metabolic process—for example, raw materials or fossil fuel. Their purpose is to configure and supply the "funds" constructed by societies to generate goods and services, and to compensate for the law of entropy by generating order. This is what allows us to understand the relationship between society and its environment (nature) as a metabolic relationship (Ayres and Simonis, 1994; Fischer-Kowalski, 1998, 2003; Fisher-Kowalski and Huttler, 1999; Giampietro et al., 2012; González de Molina and Toledo, 2011, 2014).

Following Giampietro et al., "...funds refer to agents that are responsible for energy transformations and are able to preserve their identity over the duration of representation (time horizon of the analysis). They are the ones transforming input flows into output flows on the time scale of representation" (2012, p. 184). On the other hand, "... flows refer to elements disappearing and/or appearing over the duration of representation, that enter but do not exit or that exit but without having entered... Hence, flows include matter and energy in situ, controlled matter and energy, and dissipated matter and energy" (Giampietro et al., 2012, p. 184). In brief, fund is "what the system is and what has to be sustained," and flow is "what the system does in its interaction with the context" (Giampietro et al., 2012, p. 185). Restating the above in biological terms, these elements are analogous to the anatomy and the physiology of society in relation with nature. Later, we shall see how this distinction is applied specifically to agriculture.

In short, the application of the term *flow* to exchanges of energy, materials, and information makes explicit the dynamic, unidirectional, and irreversible nature of transferences of energy, materials, and information from one point to another within the system, or between the system and its environment (Adams, 1975). The function of such flows is to configure and feed the *funds* built by societies for generating goods and services. The configuration and maintenance of such funds require a continuous flow of energy, materials, and information to counterbalance the principle of entropy through the generation of order. Such flows, as argued by Georgescu-Roegen, maintain an *entropic balance* between the system and the environment that keeps the system away from thermodynamic equilibrium and generates order within chaos. Hence, the generation of order in a society is achieved by increasing the system's total entropy through the consumption of energy, materials, and information by its dissipative structures or fund elements. The level of order will remain constant or increase if enough quantities of energy, materials, and information are added to the system, creating new dissipative structures.

Social metabolism provides a conceptual framework for the integrative analysis of natural (ecological or biophysical) and social processes. The model of social metabolism is an idealized, abstract, and general representation of the whole of human society and nature, but with an undefined location in time and space. Therefore, it must be made specific by assigning to it a dimension, and a location in space and time. Once the abstract model of social metabolism is given a concrete expression, analyses can be made either of its entirety as a process, or of its fractions, dimensions, or scales. For example, the model may assume a totalizing dimension, or on the other hand, be focused on parts of this general process.

As Koestler stated in his seminal formulation of *holon*: "Organisms and societies are multi-leveled hierarchies of semi-autonomous sub-wholes branching into sub-wholes of lower order and so on. The term *holon* has been introduced to refer to these intermediary entities which, relative to their subordinates in the hierarchy, function as self-contained wholes; relative to their superordinates as dependent parts. This dichotomy of wholeness and partness, of autonomy and dependence, is inherent in the concept of hierarchy order" (Koestler, 1967, p. 58). Because of that, this concept has become increasingly more common in approaches to complex systems: the hierarchical structure of organization whether it is biological, mental, or social. Social metabolism, as a complex system model, can be approached from multiple angles, depending on the partition of reality made by the observer. Metabolic activity occurs within a spatiotemporal dimension, that is, it is enclosed within the territory of the planet and the time spanned by the history of the planet since the origin of the species. So, the general process of social metabolism can be analyzed at several scales. The spatial narrowness or amplitude of the approach defined, or chosen, by the analyst reveals the multiscalar character of this approach. Broadly speaking, up to six scale categories can be identified: appropriation or production unit, community, microregion (e.g., municipalities or counties), national, international, and planet. Similarly, when a historical perspective is adopted, social metabolism can be approached at different time scales identified by the analyzed time periods. In this case, different temporal extensions or time scales can be recognized: years, decades, centuries, and millennia. After all, social metabolism has existed since the rise of the human species nearly 200,000 years ago.

In short, social metabolism has been defined as the organized exchange of energy, materials, and information between society and the environment with the purpose of producing and reproducing its material means for existence. Since natural processes are irreversible and energy cannot be reused, human societies as open systems must compensate for the entropy they produce through exchange with the environment. In terms of the second law of thermodynamics, we may say that social metabolism pertains to the flow of energy, materials, and information that are exchanged by a human society with its environment for forming, maintaining, and reconstructing dissipative structures, allowing it to keep as far away as possible from the state of equilibrium. In other words, all societies generate order through the importation of energy and materials from the physical environment, and the exportation to the environment of dissipated heat and waste. The flows of energy, materials, and information feed dissipative structures and are thus vital for their maintenance and reproduction. Physical structures that consume resources—both for their building and for their functioning—have been built to provide health, education, security, food, clothing, housing, transportation, and so on. We may group all these structures or infrastructures into five metabolic processes: appropriation, transformation, distribution, consumption, and excretion.

In general terms, the evolution of human social civilization can be visualized as a continual increase in the use of energy (Figure 1.2), expressed as larger volumes of foodstuffs, the mobilization of more sophisticated and complex artifacts, and more efficient transformation of materials, commodities, and people (Adams, 1975; Smil, 1994; Debier et al., 1986). Because of this trend, the material flows from and toward nature are frequently expressed in energy terms, and quantified by means of several indexes using units of energy. Almost without exception, researchers using the concept of social metabolism commonly conduct their quantitative analyses by measuring material flows in terms of energy.

Over the last 200,000 years of existence of the human species, the main societal configurations have materialized in increasingly complex ramifications that are rooted in three main metabolic regimes: hunter–gatherers (or "cinegetic or

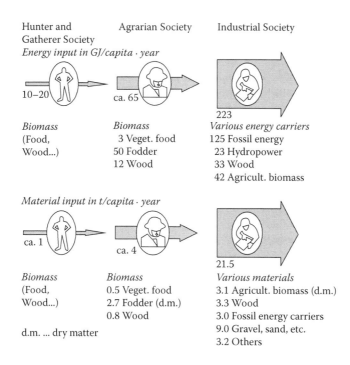

Figure 1.2 Metabolism indicators per capita and year for different regimes of social metabolism. (From Fischer-Kowalski, M. and Haberl, H., *Soc. Nat. Resour.*, 10, 70, 1997.)

extractive"), agrarian (or organic), and industrial (or fossil fuel-based). Each of them has been characterized by a certain level of energy (GJ inhab^{-1} y^{-1}) and materials (t inhab^{-1} y^{-1}) consumption, yielding a specific metabolic profile.

As represented in Figure 1.3, the main societal configurations recognized throughout history form a sequence of increasingly complex social designs interweaved with natural ecosystems and landscapes under the three modalities of social metabolism. A panorama clearly emerges formed by a sequence of socioecological stages including the different historical periods, population sizes, impacts upon and transformations of the biophysical environment, and the degrees of complexity recognized by Flannery (1972). The challenge, thus, is to decipher these and other general trends, to discover casual factors or sets of factors or changes, and to identify the patterns occurring over time.

The first consequence of increased efficiency in energy, material, and services flows was the growth of the human population. The transition between the extractive and the organic modes—taking 5000 years—multiplied the original global human population by 14, also increasing both the annual per capita consumption of energy and volume of waste excreted. But this scenario had lessened by the time of the demographic expansion that occurred during the leap from the organic to the industrial metabolism: between 1820 and 2011 the global population multiplied

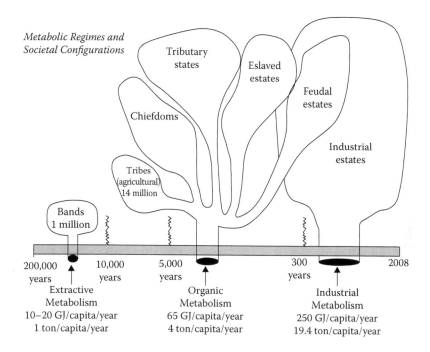

Figure 1.3 Diagrammatic representation of the three types of social metabolism and the main societary configurations suggested by Flannery (1972). The amount of energy and materials are indicated for each metabolism. (From González de Molina, M. and Toledo, V.M., *Social Metabolisms: A Theory on Socio-Ecological Transformations*, Springer, New York, p. 306, 2014.)

by seven (from one to seven billion inhabitants, and the same can be said of the annual per capita consumption of energy and material). The ten thousand years in which social metabolism was exclusively organic have a special relevance, given that it was during that period when the most significant social transformations took place. Human society changed its organization from tribal to chiefdom, finally becoming state societies. Cities appeared together with states, since the process of urbanization expresses within the territory the centralizing, hierarchical, and asymmetric character of these societies. Afterward metropolitan webs appeared, the first cosmopolitan web finally making its appearance toward the fifteenth century (McNeill and McNeill, 2004).

All these increments in social complexity were shored up by the advances made in the organic appropriation of nature. If the neolithic revolution (that allowed for the transformation from the extractive to the organic mode) was essentially an advance in human capacity for managing nature by manipulating populations of plant and animal species (originating thousands of breeds and varieties from hundreds of domesticated species), this newly conquered capability would continue its perfection and innovation during the following 7000 years.

Photograph 1.1 Field experiment on wheat varieties. A higher aboveground biomass production level can be observed for old varieties with respect to modern ones.

The available archaeological and historical analyses demonstrate such advancements. For example, productivity in Egypt is estimated to have increased from 1.3 to 1.8 inhab ha^{-1} of arable land between 2500 and 1250 BC, and during the Roman Empire, when Egypt became the breadbasket of Rome, it rose to 2.4 inhab ha^{-1} of arable land (Butzer, 1976). In China, these figures are even more striking: productivity of organic-based agriculture increased from sustaining 1–2 people per hectare during its early stages, to 2.8 in 1400, 4.8 in 1600, 5 toward 1900, and 5.5 during the 1930s (Smil, 1994, p. 63). In Mesoamerica, the lacustrine system in the Valley of Mexico sustained 4 people per hectare of arable land, including the *chinampas*— strips of land surrounded by water and fertilized by lacustrine sediments—that can sustain 13–16 people per hectare. The Inca civilization achieved similar productivity levels in elevated fields along the coastlines of Lake Titicaca known as *guaru-guaru* (Denevan, 1982). In Europe, in the more intensively cultivated regions of the Netherlands, Germany, France, and England, productivity was between 7 and 10 people per hectare of arable land (Smil, 1994).

The organic regime of social metabolism achieved other advances with the creation and perfection of two devices: ships and mills powered by wind, and the water wheel. The water wheel reached its maximum expression in Europe during medieval times when thousands of them were found throughout this territory (Basalla, 1988), becoming the leading technology for several uses including irrigation, corn milling, hide and paper presses, mining, and metallurgy, and in a way preceding the steam engine. The organic or agrarian regime as provider of food reached its limits and began to be transformed through the appearances of new engines powered by fossil energy (coal, oil, and gas), which were the product of the invention in the nineteenth century of the internal combustion engine. The advances of agrochemistry added to the transformation, and genetic science gave rise to new varieties of plants and animals.

Despite its productive limitations, the organic or agrarian metabolism is still practiced by a broad section of the human species and is the most successful and extensively used form of interplay with nature. Viewed from a wide perspective, organic metabolism is placed between the extractive metabolism, which throughout 95% of historical time maintained humans in a state of nearly total stagnation, and the industrial metabolism, in spite of its short existence for the past 200–250 years, is currently endangering not only the existence of the human species, but that of life and of the whole planetary ecosystem.

The prevalent social and ecological situation in the world today—the world-system (Wallerstein, 1974)—is the result of complex interactions taking place between the metabolisms of societies, having unequal sociopolitical complexities and historical origins, arranged as metabolic constellations in the form of networks and systems of increasingly complex webs. In general terms, the social historical process has encompassed clear trends in human population growth, higher energy flows, higher population density (sedentarization followed by urbanization), more extensive human settlements, social stratification and inequality, labor division, productive specialization, processing, storage and consumption of goods (materials and energy), technologies, knowledge, and information.

The crisis of the modern world derives from the additive accumulation of the processes of industrial metabolism. The key factor is, without doubt, a change in energy source that induced the accelerated creation of new technologies, mechanization of countless processes, and caused the change from organic appropriation to one based on fossil fuels, including uranium-based nuclear energy. This shift in energy source radically transformed practices in the use of nature, allowed for a substantial increase in surplus, and, as a consequence of increasing volumes of foodstuffs and raw materials made available, also resulted in population and industrial growth. The outcome of all this was an extraordinary intensification of exosomatic energy.

The result of all these processes was the unleashing of what McNeill (2000, p. 4) has called "...a giant, uncontrolled experiment on earth" made evident in the progressive acceleration of numerous environmental and social phenomena during the past 100 years, and in particular during the past five decades (see MEA, 2005). This qualitative leap in human transformation power first had impressive effects on the forms of appropriation of nature—agriculture, livestock breeding, fishing, management of water, forestry, and mining, among other sectors—that, in turn, potentiated the accelerated growth of human population, cities, and industry.

From a metabolic standpoint, the industrial civilization revolutionized the act of appropriation as never before in history, which propelled the circulation and transformation of products, rising consumption, but above all, increased excretion to unprecedented levels. Industrial metabolism has indeed not only substantially amplified appropriation, circulation and transformation, but has also exacerbated the excretion processes to unbelievable levels, making it the most influential process owing to our inability to controlling increasingly growing volumes of generated waste including materials, substances, gases, radiation, electromagnetic waves, and new genomes. This incapacity is mainly derived from the amount and contents of excreted waste that exceed the capacity of natural systems to assimilate and recycle them, or because such waste is intrinsically unrecyclable.

The use of energy measured in metric tons of oil equivalent is the third indicator to have been greatly accelerated during the past century, increasing sixteenfold. The energy used during the twentieth century was greater than that used by the human species throughout its history, and 10 times larger than the total energy used in the previous 1000 years (McNeill, 2000). In comparison with the former data, the use of water increased nine times, the amount of carbon dioxide (CO_2) emissions rose thirteenfold, and industrial total atmospheric pollutant emissions, by 40 times.

Likewise, the extraction and consumption of metals (copper, zinc, manganese, chrome, nickel, magnesium, tin, molybdenum, and mercury) experienced a spectacular increase over the past 100 years. Between 1900 and 2009 the consumption of resources rose from 7 to nearly 70 Gt. All types of products show a substantial increment: biomass, from 5 to 20 Gt; fossil fuels, from 1 to 13 Gt; metals, from 0.2 to 6 Gt; and building materials, from 0.7 to 28 Gt. Although the mass of biomass quadrupled, its growth has been the lowest in relative terms. In fact, the annual per capita consumption of biomass remained fairly stable during the twentieth century, while the consumption of inorganic resources went from 1 to 7 t inhab^{-1} y^{-1}. Thus, the total consumption of materials grew during the past century at a higher rate than population: whereas population multiplied by 4.4, consumption of resources increased 9.6-fold. Consequently, each current inhabitant of Earth needs 2.2 times more materials than inhabitants did at the beginning of the past century, meaning that the strong pressure exerted over resources during recent history cannot be explained by demographic causes alone since the growth of consumption has been much higher than population growth (McNeill, 2000).

1.4 THE ENERGETIC METABOLISM OF AGRICULTURE: A FIRST ATTEMPT

The theoretical and methodological approach of social metabolism could provide a very useful tool for agroecology. They share the same epistemological starting point: agroecosystems are strongly anthropized systems in which the dynamic is explained by the way societies interact with their agrarian environment. They are, therefore, socioecological constructions and, as such, they are part of nature and, at the same time, of society. Very few papers have attempted to adapt social metabolism methodology to agriculture (Risku-Norka, 1999; Risku-Norja and Mäenpää, 2007). Some authors have also attempted to estimate the metabolism of the food system (Wirsenius, 2003) and the agrofood system (Heller and Keoleian, 2003; Infante et al., 2014a) and the global and continental flows of biomass have been analyzed (Krausmann et al., 2008a; Smil, 2013a), but none of these papers have studied the role and functionality of biomass in the transition toward an industrial metabolic regime. Most studies of the socioecological transition in its historical dimension and on the scale of the nation state have analyzed social metabolism globally, with no specific analysis of agriculture (Schandl and Schultz, 2002; Krausmann et al., 2008a, 2011; Kovanda and Hak, 2011; Gierlinger and Krausmann, 2012; Singh et al., 2012; Infante et al., 2015). Only two papers, from Czechoslovakia and Spain, specifically analyzed the changes in land uses and the energy transition in agriculture (Kuskova et al., 2008; Soto et al., 2016). However, with few exceptions (Krausmann et al., 2013), there are no data regarding the environmental impacts these trends have had on agroecosystems.

Indeed, the metabolic approach is yet to be developed in this and other aspects so that it can be used to understand and analyze the specificity and complexity of agrarian activity. For example, the metabolic approach only partially incorporated fund elements and never in relation to agricultural activity, and yet productive capacity depends precisely on the reproduction of those fund elements along with the optimum provision of environmental services by agroecosystems. In fact, agrarian sustainability is increasingly associated with this provision, and environmental damage or the impact of productive activities on agriculture is linked with the degradation of fund elements, especially "natural" fund elements (land, water, biodiversity, etc.), the definition of which will be tackled later. This consideration of agroecosystems refers to their very identity, to the fact that the reproduction of such fund elements depends greatly on the generation of biomass and its distribution within agroecosystems. For that reason, it is very important to consider within the agrarian metabolism proposal all the ecosystems that make up that metabolism, regardless of their level of anthropization, and all the net primary productivity (NPP) they provide. Methodological proposals for social metabolism are usually based on the domestic extraction (DE) of biomass and barely take into account the other biomass produced, even though the latter performs vital tasks and functions for the very functioning and sustainable reproduction of the metabolism itself. Perhaps for that reason, a calculation of the amount of biomass appropriated by humans in a given territory (human appropriation of net primary productivity [HANPP]) has been proposed to evaluate the impact of the metabolism on other ecosystems and their components. However, as we will see in the following text, this methodology only takes into account a portion of net primary productivity and not all of it.

This could be due to the fact that the most widely adopted metabolic approaches assume the perspective of society and the resources it appropriates, which means they have an anthropocentric approach. Although this approach is also adequate, it leaves the structure and functioning of ecosystems to one side, even though the good ecological state of these ecosystems is crucial to the supply of goods and services received by society. To put it another way, from an agroecological perspective, the level and sustainability of DE also depends on the biomass that is not extracted and which, therefore, remains within ecosystems available to their other heterotrophic components. Consequently, the most coherent perspective with the metabolic approach is that of the agroecosystems themselves and their fund elements, since they are the reflection—like the metabolism itself—of socioecological relations. In other words, the agroecological perspective is grounded in a biocentric approach to the relationships between society and nature, the only anthropocentric approach that can be adopted from the perspective of sustainability.

The *agrarian metabolism* alludes to the exchange of energy, materials, and information between a given society and its agrarian environment. Within the general metabolism, the agrarian metabolism specializes in the *metabolic process of appropriation* carried out by all societies with regard to the products of photosynthesis. The main aim of this metabolic process was, and is, the growing and appropriation of plant biomass (net primary productivity) from the land to satisfy, directly or indirectly through livestock, the endosomatic consumption of the human species, although this has not been its sole purpose. It has also been a case of satisfying the exosomatic demand of societies with an organic metabolic regime, and it continues to be so, although to a lesser degree, in industrial societies. The satisfaction of endosomatic consumption has also signified the appropriation, growing, processing, and distribution of fluvial and marine biomass (fishing). Moreover, societies have even managed to breed fish in captivity—as if they were livestock—so we could even talk about a production process or a metabolic process for the production of fish (aquaculture). Strictly speaking, these activities are also part of the agrarian metabolism and the agrofood system. However, they will not be considered here, since they are beyond the thematic framework of agrarian activity. The remainder of this chapter shall focus on all the agrarian activities developed within a given territory to process plant or animal biomass for society, either through food for humans or animal feed, raw materials, or fuel. To this end, society colonizes or takes control of part of the available territory, establishing different levels or degrees of intervention or interference with regard to the structure, functioning, and dynamic of ecosystems, their net primary productivity, giving rise to different types of *agroecosystems*.

In the majority of research about social metabolism, biomass is considered just another source of energy and materials appropriated by society. In this book, however, agrarian activity is the focal point, the boundaries of which correspond roughly to the boundaries of the agrarian sector with any society, characterized by "producing" biomass using living organisms, a peculiarity that cannot be stated for any other production sector (with the exception of fishing). However, since the early twentieth century, the agrarian sector has also required the use of nonbiotic resources. In this respect, our methodological proposal draws a distinction between biotic and abiotic sources of energy and materials when calculating input flows that recirculate through agroecosystems. This is not only due to the need for methodological coherence, but also because not all sources of energy have the same capacity to reproduce or activate the fund elements of agroecosystems. However, for the purposes of determining the metabolic profile of the agrarian sector in each society and comparing it over time, we have to add together "apples and oranges," combining fossil energy sources with biomass, reducing them to their intrinsic energy content and the energy incorporated, adding together the input and output flows of the agrarian sector. If not, it would be impossible to capture the tremendous change that agriculture has undergone with the process of industrialization: from functioning almost exclusively through energy taken from biological sources, it has become increasingly dependent on abiotic energy and materials (fossil fuels, and metallic and nonmetallic ores) for the manufacturing and functioning of inputs. The sustainable agriculture of the future faces the challenge of minimizing the use of abiotic inputs and basing its functioning once again on renewable energy flows, particularly biological energy flows. Understanding the reasons for their progressive adoption in contemporary agriculture and their level of dependency is decisive in terms of planning the transition toward sustainable agrarian systems.

1.5 FUND ELEMENTS OF AGRARIAN METABOLISM

The metabolic exchange of energy and materials takes place by means of flows that enter and exit the boundaries of the agrarian sector, but which also circulate within it depending on the demands generated by the dissipative structures of agroecosystems, producing negentropic order; in other words, food, fuel, and raw materials for society. Describing the components of this metabolic relationship must begin, in accordance with everything stated above, by distinguishing between fund elements and the flows of energy they dissipate, according to the proposals made by Georgescu-Roegen (1971). In our view, the ultimate aim of the economy is not the production and consumption of goods and services, as predicated by neoclassical economics, but rather the reproduction and improvement of the series of processes required for the production and consumption of goods and services. This variation in the principal goals of economic activity implies, from a biophysical perspective, transferring our focus away from the flow of energy and materials onto fund elements. This shift in orientation is particularly useful when it comes to applying the theoretical and methodological proposal of social metabolism to agriculture, since it allows us to evaluate whether flows of energy and materials into and out of the agrarian sector are capable of reproducing and even improving fund elements in successive production cycles. In other words, moving the focus of attention away from the volume of production and consumption of biomass toward sustainability, and whether production and consumption can be maintained indefinitely. Consequently, the characteristics of flows are closely related with the fund from which they originate (Giampietro et al., 2014, p. 29).

Input flows involve the energy and materials consumed or dissipated by the metabolic process—for example, raw materials or fossil fuel. Their purpose is to configure and supply the "funds" constructed by societies to generate goods and services (the output flows), and to compensate for the law of entropy by generating order. Two types of factors control the rhythm of flows: external factors related to the accessibility of resources available in the environment in which metabolism takes place, and internal factors pertaining to the capacity for processing energy and materials during the process of conversion, which itself depends on the technology used and the knowledge for its management. Instead, funds are the entities or (dissipative) structures that transform input flows into output flows at a given time scale, and which, hence, remain constant throughout the dissipative process. Although the theory of social metabolism has considered and quantified above all the flows of energy and materials, it is essential to take into consideration the fund elements each society has constructed along its evolutionary trajectory. Such elements determine the nature of the flows of energy and materials and, therefore, the metabolic profile of each society.

Funds also have two particularly noteworthy characteristics: they process energy, materials, and information at a rate determined by their own structure; and they require periodic renewal or reproduction (Scheidel and Sorman, 2012). This implies that a part of the input flows needs to be devoted to constructing, maintaining, and reproducing the dissipative energies themselves, which, of course, limits their growth rate (Giampietro et al., 2008a). The human population living within the territorial limits of a given society—herein considered the processor of the energy, materials, and information (mainly endosomatically) required to produce work and residual heat—forms part of the fund elements. As discussed in the following text, the main fund elements of the agrarian metabolism are agroecosystems (land in the broad sense: soil, water, biodiversity, etc.) and domesticated livestock, which, when managed by humans, process external energy, materials, and information to produce biomass that, in turn, provides a flow that feeds other dissipative structures of social metabolism. Finally, fund elements also include what economists call *capital*, that is, the set of artifacts capable of processing energy and materials that are created by humans (Giampietro et al., 2008b). The fund elements could even be improved over time, allocating increasing amounts of energy and materials for this purpose.

Another useful distinction rescued and adapted to society by Giampietro et al. (2008b) is that of Ulanowicz (1986), differentiating between two main components of ecosystems: the hypercycle, that is, the part providing the gross energy for the whole ecosystem; and the purely dissipative part, devoted to the degradation of gross energy in the ecosystem. The hypercycle keeps the ecosystem away from thermodynamic equilibrium, while the dissipative part has important functions: it provides a control mechanism over the entire process of energy transformations, explores innovations (guaranteeing adaptability), and stabilizes the evolutionary sustainability of the whole system. In fact, an ecosystem made up of only one hypercycle cannot be stable over time. We will return to this question in agroecosystems later. Without the stabilizing effect of the dissipative part, a positive feedback "will be reflected upon itself without attenuation, and eventually the upward spiral will exceed any conceivable bounds" (Ulanowicz, 1986, p. 57). In the analogy of human societies, the hyper-cycle of the society is made up by the economic sectors generating profit and goods and services, and the purely dissipative part is the final consumption sector (Giampietro et al., 2008a, p. 3).

The metabolic relationship established by each society with its agrarian environment has access to different fund elements. We are focusing here on the elements that are critical for the reproduction of agroecosystems themselves and for the provision of their services, including the supply of biomass as the principal service. We have considered four elements (land, livestock, human labor, and traction, also known as technical capital). However, a more precise description requires a distinction to be drawn between biophysical fund elements and social fund elements that are closely interlinked, as a reflection of the socioecological relations that are at the heart of the agrarian metabolism. This interlinking of fund elements is also fundamental when it comes to explaining their dynamic.

Given the different types of fund elements considered—biophysical and social—the metric used to ascertain their entity and function is also different. The land colonized or appropriated by a society to produce useful biomass is measured in hectares and is usually divided into different uses that produce plant biomass (net primary productivity, NPP), expressing its measurement in MJ/ha per year, or its equivalent in tons of plant biomass per hectare (t/ha per year). Depending on the scale of analysis, colonized land might contain one or several interrelated agroecosystems. Livestock is used to provide services to society, fundamentally through the provision of animal biomass used also for raw materials, food, and to a much lesser degree, energy. Its entity is usually measured in standard livestock units of 500 kg (LU_{500}) (or in terms of livestock metabolizable energy requirements), and the flows it generate are expressed in kg or t of animal biomass/ha per year or LU (or MJ/ha per year or LU, if flows are expressed as energy units). Human labor, in turn, is measured in terms of average working capacity per hour or working day, and the measurement of its flows is usually expressed in energy terms (MJ/hour or MJ/year). Finally, traction, which can be animal or mechanical, depending on whether it is an industrial or preindustrial agrarian system, is measured in terms of installed capacity, expressed in kw of power or Cv, and produces labor flows expressed in terms of kw/hour or MJ/ha.

The difference in the way each fund element is measured expresses a fundamental characteristic: that these fund elements are only partially interchangeable and that the flows that reproduce them are different in terms of their identity. Their reproduction requires an amount of energy in terms of biomass and human labor, which must be dealt with in each productive process. The energy required for this can only be replaced partially with external energy, given their different nature. For example, only biomass can be used to nourish the trophic chains that sustain edaphic life and the general biodiversity of the agroecosystem. The deterioration of colonized or appropriated land can only be partially compensated by external energy and materials that are different in nature to plant biomass. For that reason, there is no optimum point capable of maximizing the NPP_{act} of crops using a supply of fuels brought in from outside the sector. However, there is a somewhat inverse relationship between fund elements: the provision of services has increased in terms of social fund elements and decreased in terms of biophysical fund elements, as we will see later.

1.6 SOCIOECOLOGICAL TRANSITIONS IN AGRICULTURE

This metabolic approach to agrarian systems is particularly useful in agro-ecology. It provides information about their physical functioning and their spatial/temporal differences. It enables differences to be shown with greater clarity, in terms of their structure and physical/biological functioning, between organic agriculture—either traditional or modern organic farming—and industrialized agriculture. It also provides information about how the industrialization of agriculture came about and how, consequently, a new transition toward a more sustainable agrarian metabolism should occur. In line with Fischer-Kowalski and Haberl (2007, p. 3), we understand the socioecological transition to be a process of change from one metabolic state to another that is qualitatively different, a process that is neither linear nor predictable.

Agricultural industrialization took place in three major waves: the first one was fostered by institutional change toward capitalism and took place within the boundaries of the agrarian sector, signifying the optimization of its possibilities by raising biomass production; the second wave was the first metamorphosis in the configuration of the agrarian sector through the injection of artificial fertilizers, that is to say, through the external subsidy of energy and materials from nonrenewable sources; and finally, the third phase was the total penetration of fossil fuels within the agrarian sector (Krausmann et al., 2008b). These three waves fit well with the canonical characterization provided by Bairoch (1973, 1999) of the history of contemporary agriculture, discerning the main transformations that, as revolutions, led to its complete industrialization.

The need to meet a growing demand for human and animal foodstuffs derived from the growing process of urbanization, and general demographic expansion was common to all countries that gradually became industrialized during the nineteenth century. Many European countries had been suffering from internal and, to a lesser extent, external pressures on their agroecosystems to increase the volume of biomass production. The population increase beginning in the eighteenth century, the process of urbanization, the elevation in consumption among the upper classes, and the different demands being generated by the newly burgeoning process of industrialization, converged in a legal–political structure that left the traditional configuration of the agrarian metabolism and the distribution of land uses unprotected. This facilitated institutional change (liberal revolutions), especially in the regime of feudal ownership, and the liberalization of agrarian markets.

The possibility offered by oil and its associated technologies of injecting large amounts of energy and materials radically changed the world's agricultural scenario during the twentieth century. Until then, coal had played a very limited role in agriculture, because of its characteristics. The main transformations took place after the Second World War in the form of the green revolution seeds, chemical fertilizers, mechanical traction, and pesticides. But in fact, the industrial metabolic regime had penetrated agriculture half a century earlier during the late nineteenth century, when chemical fertilizers manufactured by means of fossil fuels and chemical procedures made their appearance. Their introduction meant overcoming the most common limiting factor in production thus far, the lack of soil nutrients, and a break from dependence on replenishing soil fertility; in other words, reducing the land cost of soil fertilization (Guzmán Casado and González de Molina, 2009; Guzmán Casado et al., 2011). A long transition process commenced in which agrarian production shifted from a dependence on soil, that is to say on land, to a dependence on subsoil, in other words, on fossil fuels and minerals, as is the case today.

It began in this area because the critical point in terms of the resilience of the agrarian metabolism was precisely the shortage of nutrients or depletion of the soil. The successive arrangements designed in the nineteenth century to produce new essential balances in the different land uses became expensive and impracticable owing to their growing size. From the second half of the eighteenth century onward, the expansion of crops for industrial purposes or human consumption required the importation of soil/land in the form of organic matter or animal feed. But the continual increase in agricultural land area and its productive intensification aggravated the nutrient deficit to such an extent that it increasingly cost more money and effort to cover this deficit by importing organic fertilizers. This created a favorable context for the spread of land-saving technologies, especially chemical fertilizers, where the process of intensification had consumed the land's own resources, which would explain the irregular use made of this technology in the early twentieth century. In places where there were still lands with which to generate new balances there was no need to use chemical fertilizers, and that was only carried out on a partial basis. A similar pattern was observed in large extensions of land, such as the *latifundios* of some Latin American countries, or in southern Spain, Italy, and Portugal, where working livestock could be used to obtain the fertilizer required for the total or partial sowing of fallow land, thereby increasing crop intensity (see González de Molina, 2002). More intensive rotations, without fallow and with successions of crops that would have been previously impossible, were now possible stimulated by the integration of international markets for agrarian products at the end of the nineteenth century.

Saving land was the most logical option once keeping the agrosilvopastoral equilibrium was definitively impossible (Toledo, 1990; Liebowitz and Margolis, 1995; Tello, 2005). But a different course was taken where land was abundant and the equilibrium between alternate land uses was unthreatened by intensification. For example, in the United States, mechanization of farming chores preceded the introduction of chemical fertilizers. The first agricultural machines were powered by animal traction or even by steam. North American colonists had more than enough fertile land to feed their working livestock, and were not limited by scarcity of firewood for generating steam, something that was practically impossible to do in an overpopulated Europe or in most of Asia. The significant increase in labor productivity brought by such productive possibilities made U.S. farms more profitable than those of Asia or Europe. Some experts in technological change in agriculture even claimed that since technical labor-saving solutions were more effective than saving-land ones, poor countries should follow the former alternative (Hayami and Ruttan, 1971). However, due to the land costs of replacing human labor with animal power in organic agriculture, where land was scarce, labor-saving technologies could not be implemented until the arrival of motor vehicles.

In around 1904, the German chemist Fritz Haber began to experiment with the possibility of synthesizing ammonium, a form of reactive nitrogen, the shortage of which in soils poses a strong limitation to agricultural land productivity (Smil, 2001b). First, chemical nitrogen fertilizers and other agrochemical contributions, and later advances in genetics, contributed during the early twentieth century to promoting a socioecological transformation in crop fields. However, Krausmann et al. (2008b) put back the introduction of fossil fuels in agriculture until after the Second World War. But if we approach this issue from a broader perspective, which includes the Mediterranean world, the energy change in agriculture began in the

Photograph 1.2 *Dehesa* (known as *Montado* in Portugal) is a traditional silvopastoral system of the Iberian Peninsula with strong internal energy loops.

first few decades of the twentieth century, not only because synthetic chemical fertilizers entailed high energy consumption from fossil fuels, but also because these fuels were an intricate part of agrarian labor processes. In the early decades of the twentieth century, the energy change took place in Spanish irrigation systems with underground water: waterwheels and animal drawn mechanisms were replaced with systems powered by fossil fuels (irrigation water hoisting pumps powered by electric or internal combustion engines fueled by producer gas or oil). In Italy, this was even more so, bearing in mind the spread of drainage pumps powered by fossil fuels in *Bonifica* processes. The appearance and spread of these technologies were crucial to the agrarian modernization of both countries (Calatayud and Martínez Carrión, 1999; Bevilacqua and Rossi-Doria, 1984; Bevilacqua, 1989–1991; D'Attorre and De Bernardi, 1994).

However, the major leap forward came with the change in energy pattern that replaced coal with oil and natural gas, which offered higher energy densities. Associated with them, two basic innovations for the industrialization of agriculture permitted the mass subsidization of agriculture with external energy: electricity and the internal combustion engine. This began during the 1930s in the United States and reached Europe after the Second World War. It started with the mechanization of many agricultural tasks and culminated in most rich countries with the spread of the green revolution technological package at the end of the 1950s (see Table 1.1).

Table 1.1 Indicators of the Industrialization Process of Agriculture in the World

	Units	1963	1978	1993	2008
Fertilizers (N)	1,000 t	15,011	53,327	74,493	105,738
Rural population	Millions	2,106	2,656	3,134	3,385
Mechanization	1,000 tractors	12,389	20,557	26,003	–
Cereal yields	kg/ha	1,321	1,946	2,502	3,149
Food energy	Petajoules	11,027	16,075	22,393	29,060
Energy intake per capita	kcal/day	2,253	2,451	2,636	2,822

Sources: FAO, FAOSTAT, 2016, http://faostat.fao.org/; and author data.

Crop intensification had come up against new ecological conditioning factors, as had occurred in the late nineteenth century. Agricultural activity had been growing relentlessly, and livestock, the main source of traction, could not keep up in terms of traction demands or the change of diet, richer in animal proteins. Competition between the allocations of land to growing food or fodder would still be as much of an issue as ever. The presence of animal traction impeded further expansion of agriculture and intensive livestock farming. It was necessary to develop a kind of technology that would once again save land, freeing up working livestock productive areas, and a kind of technology that would replace animal traction with mechanical traction. Added to that was the convenience of saving costs to achieve a minimum threshold of profitability, situated at a lower level than the average profitability of other economic activities. The reduction of manual labor, replaced by machines or by chemical means that made certain tasks easier (e.g., weeding), was the solution. In some countries, emigration from the countryside to the city and the development of movements of paid farm laborers pushed wages up and sped up the substitution process.

Although the process of decolonization made it possible for many peripheral countries to regain sovereignty over their natural resources, in practice, control remained in the hands of the former metropolis. Many countries adopted a policy of import substitution (for a review see Bruton, 1998), which was financed by the agrarian sector that experienced a new intensification process. This was the goal of the modernization policy that accompanied the green revolution in peripheral countries. Cultivated land areas expanded, particularly dedicated to commercial crops. Permanent grassland also extended at the expense of forested areas. From 1970 to 1985 alone, the surface area of forest in Latin America and the Caribbean region fell by nearly one million square kilometers. In 1987, 80,000 km² of Amazonian forest were converted to grassland, a figure only slightly larger than that of the previous years. Beyond doubt, the case of Haiti is most relevant. By 1923, 60% of Haiti was covered by arboreal vegetation, but 60 years later that surface area was drastically reduced to below 2%, 30% of which was degraded and hence totally unproductive. The expansion of cropping and irrigating land unsuitable for agriculture also had severe consequences. The Food and Agriculture Organization of the United Nations (FAO) map of soils in 1990 included 400 million hectares of degraded soils in Latin America alone (GLASOD, 1991).

New crops depended on improved seed varieties, needed large doses of fertilizers and pesticides, and required agricultural technology that was beyond the reach of poor countries. In 1984, 20 times more fertilizers and 25 times more pesticides were used in Latin America than in 1950. From 1950 to 1972, the annual rate of average consumption of fertilizers grew by 14%. By 1980, Latin America was spending US$1.2 billion on pesticides (FAO, 2016). Hunger, poverty, and malnutrition did not disappear, but technological dependence and debt grew in an unusual way. In fact, the translation of the Western model of intensive agriculture to countries with different edaphic and climatic conditions opened up a huge market for transnational agrochemical and food corporations.

In parallel, major destruction of the agrarian subsistence sector was taking place, conditioning the loss of food self-sufficiency (Toledo et al., 1985). The expansion of the livestock industry during the postwar period is a good example of such a phenomenon. As standards of living rose in industrialized countries, the consumption of animal protein also increased, in particular of meat and dairy products. To supply the continually growing demand, peripheral zones were devoted to raising livestock or producing fodder. Global meat exports grew from 2 million tons in 1950 to 11 million tons in 1984. Many countries in Africa and Latin America converted extensive cropping areas to grazing land for cattle. In particular, large areas of forest land were converted to grasslands, while in other countries traditional crop varieties were replaced by a monoculture of forage crops. In both cases, the result was a growing production deficit of cereals and other foodstuffs formerly grown domestically (for a review, see Barkin et al., 1991). In addition, the modernization of agriculture was achieved at the expense of traditional farmers who were forced to migrate to cities to live in conditions of extreme poverty, or remain farming marginal land.

Those who remained farmers enjoyed no better living conditions. The biased distribution of property, the trend to concentrate land tenure in a few hands, and the destructuralization of rural communities brought about by modernization, forced farmers to cultivate forested and marginal lands. Many deforestation processes, overgrazing, cultivation of slopes that in some zones have accelerated erosion and desertification—such as Sahel, India, Panama, Brazil—are associated with that practice.

In short, we could state that the agrarian sector has been expelled from the energy system and has become a recipient of energy and materials from elsewhere. The nucleus of the agrarian metabolism is still DE at a national scale, but the importation (I) of energy becomes decisively important. Agriculture went from being at the heart of the metabolic process to constituting an apparently marginal segment thereof, thanks to the exploitation of fossil fuels. This metamorphosis, which occurred at an accelerating pace, began in England, made the leap to continental Europe, expanded toward its peripheries, was taken to the colonies and today is still spreading to every corner of the globe.

In fact, the production of biomass no longer provides the bulk of the energy that allows society to function (Figure 1.4). The DE of biomass represented between 95% and 100% of the energy consumption in organic metabolic regimes, whereas in most developed societies where the industrial metabolism has become the dominant way of organizing relations with nature, biomass only produces between

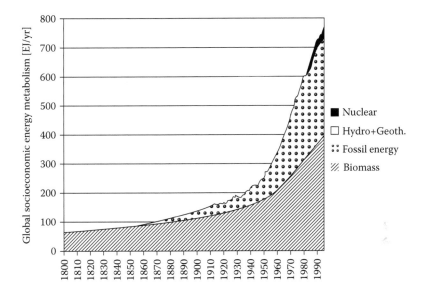

Figure 1.4 Annual global consumption of primary energy from 1800 to 1990. (From Haberl, H., *Energy*, 31, 93, 2006.)

10% and 30% (Table 1.2). Furthermore, the energy balances show that agriculture has changed from being a supplier to a demander of energy (Leach, 1976; Pimentel and Pimentel, 1979; Naredo and Campos, 1980; Carpintero and Naredo, 2006; Cussó et al., 2006; González de Molina and Guzmán Casado, 2006; Tello et al., 2015). Without the subsidy of external energy, a part of global agriculture could not function.

This major injection of energy and materials explains why yields per land unit have multiplied, offering the capability of feeding a population that has grown six-fold since the start of the nineteenth century, and giving rise to one of many paradoxes. According to Smil (2001b, p. 256), the total area of farmland in the world grew by a third during the twentieth century; however, because productivity has

Table 1.2 Weight of Biomass in Total Energy Use (%) in Organic (1750–1830) and Industrial Metabolic Regimes (2000)

	Organic Metabolism 1750/1830	Industrial Metabolism 2000
Developing countries	–	92
Developed countries	–	50
European Union—15	99	29
Austria	99	29
United Kingdom	94	12

Source: Fischer-Kowalski, M. and Haberl, H., *Socioecological Transitions and Global Change: Trajectories of Social Metabolism and Land Use*, Edward Elgar Publishing, Cheltenham, 2007, p. 231.

multiplied by four, the harvests obtained in this period multiplied by six. But as Smil himself acknowledges, this gain is partly due to the fact that the amount of energy used in farming is eight times larger (see also Pimentel and Pimentel, 1979).

It also, and particularly, explains the exponential growth registered in terms of the productivity of agrarian labor. The case studies conducted for Austria by Krausmann et al. (2003) and for Santa Fe, Spain (González de Molina and Guzmán Casado, 2006) mostly concur that the industrialization of the agrarian metabolism led to a spectacular increase in the productivity of labor, due to the mass use of new technologies and the mass input of external energy. Interestingly, both cases, built on the same methodology albeit at a different scale, coincide that this increase caused yields to augment fivefold (Guzmán Casado and González de Molina, 2008).

Agrarian activities have changed their metabolic functionality. They constitute another input in the metabolism of materials and, although the market does not reward this task, they offer essential environmental services (carbon sinks, climate regulation, water purification, maintenance of certain levels of biodiversity, etc.) for the stability of the industrial metabolism. Perhaps for that reason they have tended to become degraded through the very industrialization and commodification of agriculture (De Groot et al., 2002; Pagiola and Platais, 2002; Pagiola et al., 2004).

The socioecological transitions to the industrial metabolism regime have been accompanied by an accelerated increase in the consumption of materials, both in absolute and per capita terms, especially from abiotic materials during the second half of the twentieth century. On a global scale, it was in the late 1950s when the extraction of abiotic materials came to exceed the extraction of biomass (Kraussman et al., 2009, 2011; Singh et al., 2012; Gierlinger and Krausmann, 2012; Infante et al., 2015). In parallel, the per capita consumption of biomass has fallen in general terms. This has not been due to a complete substitution of biotic materials by abiotic materials, but only to a partial substitution caused by the replacement of biotic fuels with fossil fuels for domestic consumption. In absolute terms, the consumption of biotic materials has also increased considerably, though at a lower rate. This has been due to the changes in the functionality of biomass for social metabolism as a whole (especially as a source of domestic fuel and as "fuel" for working animals): it has gone from being the main source of energy and materials to specializing in two essential functions, the supply of food and the provision of raw materials for industry, especially wood which is difficult to substitute in many industrial processes (Infante et al., 2014b; Iriarte and Infante, 2014).

In other words, agroecosystems have gone from supplying most of the goods and services required by the world economy to specializing in foodstuffs, both animal and vegetable, and the provision of raw materials for industry. This explains why domestic extraction and, in short, productive effort have concentrated on primary crops and, to a lesser extent, on forestry production. In comparative terms, the data show different metabolic profiles of per capita biomass consumption and trends throughout the twentieth century (Figures 1.5 and 1.6). While per capita consumption varies between 3 and 4 t in Spain (similar figures to the world average), consumption in Japan is between 1 and 2 while in the United States there is greater variation, between 6 and 10 t. These regional differences are similar to

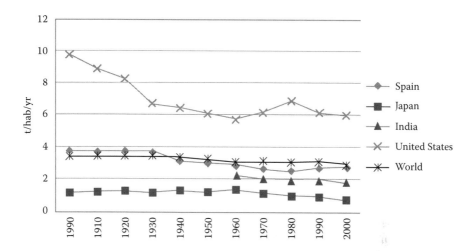

Figure 1.5 Role of biomass in social metabolism. Domestic extraction per capita of biomass in some countries in t/inhab/yr. (Spain [from Infante-Amate, J. et al., *J. Ind. Ecol.*, 19(5), 866–876, 2015]; Japan [from Krausmann, F. et al., *J. Ind. Ecol.*, 15(6), 877–892, 2011]; India [from Singh, S.J. et al., *Ecol. Econ.*, 76, 60–69, 2012]; United States [from Gierlinger, S. and Krausmann, F., *J. Ind. Ecol.*, 16(3), 365–377, 2012]; and World [from Krausmann, F. et al., *J. Land Use Sci.*, 4, 15–33, 2009].)

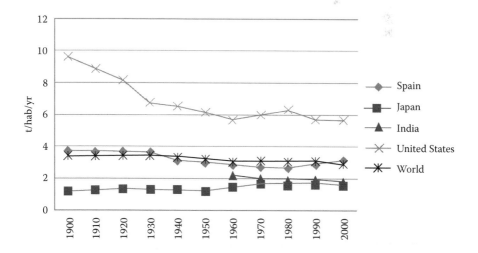

Figure 1.6 Role of biomass in social metabolism. Domestic material consumption per capita of biomass in some countries in t/inhab/yr. (Spain [from Infante-Amate, J. et al., *J. Ind. Ecol.*, 19(5),866–876, 2015]; Japan [from Krausmann, F. et al., *J. Ind. Ecol.*, 15(6), 877–892, 2011]; India [from Singh, S.J. et al., *Ecol. Econ.*, 76, 60–69, 2012]; United States [from Gierlinger, S., Krausmann, F., *J. Ind. Ecol.*, 16(3), 365–377, 2012]; and World [from Krausmann, F. et al., *J. Land Use Sci.*, 4, 15–33, 2009].)

those detected in current studies of biomass consumption levels on a regional scale. Among the explanations put forward are the availability of land, the productivity of the land, livestock and population density, trade, and income (Krausmann et al., 2008a). However, the data also demonstrate the growing importance of international trade in domestic consumption patterns in recent decades. Many developed countries, net biomass importers, have increasingly sustained their consumption thanks to trade. The case of Japan is even more evident, as it has increased its per capita domestic material consumption (DMC) for biomass, even though it started from very low consumption levels. In effect, an increasingly significant portion of biomass consumption since 1970 has taken place through imports, and so there is a progressive decoupling of production and biomass consumption (Wurtenberger et al., 2006; Erb et al., 2009; Witzke and Noleppa, 2010; Dittirch and Bringezu, 2010; Dittrich et al., 2012; Lassaletta et al., 2013). This means that the pressure on the agroecosystems of developed countries has been partly transferred to other agroecosystems. These changes have been possible as a result of factors that are exogenous to the agroecosystems and, fundamentally, as a result of the growing application of abiotic inputs into agrarian production that came with the industrialization of agriculture: fossil fuels for machinery and irrigation, chemical fertilizers, and plant health products (Infante et al., 2014a).

But perhaps the most decisive change, owing to its impact on the species itself, has been the change in diet. Rich countries increasingly consume more meat and livestock products such as milk and its derivatives, causing livestock numbers to grow to surprising levels. To feed these animals, land has been taken away from growing food for human consumption, or part of it has been dedicated to growing feed to fatten livestock. According to Krausmann et al. (2008a, p. 471), the global appropriation of land biomass in the year 2000 reached 18,700 million tons of dry matter per year, 16% of the world's net primary productivity of which 6,600 million were indirect flows. Of this amount, only 12% of the vegetable biomass went directly on human food; 58% was used to feed livestock; a further 20% as raw material for industry, and the remaining 10% continued to be used as fuel.

The importance acquired by importations of energy and materials has led agriculture to become partially decoupled from the agroecosystems that sustain it and its spatial configuration to become radically different, being based on simplified landscapes, single crops, the loss of spatial heterogeneity and biodiversity. Basic functions that in another time were fulfilled by the land (production of fuels, food for livestock, basic foodstuffs for the human diet, etc.), and to which a fairly large portion of land was dedicated, have disappeared, giving rise to a specialized landscape, peppered with constructions and areas used for urban-industrial properties (Agnoletti, 2006; Cussó et al., 2006; Tello et al., 2008; González de Molina and Guzmán Casado, 2006; Guzmán Casado and González de Molina, 2008).

CHAPTER 2

How to Measure Energy Efficiency in Agroecosystems

CONTENTS

2.1 INTRODUCTION

It may be asked whether all of the transformations that have been seen in farming have improved or, on the other hand, worsened energy efficiency in agrarian systems. This is an essential question in a panorama of growing economic and environmental difficulties facing agriculture, which is increasingly dependent on fossil fuel and under greater threat from the effects of climate change. The purpose of this chapter is to discuss the most appropriate means of measuring the energy efficiency of the management of agroecosystems. To do so, it is essential to change the traditional focus that confuses agroecosystems with cultivated land, leaving aside other spaces in the territory which are fundamental, and segment the energy flows within the territory to an extent that makes their analysis impossible. In this chapter, then, we first offer a description of how energy flows circulate through agroecosystems. As a combination of biotic and abiotic components, they show evolutionary features that bring thermodynamic peculiarities that must be studied to understand how they operate. Second, this chapter considers the reflection in the territory of the circulation of those energy flows, composing landscape arrangements in which the integration of different land uses is fundamental. This leads us to broadly characterize the very concept of an agroecosystem, where there may be different degrees of human intervention in the territory, with different productive functionalities. In accordance with this, a proposal is then made, adapting the methodology of material and energy flow accounting (MEFA) (Schandl et al., 2002), to the peculiarities of agroecosystems, measuring all of the energy flows that circulate within them on the basis of the breakdown into different categories of net primary productivity (NPP).

Having described the way in which energy flows within agroecosystems and how to quantify those flows, this chapter addresses how the efficiency of energy use should be measured. First, there is a brief review of the indicators that have been proposed to measure energy efficiency and, especially, energy return on investments (EROIs). There is then a discussion regarding the traditional focus that they have been given, very similar to that given in the economy to monetary investments in economic activities. While recognizing the usefulness of this kind of focus, a complementary method for measuring efficiency is proposed, bringing in a long-term perspective that, therefore, contemplates the sustainable functioning of agroecosystems, of their fund elements. Finally, several indicators of agroecological efficiency are presented and a formula is given to calculate them.

2.2 ENERGY THROUGH AGROECOSYSTEMS

All agroecosystems have fund elements whose reproduction and maintenance depend on their functioning correctly (Chapter 1). These funds are fed by energy flows in the form of biomass. The more complex and biodiverse the agroecosystem, the greater its capacity to host such flows within it. Ho (1998) long ago suggested that a system is more sustainable when it maximizes cyclical or circular flows of energy and minimizes dissipative flows, increasing the capacity to store energy and, therefore, the capacity to sustain the system, the number of cycles in the system, the efficiency of energy use and the space–time differentiation, expressed in levels of biodiversity, and so on, that is, minimizing the production of entropy. As it is well-known, ecosystems are an arrangement of biotic and abiotic components in which living systems with evolutionary thermodynamic specificities predominate. These specificities have been highlighted by the "thermodynamics of organized complexity," which represents progress over previous attempts to apply thermodynamics to living systems (Ho and Ulanowicz, 2005, pp. 41, 45). In accordance with this idea, which goes beyond those offered by Prigogine (1955), an ecosystem can be "far from thermodynamic equilibrium on account of the enormous amount of stored, coherent energy mobilized within the system, but also that this macroscopically nonequilibrium regime is made up of a nested dynamic structure that allows both equilibrium and nonequilibrium approximations to be simultaneously satisfied at different levels." This is possible thanks to the fact that it contains cycles or loops which, as held by Ulanowicz (1983), make "thermodynamic sense": "Cycles enable the activities to be coupled, or linked together, so that those yielding energy can transfer the energy directly to those requiring energy, and the direction can be reversed when the need arises. These symmetrical, reciprocal relationships are most important for sustaining the system" (Ho and Ulanowicz, 2005, p. 43). These cycles allow the entropy generated in one part of the ecosystem to be compensated by the negative entropy generated in another over a certain period of time (Figure 2.1). What is really decisive for living systems is not just their capacity to capture energy and material flows that keep them far from thermodynamic equilibrium, but their capacity to store the energy that circulates within the system and to transfer it between the different components.

The same is true of agroecosystems. Unlike ecosystems, which still retain their capacity for self-maintenance, self-repair, and self-reproduction, agroecosystems are unstable and require external energy for their maintenance, repair, and reproduction (Toledo, 1993; Pimentel and Pimentel, 1979; Gliessman, 1998). This energy is added through a series of tasks or operations aiming to ensure the production of biomass over successive cultivation cycles, modifying the carbon and nutrient cycles, the water cycle, and biotic regulation mechanisms. From the thermodynamic perspective, we must consider agroecosystems as complex adaptive systems that dissipate energy to compensate for the law of entropy (Prigogine, 1978; Jørgensen and Fath, 2004). To do so, they exchange flows of energy and materials with their environment (Fath et al., 2004; Jørgensen et al., 2007; Swannack and Grant, 2008; Ulanowicz, 2004). As it is well-known, the sustainable management of an agroecosystem depends on its level of biodiversity, wealth of organic material, or appropriate replenishment of soil fertility, and so on, closing biogeochemical cycles on a local scale. This represents the cost as a significant part of the biomass generated that must recirculate in order to perform basic productive and reproductive functions of the agroecosystem: seeds, animal labor, organic soil matter, functional biodiversity, and so on. In accordance with the proposals of Ho and Ulanowicz (2005) and later of Ho (2013), the sustainability of agroecosystems, therefore, correlates positively with the quantity and quality of its internal loops or cycles and, to that extent, with the energy flows that circulate within it and whose function is to reproduce the fund elements.

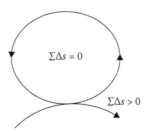

Figure 2.1 Dynamic balance of cyclic processes coupled to energy flows. (From Ho, M.-W. and Ulanowicz, R., *BioSystems,* 82, 45, 2005.)

It is generally accepted in agroecology that the more similar the organization and functioning of an agroecosystem is to a natural ecosystem, the more sustainable it will be (Gliessman, 1998; Guzmán et al., 2000). This is due to the fact that "the agroecosystem as a natural–anthropogenic system has its own biogeocenotic and biogeochemical mechanisms and self-regulation structures, which should be used to reduce anthropogenic energy costs" (Bulatkin, 2012, p. 732). The internal loops generate complex circuits that feedback in such a way that the outputs of some are the inputs of others, reducing the entropy of the system (Ho, 2013). To this extent, an agroecosystem with fund elements that require the dissipation of low levels of energy for its maintenance and reproduction by means of those recirculation processes, in turn generates low entropy in its environment and minimizes the flows of external energy. In effect, if the low-entropy energy required for the functioning of the system is provided by the available internal loops, then the external energy requirements will be lower and total entropy will fall. Systems that operate in this way are, undoubtedly, low-entropy systems that are much more durable and sustainable on a human scale. In contrast, when the internal complexity of an agroecosystem is substantially reduced, diminishing its internal loops, it needs to generate internal order through the import of significant amounts of energy. In these cases, total entropy also increases significantly, and we find ourselves before a high-entropy agroecosystem whose sustainability is seriously compromised.

In other words, the energy flows that enter agroecosystems are directly proportional to the degree of human intervention in those systems. When the intervention is minimal and generally respects the dynamics and functioning of the ecosystems (with a high density of internal loops), the imported or external flow of energy is also minimal. At the other extreme, when a complex ecosystem is simplified to the point that it hosts a monoculture, it must reduce diversity, limit interference, and modify the physical–chemical conditions to maintain optimum growth and the proper development of the crops. In this case, external energy flows must be increased significantly (Gliessman, 1998, p. 276).

But substitution is not always possible. From a thermodynamic perspective, an agroecosystem has a set of dissipative structures, which constitute its fund elements. Their reproduction requires a certain amount of energy in the form of biomass, which must be provided in each productive process. The energy required can only be partially replaced by external energy, given its varying nature. For example, the food chains that sustain both life in the soil and biodiversity, in general, within the agroecosystem can only be fed with biomass. The deterioration of a fund element cannot always, then, be replaced by external energy. For this reason, there is no optimum point that allows the net primary productivity of crops to be maximized through the supply of external energy. Substitution may allow the system to function, with a certain increase in total entropy and increasing commercial biomass, but this may be at the cost of not reproducing fund elements and, therefore, reducing the sustainability of the agroecosystem. In short, the maintenance of internal loops in agroecosystems is directly related to the use of a significant part of net primary production to fuel them.

2.3 INTERNAL LOOPS AND TERRITORIAL COSTS

The biomass production requires the appropriation of a certain amount of land for the purposes of photosynthesis. This piece of land can be more or less extensive according to the specific soil and climate conditions of each agroecosystem, the capacity of the plants used to harness solar energy, and the type of land management. Thus each way of organizing the agroecosystem requires a specific amount of land and, at the same time, leaves its distinctive physical imprint upon it, shaping specific landscapes (Guzmán and González de Molina, 2009; Guzmán et al., 2011). This cost is higher when the energy and material flows come from its own net primary productivity, something that used to occur in traditional farming and occurs partially today in organic farming. As domestic flows of energy and materials have been gradually replaced by imported flows, the land cost of modern farming has been reduced.

In any case, all production of biomass has a cost on land since the capture of solar incident energy by biological converters (photosynthesis) requires a piece of land. This cost has two components, one quantitative and the other qualitative. The quantitative dimension offers information regarding the amount of land needed to produce a specific quantity of biomass, depending on the edaphic, climatic, and technological conditions at the time (*land requirement*), whereas the qualitative dimension (*land functionality*) refers to the way in which that amount of land should be organized. It is not enough to simply have a certain amount of land; it is essential to give it structure, organizing the different components to fulfill their tasks. Each metabolic arrangement configures a particular landscape structure that conditions the ecological processes (energy and material flows, natural population regulation, etc.) in the agroecosystem. Landscape ecologists have used the term "functional landscape" to summarize the effects of landscape structure (spatial and temporal configuration) on ecological processes (Poiani et al., 2000; Adriaensen et al., 2003; Murphy and Lovett-Doust, 2004). So the functional land of (or forming part of) an agroecosystem is considered as that which possesses the necessary structure to sustain ecological processes (energy and material flows, and regulation of pests and diseases), within appropriate limits of variability, thus doting the whole agroecosystem with high levels of resilience and acceptable levels of productivity; in other words, giving it sustainability.

The complexity of the ecological processes is related to the density and connectivity of the internal loops in an agroecosystem and their capacity to store energy and feed the food chains, thereby sustaining biodiversity. This relationship between the energy available in an agroecosystem and the level of biodiversity was noted by Gaston (2000) and has served as the basis on which Ho and Ulanowicz suggested the existence of a strong relationship between the productivity of agroecosystems (measured, of course, in terms of total biomass, not just the marketable part of the crops) and biodiversity levels (Ho and Ulanowicz, 2005, p. 48). Biodiversity expresses this link (complex food chains) between low entropy and dissipative structures: some types of biomass feed others and vice versa. As we shall see in the following text, the measurement of energy efficiency through indicators such as EROIs can, if done appropriately, reflect the density and interconnection of these internal cycles and, therefore, indicate whether they are low-entropy dissipative structures, both with regard to imports and outputs, that is, whether they are more or less sustainable.

In accordance with this, each specific arrangement of the agroecosystem has a cost in terms of the territory, depending on the complexity and connectivity of the energy flows that maintain and reproduce its fund elements, that is, the complexity and connectivity of its internal loops. To that extent, each specific arrangement of the agroecosystem is reflected in a specific organization of the landscape, imposing its *particular footprint* on the territory. For example, in organic or agrarian metabolic regimes (González de Molina and Toledo, 2011, 2014), agroecosystems function in an integrated manner in such a way that the internal loops clearly extend beyond the cultivated land and cover wide stretches of the territory.

The additional energy input that allowed preindustrial farming to function had to come from biological sources: human labor and animal labor, which in turn depended on the capacity of the agroecosystem to produce biomass (Gliessman, 1998) and therefore on the amount of land available. A strict dependence was thus maintained on the land availability and on edaphic and climatic conditions (Sieferle, 2001; Toledo and González de Molina, 2007). The vast majority of energy and materials came from domestic extraction (DE) and very little was imported, since means of transport were as yet relatively underdeveloped. In other words, the virtual impossibility of importing significant amounts of external energy into managed ecosystems meant that internal needs and external demand had to be met from the territory available, fragmenting it for alternative uses. For this reason, agroecosystems had to maintain a strict balance between the different uses of the land. The increase in entropy that came about in the more intensely cultivated areas was usually compensated by the import of nutrients, generally, not only through livestock (manure), but also from other low-entropy areas such as pastureland or woodland. The result was a metabolic regime that also showed low entropy. The heterogenous space and agrosilvopastoral integration were the keys to the structuring of the different loops that captured, stored, and transferred energy.

In the Mediterranean world, for example, with its scarce rainfall and high temperatures, farmland was given over to human foodstuffs or the production of fibers and other raw materials. Pastureland was allocated to feeding animals and, finally, forested land to the production of fuel (firewood) and building materials (timber). When one of these uses failed to produce enough to meet demands, attempts would be made to compensate with others. For example, when the stock of working animals grew and exceeded the pastureland's capacity to feed it, agricultural areas had to allocate part of their production to cereals and leguminous crops for animal feed.

The three main alternative uses of the territory could certainly be found together on a single farm, combining different crops and activities (e.g., agroforestry systems), but their viability depended on the edaphic and climatic conditions in each ecosystem and its productive capacity. In climates where primary production was low, due to the lack of rainfall or nutrients, the territorial costs of the production of biomass were greater than in areas where these factors were abundant. In some dry, semiarid, and arid regions that suffered water shortages, land uses could even compete with each other and be practically exclusive, making a high consumption of territory inevitable. The useful crop area was therefore divided, in line with the activities it supported, into agricultural, livestock, and forestry land, whose degree of incompatibility depended on the aptitudes

Photograph 2.1 Agrosilvopastoral integration in the territory allows for the strengthening the internal energy loops and is expressed as a complex landscape matrix.

of the agroecosystem. Even in semidesert or arid areas, where natural productivity was low and the territory being appropriated had to be very extensive, the best option was pastoralism and nomadism (Giampietro et al., 1997, p. 155). In short, the distribution of different land uses in the territory, that is, its spatial heterogeneity, was a way of imitating the dynamics of natural ecosystems and thereby achieving maximum stability.

In contrast with this way of working, industrialized agriculture has made savings in terms of land due to the injection of growing quantities of energy and nutrients from fossil and mineral sources, mainly taken from outside the agroecosystems. The integration of forestland, pastureland, and diverse agricultural uses, which in the past ensured the diversity required for the stability of agroecosystems, has been lost and moreover, many uses of the land have been sacrificed to expand monocultures or to use exclusively for livestock. Agrarian diversity has deteriorated significantly. In this regard, the landscapes of industrialized agriculture are simplified to the same extent that the internal loops within their agroecosystems are reduced. They therefore constituted high-entropy dissipative structures. The result of all this is a considerable loss of sustainability.

The territorial arrangement of solar energy-based agriculture has changed over time and its land cost has been modified as a function of numerous variables (the supply of land, available technology, requirements of the population, etc.). In other words, the correct design of the internal loops in an agroecosystem can appreciably reduce the territorial cost that all biomass production involves, generating more biomass at a minimum cost in terms of territory. A clear example of this is the polyculture developed by traditional agriculture, whose success was based precisely on their ability to reduce the land cost (land equivalent ratio [LER]) (Gliessman, 1998; Vandermeer, 1990). This is due to the fact that the relationship between the two dimensions of land cost or biomass production, its land requirement and its land functionality, is not necessarily a direct one. When land takes on ecological functions, there is not always a parallel increase in land cost. With the correct management of agroecosystems, the land can perform the same functions, or more, without increasing the land cost. This has occurred on occasions in traditional agriculture (Guzmán and González de Molina, 2009); and currently occurs in organic production (Guzmán et al., 2011).

2.4 METABOLIC PROCESS OF APPROPRIATION: A WIDE DEFINITION OF AGROECOSYSTEM

In accordance with the previous discussion, the production of biomass requires the colonization of the ecosystems and the appropriation of part of its net primary production (Haberl et al., 2007). This process, which is central to the metabolic relationship in agriculture, could be defined as "the process by which the members of all societies appropriate and transform ecosystems in order to satisfy their needs and desires" (Cook, 1973). During this appropriation process, humans undertake three basic types of intervention on the territory, directly affecting ecosystems (Figure 2.2). The first

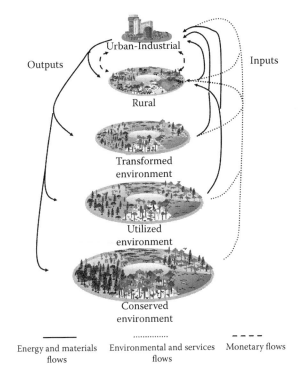

Figure 2.2 Different levels of human intervention in ecosystems. (From González de Molina, M. and Toledo, V.M., *Social Metabolisms: A Theory on Socio-Ecological Transformations,* Springer, New York, 2014. With permission.)

does not cause substantial changes in the structure, architecture, or dynamics of the ecosystem. It includes all of the hunting, fishing, and gathering activities, as well as some forms of extraction and livestock farming by foraging in the original vegetation. The second type of appropriation is the disarticulation or disorganization of the ecosystems to introduce species that have been or are being domesticated, as occurs in all forms of agriculture, livestock farming, silviculture, and aquaculture. In the first type of appropriation, the intrinsic or natural capacity of ecosystems to maintain, repair, and reproduce themselves is not affected, as we have already said. In contrast, in the second type of appropriation, ecosystems lose these abilities and require external human, animal or fossil energy, materials, and information to be maintained. In recent decades, a third form of appropriation has emerged: the conservation actions of public administrations or nongovernmental organizations. It seeks the preservation or protection of natural areas or areas undergoing regeneration processes and the provision of ecosystemic services. This latter type of appropriation, which is increasingly frequent, has been described by Toledo (1993) and called the "conserved environment," but we could expand it to cover those parts of the territory in which there is no apparent human intervention, that is, those parts of the territory that are not managed directly through farm working or indirectly through livestock or the improvement of pastureland—parts of the territory, which, for example, have been abandoned and are found very often today in developed countries.

The agroecosystems have been defined as ecosystems manipulated and artificialized by human activity to capture and transform solar energy into a specific form of biomass that can be used as food, medicine, fiber, or fuel (Margalef, 1979; Altieri, 1989). An agroecosystem is, then, that piece of nature, which can be reduced to a single unit with its own architecture, composition, and functioning and which possesses a recognizable theoretical limit, from an agronomic perspective, for its adequate appropriation by humans. By this, we are referring to the specific articulation given by humans with respect to the natural resources: water, land, solar energy, plant species, and the rest of the animal species. They are often confused with the farm, that is, with the crop area, and only in agroforest systems are other noncultivated silvopastoral spaces considered. However, agroecosystems are coherent units through which biogeochemical flows circulate, with human appropriation thus giving rise to different degrees of intervention (Guzmán et al., 2000; González de Molina and Toledo, 2011, 2014). This requires the colonization of specific ecosystems and the appropriation of part of net primary production (Haberl et al., 2007). The same can be said of the plants that inhabit agroecosystems, since very often only cultivated plants are taken into account and, among these, the aerial part of the plants, while root biomass, crop residues, and weeds are often ignored. An agroecological approach should take into account all the biomass produced within the limits of the agroecosystem, that is, the net primary productivity. The reason for this is that the reproduction of the fund elements of agroecosystems depends directly or indirectly on the total biomass produced, not just that which is harvested.

In coherence with the earlier discussion, we adopt here a broad definition of *agroecosystem*, recognizing that an agroecosystem can also include simply appropriated spaces where the level of manipulation or intervention is minimal but which form an indissoluble part of the territorial arrangement. Different units of biomass appropriation can cohabit in it, some by hunting and gathering and others by the management of plants. This is more evident when we look beyond the area of the plot and take in the territorial arrangements that must necessarily make up an agroecosystem from the point of view of different land uses. Consequently, the supply of colonized land used by society to develop its agrarian metabolism usually becomes fragmented into different categories or land uses: cropland, pastureland, and forestland, depending on the degree and type of work that is done in each one. All of these can be subdivided, in turn, into different categories, depending on their specific or multiple exploitations and the intensity of the farmwork done on them. This characteristic, to which we shall return later, reflects the functioning of the living beings and is an essential element for sustainability. In short, an agroecosystem can contain very heterogenous units of landscape or territorial arrangements with different degrees of artificialization.

But agroecosystems not only provide biomass to meet the endosomatic metabolism of humans and livestock. As ecosystems, also provide ecosystem services. The fund elements of agroecosystems generate flows of ecosystem services, part of which are used for their own renovation (de Groot et al., 2003; Ekins et al., 2003; Millennium Ecosystem Assessment, 2005; Folke et al., 2011). According to Schröter et al. (2014), every agroecosystem has a specific capacity to provide these services, depending on soil and climate conditions. Since agroecosystems are dependent on human management, the quantity and quality of fund elements and the rate at which they provide services depends on how they are managed. An adequate provision of services will depend on the state of health of the agroecosystem, that is to say, on the sustainability of its fund elements (Cornell, 2010; Costanza, 2012). Conversely, the degradation of the fund elements of an agroecosystem can lead to the reduction of ecosystem service supply (Burkhard et al., 2011) (Table 2.1).

These services are usually grouped into four categories: supply, regulation, support, and cultural services. Supply includes the extraction of goods (timber, firewood, foodstuffs, and fibers); regulating services help to modulate ecosystem processes (carbon sequestration, climate regulation, pest and disease control, and waste recycling); support services contribute to the provision of all the other categories (photosynthesis, soil formation, and nutrient recycling); while cultural services contribute to spiritual well-being (recreation, religion, spiritual, and aesthetic values) (de Groot et al., 2010).

Table 2.1 Environmental Services of Agroecosystems

Supply	Regulation	Auxiliary Services
Products Obtained	Benefits Obtained	Services Necessary for the Production of the Others
Food, freshwater, firewood, fiber, biochemical products, genetic resources, so on	Climate regulation, carbon sequestration, disease control, regulation of water, purification of water, pollination, etc.	Soil formation, nutrient cycle, primary production, biodiversity, etc.

Source: Adapted from FAO (SOFA, 2007).

Photograph 2.2 Sugarcane cultivation in the north-east region of Brazil. The systematic burning of aboveground biomass generates a strong degradation of fund elements in these industrialized agroecosystems.

2.5 METABOLIC APPROACH TO ENERGY ANALYSIS OF AGROECOSYSTEMS

As stated earlier, the limits of the metabolic analysis of agrarian activity are identified with those of the agrarian sector itself and they therefore leave out all of the processes (transformation, distribution, consumption, etc.), which take place after harvesting. Our calculations end, then, at the farm gate. Our main challenge, though, is the application of metabolic analysis to agrarian activity, adapting the usual methodology of the social metabolism approach and, specifically, that of MEFA (Schandl et al., 2002; Haberl et al., 2004) to the energy analysis of agroecosystems. This methodology describes and quantifies the energy flows that enter and leave the agroecosystem, but do not reveal their function within the system. The most that it allows us to ascertain is whether a part of those flows accumulates and maintains a given compartment of the system. In accordance with MEFA methodology, the most important stocks or compartments of the social metabolism are the population, the built environment (buildings and infrastructure that only dissipate energy and materials in their maintenance, not in their operation), livestock, and other domestic animals. The environment as such is not considered a stock, since it is no more than the opposite pole of the metabolic exchange. In this way, its dynamics and functioning remain outside the socioecological relationship. From an agroecological point of view, agroecosystems are, however, anthropized ecosystems and, therefore, they constitute the center of metabolic activity, the concrete expression of the exchange of energy and materials between society and its agrarian environment. They are, consequently, the expression of the socioecological relationship in agriculture.

It is crucial, then, to ascertain and assess the role that energy flows play in sustaining and reproducing the fund elements of agroecosystems. Apparently, the basis for an analysis of this type is given in the MEFA methodology itself (Figure 2.3), but this methodological proposal has still not developed this aspect of social metabolism. As far as we are aware, there have been only a few contributions that attempt to quantify the variations seen in stocks (e.g., in infrastructure) in societies with an industrial metabolism (Fishman et al., 2014). However, this methodological approach does not usually take into account the differentiated role, from the thermodynamic point of view, of some stocks with respect to others, or the identity and quality of the flows that feed them. This has rightly been criticized by Giampietro et al. (2014) and it is particularly serious when we speak of agrarian metabolism and agroecosystems, where part of those stocks are, in reality, fund elements, made up of a specific combination of living beings in interaction with their abiotic environment.

The adaptation of metabolic methodologies to the energy analysis of agroecosystems should, in consequence with the aforementioned, distinguish between stocks and fund elements. Stocks do not necessarily require a continuous flow of energy

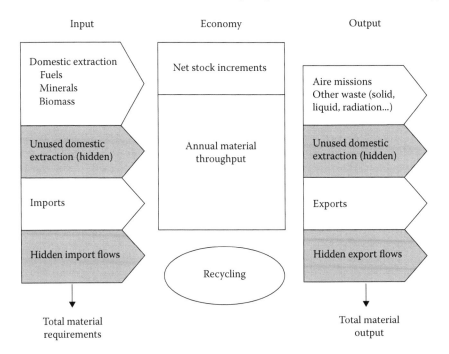

Figure 2.3 Material and energy flow accounting (MEFA). (Adapted from Schandl, H. et al., *Handbook of Physical Accounting. Measuring bio-physical dimensions of socio-economic activities MFA–EFA–HANPP, Social Ecology Working Paper 73*, Institute for Interdisciplinary Studies of Austrian Universities [IFF], Vienna, 2002.)

and materials for their reproduction; they can provide identical services until they are completely exhausted, while the deterioration of the fund elements affects their capacity to provide services from the very first moment, that is, the capacity to reverse the entropic process, generating order, or their own status or identity as a dissipative structure. The reproduction of a fund element is a process in which certain amounts of energy and materials are invested and these must be deducted from the flows that are invested in their own functioning. Consequently, it is necessary to distinguish between productive flows that drive the functioning of the fund elements and reproductive flows that are crucial for their maintenance over time. This allows us to introduce a key factor for environmental analysis, sustainability; that is, whether a given dissipative structure can maintain itself indefinitely over time. This distinction or breakdown of energy and material flows into two parts or sections, productive and reproductive, is a distinctive feature of our proposal and the central objective of the energy analysis. In other words, whether the energy and material flows that circulate in agroecosystems are able to reproduce, that is, maintain and even improve the fund elements that they contain.

In short, MEFA methodology only takes into account the entry and exit flows, and it is impossible to appreciate the decisive processes that take place within agroecosystems, which are like black boxes. In contrast, our proposed methodology, which is also based on the MEFA proposal, takes into account the flows that circulate within agroecosystems, between some fund elements and others, and whose importance is decisive to the long-term maintenance of their structure and functioning. These internal flows must be appropriately characterized and quantified. In fact, as we have seen, it is the interrelationship between these flows and the fund elements that explains the metabolic dynamics and the degree of sustainability, as will be seen in the following section.

2.6 BREAKING DOWN THE BIOMASS FLOWS IN AGROECOSYSTEMS

As mentioned earlier, our proposal incorporates the distinction between flows and funds into MEFA methodology, that is, the existence of fund elements in agroecosystems that require energy flows for their maintenance and reproduction. These energy flows, as we have also seen, are flows of biomass whose substitution by flows of external energy in different forms is not always possible. Thus, it is useful to know the destination of the biomass flows that circulate within agroecosystems. To this end, it is necessary to break down the net primary productivity into the different categories of the final destination of the biomass flow. The study of these flows and their final destination allows us to ascertain whether or not they are of sufficient quantity and quality for the functioning, maintenance, and reproduction of the fund elements of the agroecosystem.

The metabolic relationship between a society and its agrarian environment is expressed through the exchange of energy, material, and information (see Figure 1.1), or, in energy terms, it is expressed through the appropriation of certain amounts of low-entropy energy from the environment and the return to the environment of dissipated energy. Agroecosystems are at the heart of this exchange. Biomass is extracted from the agroecosystem and part of that biomass is socialized to satisfy human requirements for food, raw materials, and fuel, while the other part is recycled (animal feed, seeds, or the reproduction of the wild food chains) or simply accumulated. The metabolic accounts can, therefore, be broken down into the input of energy from outside the agroecosystem, the extraction of biomass within its boundaries, and the output of biomass destined for use by society. But let us examine this in a little more detail.

2.6.1 The Input Side

On the input side, we must first consider energy *Imports* (I), which enter the agrarian metabolism, measured in MJ/year. Both the amount of imported biomass and of other types of material (fossil fuels, metallic and nonmetallic minerals, and construction materials) and the capital goods and fuel used by agriculture must be included under this item, whether they originate from abroad or from within the country, but from another sector (any nonagrarian sector) of the economy. It is useful to distinguish these flows by their nature, that is, whether they are biotic or abiotic.

Imported biomass must be included among the biotic ones, that is, the biomass that is not produced in the agroecosystem being studied and that is destined for livestock; the germplasm that is inputted into the agroecosystem or the imported organic fertilizer. Among the abiotic ones, we should include the imported inorganic inputs necessary for production. These include chemical fertilizers, machinery, plant health products, and so on. These imports incorporate energy costs (embodied energy), which must be taken into account not to distort the real energy costs of imports. We shall discuss in detail of these in Chapter 4.

2.6.2 Inside Agroecosystems

The appropriation of biomass, or the *domestic extraction* of biomass, occurs within the agroecosystem. In accordance with the foregoing, it is an error from an agroecological point of view to consider only the extracted biomass, leaving aside the unharvested and the accumulated biomass (*AB*). In the same way, taking the extracted biomass as a whole, without breaking it down into its different functions and ends, is also an error as it conceals the internal functioning of the agroecosystem and its capacity to maintain itself over time. To express the different basic functions that biomass performs in the reproduction of the fund elements of agroecosystems, it is useful to divide it into different categories, distinguishing mainly between their use by humans, animals, or the agroecosystem. We therefore believe that it is necessary to break down the net primary productivity (*NPP*) of agroecosystems into several different categories (Figure 2.4).

Socialized vegetable biomass (*SVB*): This is the vegetable biomass (timber, firewood, cereal grain, olives, etc.) that is directly appropriated by human society, considered as it is extracted from the agroecosystem, that is, prior to its industrial processing, if any (transformation into flour, oil, etc.). In this way, we avoid the effect of changes in agroindustrial efficiency on the comparison of the agroecosystems studied.

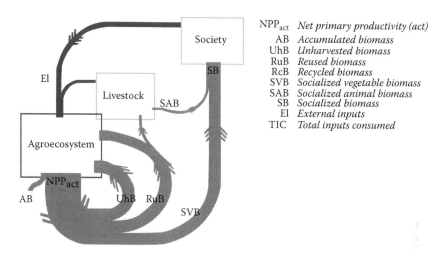

NPP_{act}	*Net primary productivity (act)*
AB	*Accumulated biomass*
UhB	*Unharvested biomass*
RuB	*Reused biomass*
RcB	*Recycled biomass*
SVB	*Socialized vegetable biomass*
SAB	*Socialized animal biomass*
SB	*Socialized biomass*
EI	*External inputs*
TIC	*Total inputs consumed*

Figure 2.4 Biomass flows in agroecosystems. (Author data.)

In the same way, socialized animal biomass (*SAB*) is the animal biomass at farm gate (animal, milk, wool, eggs, etc.) that is appropriated directly by society, considered as it is extracted from the livestock, before industrial processing, if any. The sum of *SVB* and *SAB* gives the socialized biomass (*SB*), which is the total biomass appropriated by society.

The concept of socialized biomass does not imply the existence of an economic exchange in monetary terms. That is to say, socialized biomass includes all of the biomass (food, fiber, timber, firewood, etc.) that is self-consumed or exchanged by barter. There may also be biomass outputs from an agroecosystem involving monetary exchange but not considered socialized biomass. This would be the case of that biomass, which leaves the agroecosystem but which is not destined for society, but to sustain the functions of another agroecosystem. For example, hay sold as feed for the livestock of another producer, the sale of working animals, and so on.

Recycled biomass (*RcB*) is the biomass that is recycled through the agroecosystem, whether intentionally or not. The seeds and reproductive organs of plants form part of this recycled biomass. However, most of the recycled biomass is that which is recycled by livestock or through wild heterotrophic organisms. The recycled biomass can, in turn, be divided into two portions from the perspective of society: (1) *Reused biomass* (*RuB*) is that part which is returned to the agroecosystem intentionally by farmer. This means that the reincorporation into the agroecosystem of this plant biomass is done through human labor and has a agronomic purpose that is recognized by farmer, for example, to obtain a product or a service (animal feed for the supply of meat or milk). This category includes the biomass that is destroyed by fire (e.g., stubble burning) since it involves conscious work and has an agronomic purpose, and (2) *Unharvested biomass* (*UhB*), which is that part that is simply abandoned and allowed to return to the agroecosystem, for no specific purpose. Its return to the system does not involve any human labor. This would be the case of crop residue that is neither used nor burned, the portion of pasture that is not consumed by the livestock, and the woodland waste and most of the root systems that are not harvested by society and that are recycled by wild heterotrophic organisms. This is the logical consequence of taking into account, as proposed earlier, both the root and the aerial biomass. The *UhB* can be divided into aboveground unharvested biomass (AUhB) and belowground unharvested biomass (BUhB), depending on the location of this biomass when it is abandoned.

Accumulated biomass (AB) in agroecosystems with perennial species—in addition to the biomass extracted by society and that recirculates every year, there is another portion of biomass that accumulates annually in the aerial structure (stem and crown) and roots.

Accordingly, the actual NPP (NPP$_{act}$) would be the result of the sum of the vegetable biomass appropriated directly by society (socialized vegetable biomass), the biomass that is recycled through the agroecosystem, whether by intentional reincorporation (reused biomass) or by simple abandonment (unharvested biomass), and the biomass that accumulates annually (accumulated biomass) in the aerial structure (stem and crown) and roots of perennial species on pastureland, forestland, and cropland. In this way, the domestic extraction of the agroecosystems would be the equivalent of the sum of the socialized vegetable biomass and the reused biomass, obtained through the intentional management of the agroecosystems.

$$NPP_{act} = SVB + RuB + UhB + AB$$

$$SB = SVB + SAB$$

$$RcB = RuB + UhB$$

$$ED = SVB + RuB$$

2.6.3 The Output Side

In this section, we consider *exports* (*E*), also measured as gross energy contained in the biomass exported each year. That is to say, the exports of the agrarian metabolism are the total amount of plant and animal biomass that crosses the farm gate and is destined for human consumption, the agrifood industry and industry in general.

$$E = BS = SVB + SAB$$

2.7 EROIs: BEYOND THE ENERGY EFFICIENCY

Energy return on investments (EROIs) are indicators (Gupta and Hall, 2011, p. 28; Pervanchon et al., 2002, p. 150) that aim to measure the efficiency of energy use and, in doing so, provide information about decision making on this vital aspect of the operation of productive activities. This important tool for "energy analysis" or "net energy analysis" (Hall et al., 2009, p. 26) is strictly economic in origin and is based on the same valuation criteria as monetary investments, that is to say, on cost–benefit analysis. It has been used for some time for the conversion of oil and other primary energy sources into fuel and other energy products, when attempting to measure the efficiency of the process (Cleveland et al., 1984; Hall et al., 1986, 2008, 2009; Cleveland, 1992; Mulder and Hagens, 2008; Giampietro et al., 2010; Hall, 2011, pp. 2–3). It provides a numerical indicator that can be quickly and easily used for comparison with other similar energy processes in both space and time (Murphy et al., 2011, p. 8).

When applied to agriculture, it measures the amount of energy invested to obtain a unit of energy in the form of biomass. Put more simply, we could say that in agriculture an EROI measures the "energy cost" (Scheidel and Sorman, 2012, p. 3) of net biomass produced to be used by society (Martinez Alier, 2011), whether in the form of foodstuffs, raw materials, or biofuels. This indicator is particularly important in the context of the current energy crisis, especially in the context of industrialized agriculture that uses large amounts of external energy, both directly and indirectly, and is facing the challenge of reducing energy costs and greenhouse gas (GHG) emissions. Given that the endosomatic metabolism of people and the production of raw materials that are difficult to produce synthetically can only be satisfied by producing biomass, the efficiency of energy use in agriculture has become a basic issue (Tello et al., 2015, p. 9).

However, using EROI as the sole indicator can produce contradictory results, depending on the system boundaries chosen. Furthermore, it is a prime example of the need for multiple criteria in analysis. As highlighted by Giampietro et al. (2010), energy efficiency cannot be reduced to a single figure or a single criterion for analysis, especially when applied to agriculture. In addition to the advisability of using several EROIs, an analysis should be carried out from different perspectives to aid coherent decision making on energy use. This spirit has guided proposals made elsewhere for the use of several EROIs in the study of agriculture and its history (Tello et al., 2015). This social perspective addresses the profitability of the investment in energy for the production of net biomass for the farmer (microeconomic perspective) or for society as a whole (macroeconomic perspective).

2.8 EROIs FROM THE POINT OF VIEW OF SOCIETY: KEY INDICATORS AND MEANINGS

These EROIs inform us of the return on energy intentionally invested by society in agroecosystems. In traditional agroecosystems, the energy investment was fundamentally the energy of the RuB, while the external inputs (EI) invested were minimal. The EI include human labor, as well as all of the inputs (fertilizer, pesticides, machinery, plastic, feed, etc.) that originate outside the agroecosystem. In industrialized agriculture, the EI have shown strong growth, which would indicate a parallel reduction in investment in RuB which, apparently, was now not so necessary for the functioning of the agroecosystem.

The proposed EROIs, from this point of view are as follows:

$$\text{Final EROI (FEROI)} = SB/(RuB + EI)$$

where socialized biomass (SB) $= SVB + SAB$.

This explains the return on the energy investment made by society. SB is a net supply of energy carriers able to be consumed by the local population or for use in other socioeconomic systems (Fluck and Baird, 1980; Pracha and Volk, 2011). It should be noted that this return is not strictly related to the productive capacity of the agroecosystem. For example, two agroecosystems with the same productivity (NPP_{act} and SVB) and similar external energy investment can give rise to a different *FEROI*, depending on the amount of biomass used as animal feed. In this regard, the diet of society has a strong impact on this EROI, as does the need for animal traction or manure, due to the low efficiency of livestock as an energy converter. The case studies analyzed in the second part of the book (Chapters 5 and 6) will give examples of this impact.

The FEROI can be broken down, in turn, into two elements: *external final EROI* and *internal final EROI* (Tello et al., 2015).

$$\text{External final EROI (EFEROI)} = SB/EI$$

EFEROI relates *EI* to the final output crossing the agroecosystem boundaries (Carpintero and Naredo, 2006; Pracha and Volk, 2011). In the academic literature, it has frequently been called "net efficiency," and it is one of the indicators most commonly used to evaluate agriculture from the energy perspective (Guzmán and Alonso, 2008). This ratio links the agrarian sector with the rest of the energy system of a society—and thus assesses to what extent the agroecosystem analyzed becomes a net provider or rather a net consumer of energy.

$$\text{Internal final EROI (IFEROI)} = SB/RuB$$

This explains the efficiency with which the biomass that is intentionally returned to the agroecosystem is transformed into a product that is useful to society. This indicator has not habitually been used but its usefulness is growing since this biomass can have alternative uses (e.g., biofuels), since poor management can cause environmental problems (e.g., pig slurry pollution) or due to the ecosystem services it can provide (e.g., soil carbon sequestration).

The interrelationship among FEROI, EFEROI, and IFEROI is interesting as well as controversial. Tello et al. (2015) developed a mathematical expression of this relationship and suggested that potential improvements are higher if FEROI is lower and/or when the combination of EFEROI and IFEROI is unbalanced—that is, when the EI:RuB ratio is far from one. If this were so, an increase would be seen in FEROI at the beginning of the modernization process and also today, when industrialized agriculture becomes organic, on substituting part of the EI with internal biomass flows. However, this article also warns that "the function relating SB [called "final product" in that article] with RuB and EI is too complex to be determined, due to in the agroecosystems any internal or external biophysical flow interacts with a set of funds, which can only bring about a socialized biomass within a limited range of variation in yields and in a discontinuous manner. What really matters are the emerging properties arising out of the whole network of synergistic links of flows established among a myriad of fund components of subsystems working together to attain a joint outcome—and that is the main focus of agroecology as a science." The case studies analyzed in this book allow us to offer some conclusions with respect to this debate.

Other EROIs developed from an economic point of view are the *fossil final EROIs* (*SB/fossil energy*) or the return on investment for other factors of production, such as labor. Since they are widely used, they are not applied to the case studies included in this book and therefore they will not be explained in this chapter.

"Economic" EROI indicators offer a measurement of efficiency only in the short term. These EROIs show whether society receives a sufficient flow of useful biomass in exchange for the flow of energy invested, but say nothing about durability. From this point of view, it is licit even to consider whether it is convenient to maximize the flow of socialized biomass in relation to the energy invested, even though that would mean the incorporation of a certain amount of external energy. In such a case, an "optimum" balance could be sought between external and internal inputs that would allow the maximization of socialized biomass and raise the final EROI as proposed by Tello et al. (2015). However, the use of this type of "economic" indicator does not allow us to ascertain whether this increase in efficiency has been achieved at the cost of the deterioration of the fund elements of the agroecosystem, reducing the medium- and long-term supply of useful biomass to society. In other words, it is of interest not only if society receives an adequate flow to maintain and reproduce itself, but also if it can do so indefinitely, in line with the concept of the (sustainable) economy, as stated by Georgescu-Roegen. This focus is also economic, but it is long term and is the same type of reasoning as that applied in ecological economics when it calls for a focus on efficiency, which goes beyond a simple cost–benefit analysis (typical of a short-term focus), and which looks at the long term, that is, that focuses on sustainability. For this reason, this book highlights the need to combine these "economic" EROIs with others of an agroecological nature.

2.9 EROIs FROM AN AGROECOLOGICAL POINT OF VIEW: KEY INDICATORS AND MEANINGS

As we have seen in Sections 2.7 and 2.8, the EROIs have traditionally been calculated on the sole basis of supply services (Pérez-Soba et al., 2015, p. 6), ignoring the fact that regulation and maintenance services are essential for supply services to survive over time. They are not maintained and they do not reproduce themselves independently, but through human intervention. This book considers EROIs that go beyond the social benefits offered by increasing investment of energy in agriculture. This requires us to recognize that not only is it necessary to invest energy in the production of biomass useful to society or to farmers, but also to invest energy in maintaining the agroecosystem so that it can continue to produce biomass under the best possible conditions. EROIs can, in effect, become a measurement of metabolic efficiency, that is to say, of the exchange of energy between agrarian systems and the environment, to establish whether this metabolic exchange is sustainable over time (Schramski et al., 2013). In this sense, we have to consider not only the energy cost of the production of socially useful biomass, but also the maintenance cost of the ecosystem services provided by an agroecosystem: this cost does not end with the reuse of seeds, green manure, or the production of animal feed (which corresponds only to the supply services provided by agroecosystems), but also extends to the maintenance of the remaining ecosystem services (Cornell, 2010; Costanza, 2012; Burkhard et al., 2011, 2012; a review in Häyhää and Franzese, 2014). It is therefore necessary to adopt an agroecological focus. As in the case of ecosystems, maintenance and reproduction depend on an adequate supply of energy in the form of biomass.

Therefore, the objective of agroecological EROIs is to ascertain whether a given agroecosystem is capable of maintaining its biomass production and ecosystem services or whether it degrades them, requiring increasing amounts of external energy to compensate for the loss only partially. As we have seen in Section 2.2, low-entropy systems can be obtained by recirculating energy within the agroecosystem, since the agroecosystems in which internal recirculation processes have been simplified require large amounts of external energy, and are thereby converted into highly entropic agroecosystems.

As it is known, whether an agroecosystem is more or less sustainable depends on its level of biodiversity, its wealth of organic material, on whether fertility replenishment is performed on the scale of the agroecosystem, and so on. Therefore, most biogeochemical cycles are closed on a local scale. This implies an internal energy cost (generation of biomass) or a territorial cost, understood as the functional land in an agroecosystem that possesses the necessary structure to sustain ecological processes (energy and material flows, and pest and disease regulation) within appropriate limits of variability, thus making the whole agroecosystem highly resilient with acceptable levels of productivity (Guzmán and González de Molina, 2009; Guzmán et al., 2011). This represents the cost as a significant part of the biomass generated that must recirculate to perform basic productive and reproductive functions of the agroecosystem: seeds, animal labor, organic soil matter, functional biodiversity, and so on.

So the fund elements of an agroecosystem require a specific amount of energy for reproduction and maintenance that can only partially be substituted by external energy. For instance, only biomass can feed the food chains that sustain both the life in the soil and the general biodiversity of the agroecosystem. In this regard, we should note again the idea expressed by ecological economists that natural capital cannot be replaced by manufactured capital (Ayres, 2007; Häyhää and Franzese, 2014, p. 125), in the same way that not all types of energy have the same use or are interchangeable (Giampietro et al., 2010). The fund elements of agroecosystems cannot be sustained by oil or coal or their fuel derivatives.

These peculiarities of the throughput of energy in agroecosystems can be captured by an EROI if it is designed following agroecological criteria, taking into account the flows of biomass used for the appropriate maintenance of the fund elements of the agroecosystem, that is to say, to subsidize the production of ecosystem services such as nutrient recycling, biological pest control, soil conservation, and so on. This task is performed not only by the reused biomass but also by the NPP as a whole. An EROI of this type could be a means of measuring the state of the agroecosystem and its sustainability (Murphy et al., 2011, p. 8). This sustainability means that an agroecosystem could provide an optimal level of biomass production over time without deteriorating the basis of its fund elements while maintaining an optimal provision of ecosystem services. So, agroecological EROIs inform us of the real productivity of the agroecosystem, not just the part that is socialized. Furthermore, they inform us of the balance between the uses to which the biomass is put. The interests of society often center on socializing the greater part of the biomass produced. However, this use should be limited, since there should be a reinvestment in the fund elements, that is, in the structure of the agroecosystem (e.g., biodiversity, spatial heterogeneity, and the complexity of agroforest landscapes or soil quality) to sustain basic ecosystemic services. We propose four different EROI indicators that are the following:

$$NPP_{act} \; EROI = NPP_{act}/\text{total inputs consumed}$$

Total inputs consumed (*TIC*) being $= RcB + EI = RuB + UhB + EI$.

NPP_{act} EROI explains the real productive capacity of the agroecosystem, whatever the origin of the energy it receives (solar for the biomass or fossil for an important portion of the EI). We speak of "real productivity" because it considers all of the vegetable biomass produced, not just what is socialized, and because it is independent of other factors such as the transformation of the biomass through livestock farming that influences the FEROI. The degradation processes affecting natural resources, such as soil salinization or erosion, genetic erosion, and so on, must be compensated by the incorporation of increasing amounts of energy to palliate the loss of productive capacity of the agroecosystems. Falling NPP_{act} EROI values in an agroecosystem over time indicate degradation of productive capacity.

$$\text{Agroecological final EROI (AE-FEROI)} = SB/TIC$$

From an ecological point of view, the *SB* is the result not only of the energy expressly invested by society in the operation of the agroecosystem, but also what is really recycled without human intervention. This EROI gives a more exact idea of the energy investment required to obtain it. From an agroecological point of view, the relationship between this indicator and the final EROI is of great interest:

$$\text{Biodiversity EROI} = 1 - \frac{\text{AE-FEROI}}{\text{FEROI}} = UhB/TIC$$

It can reach a minimum of 0, when all of the recycled biomass is reused, indicating agroecosystems with very significant human intervention, which could even be organic, but in which no biomass is left for wild heterotrophic species. It has a maximum value of 1 when there are no external inputs and no biomass is reused by society. This would be the case in natural ecosystems without human intervention. By the very nature of agroecosystems, a scenario with a value of "1" is not possible, but agroecology considers the need to leave biomass available for other species that will allow the generation of complex food chains to guarantee ecosystemic functions. At the same time, doing this at the expense of *RuB* might also entail reducing the need for integrated land-use management. Thus getting rid of *RuB* per unit of *TIC* might lead to a decrease in the spatial heterogeneity and complexity of agroforest landscapes, and a reduction in the species richness they can shelter (Gliessman, 1998; Guzmán and González de Molina, 2009; Perfecto and Vandermeer, 2010; Marull et al., 2015). Furthermore, a drastic reduction in *RuB* would lead to an increase in the use of *EI* for the functioning of the agroecosystems and, consequently, of fossil fuels. Therefore, from an agroecological point of view, balance is needed between these two uses of the biomass and the value obtained should be analyzed from this perspective.

Furthermore, this indicator allows us to explore the hypothesis of land sparing versus land sharing from the perspective of energy, since it links the productivity of the system with the biomass available for wild heterotrophic species. The availability of phytomass is necessary to sustain complex food chains of wild heterotrophic species, but on its own it is not sufficient. Other factors, such as the absence of biocides and the presence of a diverse territorial matrix, are also pillars that sustain biodiversity in agroecosystems. The absence of biocides is an inherent characteristic of traditional agriculture and, to a large extent, of certified organic agriculture. Likewise, we have shown in Section 2.2 that traditional agriculture and, to some extent, organic farming necessarily generate complex territorial organizational matrices. It, therefore, remains for us to ascertain whether or not these types of agriculture are able to liberate greater proportions of phytomass than industrialized agriculture, and this is what the biodiversity EROI allows us to do. It should be noted that the investment in external energy in agroecosystems has been considered to be a means of intensifying agricultural production that will allow territory to be freed (land sparing) for the recuperation of wild biodiversity (Phalan et al., 2011). This theory has been discussed by other authors such as Perfecto and Vandermeer (2010), Phelps et al. (2013), and Tscharntke et al. (2011).

Land sparing for biodiversity can have several meanings. It can be understood to mean the liberation of phytomass, as we have proposed with the biodiversity EROI indicator, but it can also be understood to mean the liberation of physical space, for example, through the conversion of cropland or pastureland to woodland. The application of the following EROI (woodening EROI), allows us to look in greater depth at this aspect.

$$\text{Woodening EROI} = AB/TIC$$

This agroecological EROI tells us whether the energy added to the system (*TIC*) is being stored in the form of accumulated biomass (*AB*). Accumulated biomass can be considered a fund element, insofar that it can be related with the ecosystemic services provided by forests (when that biomass is increasingly accumulated in woodland spaces) and/or with the environmental benefits of agroforestry systems (e.g., the development of wooded cover in agricultural areas: hedges, shade trees, etc., which provide ecosystemic services to agrarian activities, carbon sequestration, etc.). From the agroecological point of view, investment in *AB* is a desirable scenario.

Photograph 2.3 Agroecological rural extension develops complex agroforestry systems with peasant communities in Brazil. High biomass productivity is allowing for the recovery of fund elements in these agroecosystems, formerly devoted to sugarcane cultivation. View of Chico Mendes III settlement.

CHAPTER 3

The Output Side
Calculating the Net Primary Productivity and Its Components

CONTENTS

3.1 INTRODUCTION

The contents of this chapter and Chapter 4 are basically methodological. They show how to make standard calculations of agricultural production outputs and the inputs used in such a production. This information is required to calculate the energy return on investments (EROIs). However, before showing how inputs and outputs are calculated, it is necessary to accurately define the scale of the analysis, carefully set the boundaries of the study and, finally, select and collect data from the most suitable information sources.

The analyzed agroecosystems must be placed in a concrete dimension in space and time. As a complex system model, agroecosystems can be approached from multiple angles, depending on the partition of reality made by the observer. Such a partition is framed along at least two axes. First, the spatial dimension represented by the territory, and the second, the time dimension represented, for example, by the history of agroecosystems. This two-dimensional framework at the same time reveals the narrow relation among dimension, scale, and time as aspects functioning in permanent reciprocity. For example, an aggregated study of agrarian metabolism at national scale probably need to sacrifice its resolution and be based on a gross scale space and a shallow historical horizon. Likewise, an analysis of a specific agroecosystem will have a finer spatial resolution, and be appropriate for making more detailed analyses. Adoption of a given scale or time will depend not only on the skills of the analyst, but also on the availability of evidence, data, and sources of information.

When a historical perspective is adopted, agrarian metabolism or a specific agroecosystem can be approached at different time scales identified by the analyzed time periods: years, decades, centuries, and millennia. Similarly, the study requires an analysis at four different spatial scales closely interlinked. The first of them is the *crop scale* where the energy flows and energy efficiency changed from past organic agricultures to nowadays. In this sphere, farmers have aimed to maximize the harvestable part of the plant and especially the part that offers the greatest commercial value or the livestock species or breed with the greatest economic yield. The case of the olive tree is paradigmatic: from a tree that produced wood, fodder and the skins and stones used to feed cattle, domestic lighting and edible oil, they are now used almost exclusively to produce oil, bringing about changes in their management and morphology (Infante and González de Molina, 2013).

The second level of analysis focuses on the farming estate. Thus, this analysis addresses farm management practices such as rotations, types of crops, farming activities, soil fertility, and so on. From heterogeneity in terms of crops and plants and their layout, we have moved toward single crops, significantly reducing genetic, structural, and functional diversity (Gliessman, 1998). The progressive simplification of rotations, reducing the presence of fallow and/or making the insertion of more commercial crops more frequent in rotations have been the main transformations at this scale.

The third level of analysis corresponds to the *organization of the agroecosystem.* During the transition to industrial agriculture, there has been a growing segregation in the uses of land and the loss of productive and functional synergies generated by agroforestry and pastoral integration. The progressive trend toward productive specialization has been an ever-increasing demand that has tended to impose specialist land uses in accordance with market demands and the aptitudes of the lands and the provision of natural resources. The result has been the loss of geodiversity and spatial heterogeneity. With this, flows of energy and materials that tended to be local and closed (renewable) have become global and based on fossil fuels. If, at the scale of agroecosystems, this phenomenon provoked a progressive decline in the agroecosystems' capacity to replenish their fertility autonomously, at the scale of individual estates, it provoked a considerable increase in the relative demand for fertilizers.

The fourth and last level refers to the "greater society," in other words, to the nation-state, first, and to the different stages in the process of globalization.

The industrialization of agriculture has favored the integration of agroecosystems in a broader geographical scope, boosting the specialization of each country according to its comparative advantages and building a global agrarian market and a single global agrofood system.

On the other hand, the choice of a suitable unit of analysis (crop, farm, agroecosystem, country, global, etc.) is a critical decision that influences the results. The system boundaries should be clearly defined so that no concern should arise. Thus, there should be no doubt with regard to the input, output, and throughput flows of energy of the unit of analysis. Accordingly, it is reasonable to assume that the results may differ depending on what flows are considered input, output, and throughput. Chapter 2 shows how different the results may be if social criteria are used to define the system output flows or, on the other hand, if agroecological criteria based on biomass functionality are used. In this sense, the specific units of analysis are defined with respect to other crops, farms, agroecosystems, countries, or other economic activities. They are defined because they cover a particular piece of land in which they can photosynthesize. Such a definition is made according to environmental, social, economic, and political criteria of different nature. The boundaries are very often arranged by agreements or political conventions in the case of countries, regions, and municipalities. In other cases, such boundaries are arranged by cultural habits that feature the territories with specific identities and territorial limits. However, the choice of systems boundaries should not be made at random or according to the researcher's interest.

Finally, the choice of information sources is paramount for a correct development of the calculations. The scale and timing of the analysis determine the type of sources used. In the case of integrated scales (national, regional, etc.), primary sources of a statistical nature created by the appropriate governments are used. The primary and secondary information provided by the resources of Food and Agriculture Organization of the United Nations (FAO) and its Statistical Department (FAOSTAT) are extremely interesting in this sense. Such information is available online (since 1961) or in the library of the organization's headquarters in Rome (Italy), which keeps information from the early twentieth century. It is more difficult to have statistical sources addressed to local case studies or when it comes to analyzing agroecosystems, crops, or specific farms. In such a case, for current case studies, the detailed information required should be collected by means of surveys, questionnaires, and use of a wealth of secondary information. On the other hand, with regard to the case studies of historical content, the information sources, frequently not in series, are usually more abundant at a regional or local scale, as they are kept in archives. The information required for specific times of the past is not always found, so this limits the historical depth of the analysis. However, there are methods of estimation or modeling, based on abundant studies on organic agriculture and stockbreeding, which can provide some substitute information, given that the traditional organic and modern agriculture share some similarities.

From a metabolic standpoint, agroecosystem outputs always consist of biomass; however, this differs greatly in the case of inputs. Almost since the early twentieth century, agriculture has operated with increasing flows of fossil fuels for both direct use and input manufacturing. Accordingly, this chapter addresses primarily how to calculate the output of biomass and energy content (output side), both from the point of view of the biomass produced (NPP), and from the point of view of animal biomass produced by livestock. Chapter 4 addresses the way in which the inputs used in agricultural production (input side) should be calculated, keeping in mind the changes undergone by manufacturing and operation costs over time. The appendices located at the end of this book are referred to, where the reader will observe the energy contents and manufacturing costs of inputs, as well as the energy contents of all the types of biomass produced.

3.2 NET PRIMARY PRODUCTIVITY (NPP) OF AGROECOSYSTEMS

As seen in Chapters 1 and 2, the metabolic process of appropriation is at the core of agricultural metabolism. This process is also called ecosystem *colonization*. In the process of colonization, the humans appropriate both ecosystem goods and services, whose main characteristic is that they are the direct or indirect result of photosynthesis, in other words, of net primary productivity (NPP). The human colonization of ecosystems implies the total or partial territorial colonization with different levels of human intervention with the purpose of taking total or partial control of its net primary productivity.

In ecology, primary productivity is the term given to the production of organic material (biomass) or the accumulation of energy by autotrophic organisms through the processes of photosynthesis or chemosynthesis using inorganic material. Chemosynthesis is relevant in certain, very specific ecosystems (ocean bed, hydrothermal vents, etc.) and, therefore, it is not of interest when we consider agrarian metabolism. In terrestrial ecosystems, the main primary producers are plants, with a small contribution from algae. In the oceans, the primary producers are, above all, algae, mainly phytoplankton. Terrestrial primary productivity by plants is the basis of agrarian metabolism. However, in flooded agroecosystems such as rice fields or in those where marine algae are used as fertilizer, the primary productivity of algae may be relevant.

Primary productivity is divided into gross productivity and net productivity. The former includes that part of solar energy that is captured by photosynthesis but which is not accumulated as biomass since it is lost in the process of respiration. Net primary productivity (NPP) is the amount of energy really incorporated into plant tissues (increase in accumulated biomass) and is the result of the opposed processes of photosynthesis and respiration. Net primary productivity is expressed in terms of energy accumulated (joules/hectare/year) or in terms of the organic material synthesized (grams/meter^2/day, kilogram/hectare/year). NPP measures an annual flow and is therefore not equal the amount of standing biomass per unit of area that measures a stock at a certain point in time. The stock or perennial plants can therefore be much larger than annual NPP. This needs to be considered when biomass from perennial plants is harvested.

With regard to agrarian metabolism, it is the net primary productivity that is of interest, since this is the basis on which the food chain is built. That is to say, the NPP establishes the limits of the capacity for the maintenance of heterotrophic populations: all of the members of the animal kingdom (human population, domesticated animals, and wild fauna), fungi, and a large part of the bacteria and the archaea. From this derives the fact that the appropriation of the NPP by human society affects the maintenance of the rest of the populations of heterotrophic organisms that depend on the same resources (Wright, 1990).

Several methods were proposed to measure this. The most widely used method, within the scope of studies of social metabolism, is the so-called human appropriation of net primary production (HANPP) (Vitousek et al., 1986; Haberl et al., 2007, 2014). According to Schandl et al. (2002, p. 49), human beings appropriate certain quantities of biomass produced annually by plants in a given territory, disrupting the natural flow of energy with agricultural, stockbreeding, or forestry activities and reducing the amount of biomass remaining in food chains. Thus, HANPP reflects the degree of human colonization of ecosystems and measures the integrated effect of land uses on the net primary productivity of ecosystems. HANPP is considered the point of contact between the social metabolism and land uses and it is defined as the difference between the energy flow (NPP) of the potential vegetation and the amount of energy (biomass) remaining in ecological cycles after subtracting the appropriated biomass (human harvest).

To calculate this, the following parameters should be previously calculated: annual productivity of the potential vegetation, annual productivity of the prevailing actual vegetation, usually divided into various kinds of land use and land cover, and the amount of biomass harvested annually. Thus, HANPP is calculated as follows:

$$HANPP = NPP_0 - NPP_t \qquad (3.1)$$

where NPP_0 is the productivity of the potential vegetation, and NPP_t is the productivity remaining in the ecosystems after human harvesting took place. In turn, NPP_t may be calculated as follows:

$$NPP_t = NPP_{act} - NPP_h \qquad (3.2)$$

where NPP_{act} is the NPP of real existing vegetation and NPP_h is the NPP appropriated by society.

According to these authors, NPP_{act} may range between 0% (clear areas or without vegetation and built-up areas) and more than 100% (fields with many inputs) of potential NPP, depending on land uses. Therefore, keeping in mind the land uses of the case study and the NPP_{act} of each area, the total value of NPP_{act} is obtained.

The concept of potential vegetation is a dubious one from the ecological point of view, because as there is no ecological balance one cannot imagine the existence of a potential or pristine vegetation either. Ecosystems change and their evolutionary dynamics actually go through periods with no change but also through periods of intense change, so it is impossible to think of a stable and ideal vegetation. Moreover, from a long-term ecological point of view, agroecosystems and the vegetation associated with them are as "natural" as the potential vegetation.

On the other hand, HANPP actually measures land uses but, paradoxically, it does not "assess" biomass production intensity. Agricultural intensification (e.g., achieved with improved genetic material that alters the grain/straw ratio, and the use of synthetic fertilizers to increase yields per unit area) can mask HANPP. An example is the use of fertilizers that produces larger harvests, but which may cause an apparent decrease in HANNP. In that case, the intensification of production seems to produce beneficial effects on all the ecosystems, as it focuses the pressure over a territory and eliminates or reduces pressure on another (for instance, on protected areas, the basis of the reasoning behind the proposed "land sparing"). Following the same reasoning, any reduction of fallow, addition of water or nutrients in large amounts result in reduced HANNP; although this statement is arguable as a concept and in terms of sustainability.

This leads to another relevant topic. The idea of HANPP itself and its methodological development are more suitable to climates with no hydrological stress. In semiarid climates such as the one dominant in the Mediterranean region, the addition of sufficient amounts of water through irrigation may mean that the actual vegetation level is higher that its potential. In such cases, HANPP does not reflect the degree of agroecosystem artificialization either in qualitative or quantitative terms.

Similarly, it cannot be used to measure, even indirectly, the impact on agroecosystems, since it does not take into account the entity and state of the biomass parts on which the reproduction of fund elements depends. In this sense, the usual calculation of HANPP only takes into account the vegetation that exists on the ground, that is the aerial vegetation, and in no case the belowground biomass. Indeed, the biomass produced in agroecosystems through the transformation of flows of energy (solar and, currently, fossil) and the mobilization of nutrients and water are the basis of the operation of traditional societies and, to a certain extent, of industrialized societies. However, only that biomass that has a use value to society and often only the fraction that has been given a monetary exchange value is quantified. This focus ignored a significant part of the biomass produced, whose recirculation in agroecosystems is fundamental to their functioning and to the maintenance of numerous populations of heterotrophic organisms that inhabit the planet. From this point of view, the need to quantify all biomass produced by agroecosystems becomes more acute, as a response not only to the flows of imported energy and materials, but also to those that recirculate within the limits of the system. The same can be said of the need to evaluate the magnitude of the human appropriation of biomass, which characterizes the different metabolic arrangements.

Accordingly, this book considers actual net primary productivity (NPP_{act}) existing at all times and it is calculated taking into account all types of biomass present in the ecosystems, both root and aerial and both harvested and nonharvested. This requires its breakdown into several categories and the suggestion of the calculation method described in Section 3.3, on which all the subsequent calculations in this work are based.

3.3 HOW TO ASSESS THE NET PRIMARY
PRODUCTIVITY (NPP) OF AGROECOSYSTEMS

The NPP calculation focuses on the assessment of the actual NPP in agroeco-systems, that is to say, the amount of NPP harvested and used by humans and the amount of NPP remaining in ecosystems for other species. To calculate it correctly we must consider the productivity both of cropland and of areas devoted to pasture and forestry. That is, we consider the productivity of all those spaces from which the human society under study extracts biomass to meet the needs of its own metabolism.

Since not all accumulated biomass is of equal interest or may not be appropriated with equal ease by human populations but still has important ecosystemic functions, we propose to distinguish different fractions of NPP. The first division is the position on or below the soil of the biomass accumulated by plants. With the exception of harvested roots and tubers, belowground NPP is typically not considered in metabolic studies, since most of it is not harvested and since it is difficult to quantify or measure. Its absence from the quantification of material and energy flows also indicates a certain disregard for or ignorance of its ecosystemic functions both in relation to the maintenance of food chains (edaphic biodiversity has only recently attracted interest with respect to the sustainability of agriculture), and also in relation to its role in the storage of nutrients and carbon in the soil. This latter function, which is useful for mitigating climate change, has led to studies that quantify the biomass of the root systems of plants either by direct measurement or through models.

To facilitate the calculation of total biomass production, we have built a database (Appendix I) with conversion factors that allow the user: (1) to calculate the total biomass produced in the agroecosystem on cropland based on information on harvested biomass (e.g., crop production), which is the most commonly available data, in particular for historical sources. A list is included with over 100 crops to calculate the total aerial biomass and more than 30 to calculate belowground biomass, (2) to convert the fresh biomass into dry biomass and vice versa, and (3) to convert the biomass into gross energy (GE). The conversion of biomass into gross energy is essential in the study of the energy efficiency of agroecosystems.

Here, conversion factors are not included to calculate the total biomass produced on grasslands or woodlands from the amount of harvested biomass. The main reason is that these conversion factors are highly variable and dependent on circumstances. Typically only a fraction of the aboveground biomass production on pastures is grazed by livestock—depending on stocking density, composition of vegetation, and quality of feed. In woodlands, harvested wood can be smaller or much larger than annual aboveground biomass production—it is not straightforward to extrapolate annual biomass produced and annual biomass produced remaining in ecosystems after harvest from wood. However, for calculating the net primary productivity of agroecosystems the biomass produced in these spaces should also be accounted for. To do that, other approaches are possible: for example, experimentally recreating past conditions and carrying out direct measurements that can be extrapolated (experimental history), or using algorithms that take into account variations in vegetation and soil and climatic conditions, and so on. The latter option has been used in the example given in Section 3.5.

Most conversion factors included in Appendix I (biomass partitioning coefficients, moisture, and gross energy content of biomass) have been collected from studies performed and based on very different land use types, crops, technological, and climatic conditions. In this sense, they are globally applicable. Nevertheless, these conversion factors are influenced by the genotype of the variety, the hormonal regulation of each plant, the phenological state, and the growth conditions (climate, soil, inter- or intra-species competition, cultural practices, etc.). The variability due to the method and moment of the estimate should be added to these (Unkovich et al., 2010). Therefore, the values offered in the database must be considered approximate, being averages taken from data collected from different sources. We include the deviation from the averages in terms of standard deviations. The consulted references for each conversion factor are also available. If more precision is needed, the user can select the conversion factors provided by studies that are closer to its environmental conditions.

Only the conversion factor of "weed biomass" is explicitly referred to Mediterranean climate conditions. The application to another specific region requires using data obtained directly from it or from regions with similar environmental conditions.

Regarding their temporal application, most of the coefficients come from current literature and handbooks. For most coefficients, we do not expect large variations over time. For others, like the harvest index, which changes over time, we have also provided information for preindustrial time periods in some crops.

3.3.1 Root:Shoot Ratio

Table AI.2 (Appendix I) shows the root:shoot ratio. This includes examples of herbaceous and ligneous species, which can be used for reference. Normally, this ratio is calculated from dry biomass, but on occasion it refers to fresh matter. In this latter case, a comment has been included in the database. The number of entries on this spreadsheet is small due to the lack of reliable data found in the literature. Undoubtedly, information on this ratio for different crops will increase significantly in the coming years. With regard to the ratio between the root and the aerial parts, there are numerous edaphoclimatic, hormonal, and so on, factors (Lynch et al., 2012), which means that the value given on the spreadsheet should be taken as an approximate value. For example, in areas with a Mediterranean climate, the root:shoot ratio is usually larger than in areas of higher precipitation due to the need to spread roots over a larger area to capture sufficient water (Hilbert and Canadell, 1995).

3.3.2 Harvest Index

Other better known biomass partitioning indices are habitually used in metabolic calculations. The main one of these is the *harvest index*, which tells us the biomass of the main product harvested in relation to the sum of that crop plus the rest* of the aerial biomass at the time of harvest. The harvest index most usually studied is that for annual grain crops, mainly cereals and legumes. In this case, the harvest index is the percentage of the biomass harvested (grain) in relation to the total aerial biomass (grain + straw). It is usually calculated from the fresh material (i.e., with the moisture content at the time of harvest).

In the case of ligneous species, such as fruit trees, the harvest index contemplates in the numerator the fruit harvested annually and, in the denominator, the sum of fruit harvested plus the wood extracted in pruning. This is not strictly the harvest index, since the denominator should also include the part of biomass produced annually but which does not leave the system; for example, most of the leaves and some of the branches. As an illustration, in the case of the holm oak (*Quercus ilex*), the acorn represents 15% of the total aerial biomass produced annually, with wood from pruning representing 50% and the rest (35%) corresponding to the leaves (Almoguera, 2007). Strictly speaking, the harvest index would be 15%. However, since the denominator does not include recirculated biomass, the harvest index rises to 23%. In the case of orange trees, the fruit is 42% of the annual dry aerial biomass. Pruned firewood is 22% and the rest (34% of dry material) is the leaves and branches that remain on the ground (Roccuzzo et al., 2012). In kiwis, 46% of dry aerial biomass corresponds to the fruit, 24% to leaves and 30% to branches (Smith et al., 1988). Likewise, in these two cases, the biomass generated annually that is recirculated on the same plot has not been used to calculate the harvest index. We would draw attention to the ecosystemic functions of the recirculating biomass and the need to take it into account in metabolic studies. However, due to a lack of data, we have included the crop and residue indices in the same way as they are usually reported in the literature and we have used the example of the holm oak, orange, and kiwi to illustrate the magnitude of the biomass, which is excluded.

Only in the case of the cereals, which are most affected by scientific varietal improvements, we offer harvest index values for old varieties (prior to the 1940s) differentiated from current values. In these crops, genetic selection focused on the increase of grain production to the detriment of straw and, clearly, current varieties have an average harvest index, which is greater than that of the older varieties.

Table AI.1 shows the harvest index of numerous crops. It also gives other indices such as "kg of residue/kg of aerial biomass." This index complements the harvest index. The sum of both is 1. The third index is "kg residue/kg product." All of these indices are expressed in terms of fresh biomass, although in some cases they have been recalculated if they appeared as dry biomass in the original document. In these cases, it has taken into account the specific moisture content of the product and residues at the time of harvest, since they are usually significantly different. In the case of trees, they refer to adult specimens at peak production.

---

* In the case of sugar beets and other root crops, this refers to the ratio between the root harvested and the sum of the harvest plus the aerial biomass.

3.3.3 Weed Biomass

Part of the net primary productivity of agroecosystems is not cultivated. It is the adventitious flora that escapes the control strategies of the farmer. In modern agriculture, with the continuous use of herbicides, this biomass may be minimal but, in traditional agriculture and in today's organic farming, its biomass is relevant. Again, we underline the importance of including it in the energy and material flows of agrarian metabolism due to its ecosystemic functions. The "weed biomass" table (Table AI.4) gives examples of the magnitude of this biomass expressed as dry weight for different crops and different management methods in Mediterranean agroclimatic conditions.

3.3.4 Moisture Content of the Biomass

When studying the hydrometabolism, it is essential to ascertain the moisture content of the biomass. Furthermore, it is necessary as a conversion factor in any metabolic calculation to always refer the data to the same units. In Sections 3.3.1, 3.3.2, and 3.3.3, we presented some indices, which usually refer to fresh material and others to dry material but, within the indices, there is also variation in the way these are expressed, depending on the authors. Three different values can be found in the literature: fresh weight typically refers to the moisture content of living biomass or biomass at the time of harvest; air-dry weight refers to biomass at a standardized water content of typically 15% and dry matter refers to moisture free biomass (moisture content 0%). Care must, therefore, be taken with the databases and the precise method of calculation must be verified.

The moisture content of wood is the proportion of free and hygroscopic water expressed as a percentage with respect to the dry weight (Ruiz and Vega, 2007). The wood is not usually totally dry, but contains humidity that may vary between 15% and 60%, depending on the open-air drying time. Wood is a porous, hygroscopic material and, given its chemical–histological structure, it has two types of porosity: macroporosity, created by the cavities in the conducting vessels and the parenchymal cells that contain free water (or imbibition water), and the microporosity of the ligneous substance itself (fundamentally, cellulose, hemicellulose, and lignin), which always contains a certain amount of bound water. Wood begins to lose water from the moment at which the tree is felled. First, imbibition water evaporates from the outer part (sapwood) and, subsequently, from the internal parts (heartwood) of the trunk. At a certain point, all of the free water of the dry wood evaporates, while the bound water reaches a point of dynamic equilibrium with external humidity, falling to a value of less than 20% (Francescato et al., 2008). Tay (2007) reported that newly cut biomass may have 80%–90% moisture content and, on drying, this figure could fall to 10%–26%. Table AI.3 gives the average percentage moisture content of the wood of different fruit trees after a variable period of open-air drying, together with the standard deviation of the data.

The dry matter of green fodder varies with the phenological state of the plant. Mainly, the dry matter given in the database refers to when the fodder is at 50% of floration. The dry matter of the main fruit and vegetable products refers mainly to the whole fruit or vegetable. Normally, the dry material data for fruit and vegetables that appears in the literature refers to the edible part. In the case of some products (e.g., lettuce, spinach, etc.), the water content of the edible part is the same as that of the residue (peel, outer leaves, etc.), but in other cases (peel of Cucurbitaceae, stones of drupe fruit, shells of nuts, etc.), the moisture content is significantly different. Given that the production data that appears in agricultural statistics refers to complete fruit or vegetables, we have attempted to compile dry material data for complete fruit, which in some cases we have calculated from the dry material of the parts and of the proportion of each one in the product.

In cereals, legumes, fruit, and vegetables, we give not only the dry material of the main product, but also the dry material of the rest of the plant (straw, prunings, and plant remains) which, while it is not the main product, can also be sold, buried, burned, left on the land, and so on. Depending on the treatment given to it, this biomass is considered in different ways in agrarian metabolism. As an example, we have included data on the dry material of livestock products, some processed industrial products and large volume by-products of agroindustry.

3.3.5 Assessing NPP$_{act}$ and Its Different Categories

As a result, NPP$_{act}$ would be the result of the sum of plant biomass directly appropriated by society (socialized plant biomass); biomass already circulated through the agroecosystem either by intentional reintegration (reused biomass) or simple absence of harvesting (unharvested biomass); and biomass accumulated annually (accumulated biomass) in its aerial structure (stem and crowns) and roots of perennial species of pastures, forests, and crops. Thus, the domestic extraction of agroecosystems would be equivalent to the sum of the socialized plant biomass and reused biomass obtained through conscious management of agroecosystems. The formulae included in Chapter 2 should be restated at this time:

$$NPP_{act} = SVB + RuB + UhB + AB$$

$$SB = SVB + SAB$$

$$RcB = RuB + UhB$$

$$DE = SVB + RuB$$

Herein, the calculation of net primary productivity does not take into account the total biomass lost within the agroecosystem before harvest in different ways (exudation of compounds, herbivory, etc.) (Smil, 2013a). According to the existing literature, these losses are awarded NPP percentages that are too general, arbitrary, and uncertain. Hence, we have chosen to disregard them. However, when calculating

NPP from a historical perspective, there should be major differences between the values applicable to the past and those applicable to the present, due to the current use of pesticides, in the same way as there are major differences between farmland and pastures or warm or cold areas.

3.4 INDIRECT CALCULATION OF THE GROSS ENERGY OF BIOMASS

The *gross energy* (GE) is the energy liberated as heat when an organic substance is completely oxidized to carbon dioxide and water. In the International System, it is expressed in Joules per gram of substance. It is also common, however, to find GE expressed in calories per gram. We must take care to note whether the GE value refers to humid or dry matter to multiply it by the amount of biomass, whether humid or dry, as the case may be, whose GE is being calculated.

The GE content of an organic substance (human foodstuffs, fodder, wood, etc.) can be obtained directly by measuring the energy content of a given mass of the substance, as combustion heat in a calorimeter (bomb calorimeter), or indirectly by estimating from the chemical–bromatological composition of the substance.

It is essential to ascertain the GE of organic substances to calculate the EROIs. However, a calorimeter is only available in a few cases to make direct measurements of the GE of different products and residues from agricultural and forestry activities. In practice, we shall make a comprehensive review of the literature to obtain published GE data, such as the calculation based on chemical–bromatological composition tables of biomass. By means of this indirect calculation, we can also verify data found in the literature on GE that appears to lack credibility.

We should warn that the energy that usually appears in the tables relating to human and livestock nutrition is not gross energy, but metabolizable energy. Metabolizable energy is the result of deducting the energy of feces, urine, and gases from the gross energy. It is, therefore, useful when preparing diets but not to calculate the EROIs. The database presented in Table AI.5 (Appendix I) to facilitate the calculation of the EROIs uses both sources: literature and indirect calculation, which was performed as described in Sections 3.4.1 through 3.4.4. The database specifies, in each case, the source of the information.

The energy unit used in the database is the Joule. We have used a conversion factor to calories (thermochemical calorie) of 4.184 cal/J (FAO, 1971).

3.4.1 Calculation of the Gross Energy of Human Foodstuffs

Each pure substance that makes up organic material has its own gross energy (GE) (e.g., 17.7 kJ GE/g for starch, 15.7 kJ GE/g for glucose, 16.0 kcal/g for hemicellulose, etc.), and so if we know the composition, we can calculate the GE of the substance. To simplify the calculation, average GE values are used for proteins, lipids, and carbohydrates, since these are the compositional data of human foodstuffs that are easiest to find, being available in many tables.

To calculate gross energy in our database, we have used figures of 23.5 kJ/g for proteins, 39.5 kJ/g for lipids, and 17.5 kJ/g for carbohydrates (Flores Mengual and Rodríguez Ventura, 2013). These values are similar to those used by other authors. For example, De Masson (1997) proposed values of 5.4 kcal or 23 kJ/g for protein, 4.1 kcal or 17 kJ/g for carbohydrates, and 9.3 kcal or 39 kJ/g for lipids. Maynard et al. (1979) used 4.15 kcal/g for carbohydrates, 9.4 kcal/g for fat, and 5.65 kcal/g for proteins. Merrill and Watt (1973) also offered GE for fat, carbohydrates, and proteins from different sources.

The composition of foodstuffs has been obtained mainly from Moreiras et al. (2011). In the few cases in which the foodstuffs did not appear in this publication, we have used Mataix and Mañas (1998). These authors give the percentage of food consumed by the person (e.g., 84% of an apple) and the composition of the part consumed. Given that many foodstuffs have a part that is not consumed, we would be underestimating the gross energy of the agricultural product if we did not also consider the combustion energy of the waste. To avoid this underestimation, we have also calculated the GE of the waste, as explained in Section 3.4.3.

Therefore, the database (Table AI.5) includes the GE of the consumable foodstuff, the GE of the waste and total GE, which is the sum of both. We must, then, simply multiply the total GE of the foodstuff by the weight of the food products (kilogram of wheat, kilogram of wheat/ha, liters of milk, liters of milk/farm, etc.) to obtain the GE of the part extracted from the agroecosystem in the form of human foodstuffs. If the residue is partially or totally returned, database users can also estimate the GE returned to the agroecosystem.

3.4.2 Calculation of the GE of Livestock Feed

To calculate the GE of processed livestock feed such as silage, oil cake, or composite feedstuff, different formulae are available from the literature that uses information of the chemical composition of feedstuff and statistical relations between material characteristics and energy content. In the literature, the following formulae to calculate the GE can be found:

- For concentrates (Nehring and Haenlein, 1973 in Meineri and Peiretti, 2005): GE (kcal/kg dry matter) = 5.72 × raw protein + 9.5 × ether extract + 4.79 × raw fiber + 4.03 × N-free extract ± 0.9 (in g/100 g dry matter)
- For silage (Andrieu and Demarquilly, 1987 in Meineri and Peiretti, 2005): GE (kcal/kg organic matter) = 3910 + 2.45 × protein + 169 pH ± 84 (in g/kg organic matter, $R^2 = 0.59$)
- For alfalfa silage (Valente et al., 1991 in Meineri and Peiretti, 2005): GE (MJ/kg dry matter) = 21.54 − 0.011 × Total N − 0.011 × dry matter + 1030 pH − 0.073 × acetic acid + 0.018 × lactic acid − 0.056 × ethanol ± 0.22 (g/kg dry matter, $R^2 = 0.91$)
- For crimson clover silage (Peiretti et al., 1994 in Meineri and Peiretti, 2005): GE (MJ/kg dry matter) = 14.74 + 0.319 × methanol − 0.008 × lactic acid + 0.082 × Total N + 0.012 × acetic acid ± 0.21 (g/kg dry matter, $R^2 = 0.91$)
- Ewan Formula, 1989 (in NRC, 1998): GE (kcal/kg fresh matter) = 4143 + (56 × % ether extract) + (15 × % raw protein) − (44 × % ashes) ($R^2 = 0.98$) (% of fresh matter)

In our database (Table AI.5), the GE of livestock foodstuffs (grain, feed, and cake) is calculated using the Ewan formula (1989, in NRC, 1998), unless otherwise indicated. The composition of the foodstuff (ether extract, raw protein, and ashes) comes from the tables of "ingredients for animal feed" of the Spanish Foundation for the Development of Animal Nutrition (FEDNA, 2010). In the case of green fodder and humid fibrous by-products, for which this formula is not appropriate, the calculation has been performed as indicated in Section 3.4.3.

3.4.3 Calculation of the GE of Crop Residues, Food Waste, Green Fodder, and Fiber

The term "crop residue" refers here to the aerial biomass of herbaceous plants that are not harvested as the main crop product. It may or may not be used by society. Crop residue is the straw and stubble of cereals and legumes whose main product is the grain, although this residue is frequently used as animal feed. Crop residue also includes other herbaceous crops (sugar beet, sugarcane, horticultural, industrial crops, etc.) some of which can be used as feed or energy carrier. Food waste is the inedible part of foodstuffs as described in Section 3.4.1. Green fodder refers to the aerial parts of these crops at the moment in which they are harvested as fodder for livestock. They have not, therefore, undergone the process of haymaking or silage. Fiber refers to the product of fiber-producing crops (cotton, flax, etc.).

In these four cases, the calculation is based on the assumption that plant biomass is composed basically of carbohydrates and it has, on an average, 4200 kcal/kg of dry matter (17.57 MJ/kg dry matter) (Merrill and Watt, 1973; González González, 1993). In this regard, there is a certain variation between authors, between 4000 and 4400 kcal/kg dry matter (Campos and Naredo, 1980; NRC, 2001). In fact, since there is a slight variation in the proportion of the different carbohydrates contained in the different plant species, as well as the presence of other substances in small quantities (resins, lignin, etc.), a certain amount of variation is to be expected. Therefore, the GE of 1 kg of fresh matter of these products (Table AI.5) is obtained by multiplying the percentage of dry matter by 17.57 MJ/kg dry matter. Table AI.3 shows the percentages of dry matter GE of all products and residues.

We would calculate the GE of the biomass of weeds (adventitious flora) in the same way.

3.4.4 Gross Energy of the Wood in Forest Species and Pruning Residue of Fruit Trees

According to the FAO (1991), the gross energy of wood depends very much on the species and the part of the tree that is used, varying between 17 and 23 MJ/kg dry matter of wood. Generally, conifers have higher values than broadleaf trees, with an average value of 21 MJ/kg of dry matter for resinous wood and 19.8 MJ/kg dry matter for other woods. There is very little variation in the GE of the substance of the wood, which is 19 MJ/kg of dry matter, with the difference between species depending on the proportion of resin. Resin has a GE of 40 MJ/kg dry matter (FAO, 1991). Likewise,

Francescato et al. (2008) said that the GE of different species of wood varies within a very reduced interval, of 18.5–19 MJ/Kg dry matter. In conifers, it is 2% higher than in broadleafs. This difference is due fundamentally to the higher lignin content of conifers but also in part to their higher resin, wax, and oil content. In comparison with cellulose (17.2–17.5 MJ/kg dry matter) and hemicellulose (16 MJ/kg dry matter), lignin has a higher GE (26–27 MJ/kg dry matter) (Francescato et al., 2008).

To calculate the GE of different types of wood, we have reviewed the literature (see Table AI.5). Since this biomass is habitually used to generate energy, it is possible to find information for each species or group of species. The data are normally given for dry matter, and so we have also considered the percentage of dry matter per kilogram of fresh wood to calculate the GE per kilogram of fresh wood. Since, as we have seen, the percentage of dry matter of the wood varies with the time that has elapsed since it was cut, the storage conditions, and so on, we have standardized the moisture content for all wood on a 25%.

This decision is arbitrary and would correspond to wood that has been aired for a certain period of time, without being exposed to rain. In our case, we have considered that the wood production data that appears in historical sources refers to wood in this condition and not to newly cut wood. In other cases, if there is a suspicion that the production data refer to other conditions, the GE value may be adjusted, dividing by 0.75 and multiplying by the decimal representing the percentage of dry matter considered most appropriate in each case. In the case of pruning residue, the dry matter content is taken from a review of the literature (see Table AI.3 in Appendix I).

3.5 AN EXAMPLE FOR ASSESSING NPP AND OTHER CATEGORIES OF BIOMASS

As an example of the use of the database, we offer a case study of the municipality of Santa Fe (Granada) in the south-east of the Iberian Peninsula, in the mid-eighteenth century. This case study has been widely described in a book and several articles, which makes it possible to investigate the agrarian social metabolism of Santa Fe from the mid-eighteenth until the end of the twentieth century (González de Molina and Guzmán, 2006; Guzmán and González de Molina, 2007, 2009, 2015). In addition, Chapter 5 presents the calculation of energy efficiency indicators (EROIs) of this case study and the results are discussed.

The agricultural area and agricultural production in the municipality of Santa Fe (Granada) in 1752 are shown in Tables 3.1 and 3.2. The information about the agricultural area, crop production, and forestland comes from historical sources (see Chapter 5). The aerial production of pastureland was obtained as dry matter using models, which take into account edaphoclimatic and vegetation variables.

From this data, we can obtain an approximate figure for the real biomass production of the agroecosystem using the conversion factors in the database (Table 3.3). For example, in the case of broad beans, we would have to multiply the harvest (124,568 kg of fresh material) by the dry matter conversion factor for broad beans (0.915) to obtain the harvest of dry matter (114,021 kg of dry matter in the harvest of broad beans).

Table 3.1   Agricultural Area (Cropland and Forestland) and Harvest in Santa Fe
(Granada) in 1752

	Agricultural Area (Hectares)	Yield (kg Fresh Matter)
Broad beans	67.7	124,568
Hemp	20.4	6,780
Wheat	564.7	1,030,013
Flax	199.8	60,228
Corn	6.3	14,364
Irrigated barley	52.2	91.768
Chickpeas	3.9	600
Millet	20.1	44,300
Onions	1.5	1,074
Grass peas	7.3	600
Common beans	5.3	3,794
Safflower	2.5	600
Dryland barley	376	110,168
Olives	189	27,062
Grapes (cultivated with olives)	–	191,268
Poplars/riverbank vegetation	3.4	31,897

Table 3.2   Agricultural Area (Pastureland) and Harvest in Santa Fe (Granada) in 1752

	Agricultural Area (Hectares)	Yield (kg Dry Matter)
Fallow	1,180	2,049,660
Dehesa pastureland	366.3	331,684
Floodable pastureland	700	980,000

To obtain the aerial dry biomass of the residues generated by the broad beans harvest, we would multiply 124,568 kg of fresh matter harvested by the residue index for broad beans (1.56) and by the dry matter conversion factor for broad bean residue (0.886), giving a figure of 172,449 kg of dry matter.

To calculate the dry root biomass, we add the dry biomass of the harvest and the residue (286,470 kg of dry matter) and multiply it by the root:shoot ratio for broad beans (0.6). The root biomass would come to 172,837 kg of dry matter, an amount similar to that of the residues (straw) of the broad bean crop.

In the case of crops for which we have not found data, we have used equivalents in similar crops. For example, we have considered that flax is similar to hemp where we did not have any conversion factor available. Grass peas were compared to "other legumes" or peas, depending on the conversion factor.

The aerial biomass of the vegetation accompanying the crops was obtained by multiplying the crop area (hectares) of the broad beans with value of average dry matter production of weeds per hectare for extensive crops (873 kg dry matter/ha). The dry root biomass is obtained by multiplying the dry biomass of the aerial parts (59,125 kg dry matter) by the root:shoot ratio for pastureland (0.8), which we have

Table 3.3 Net Primary Productivity (Dry Matter) of the Santa Fe Agroecosystem in 1752

Crops	Crops Aerial Part			Root (kg)	Weeds Aerial Part (kg)	Root (kg)	Total Mg
	Harvest (kg)	Residues (kg)	Accumulated Perennial Structures (kg)				
Broad beans	114,021	172,449		172,837	59,125	47,300	565.7
Hemp	6,177	1,544		1,418	17,816	14,253	41.2
Wheat	905,381	2,257,516		2,024,254	493,174	394,539	6,074.9
Flax	55,952	12,798		12,627	174,493	139,594	395.5
Corn	12,382	15,467		6,769	5,502	4,402	44.5
Irrigated barley	81,215	149,345		147,558	45,588	36,471	460.2
Chickpeas	566	911		87	3,406	2,725	7.7
Millet	39,006	48,730		21,325	17,554	14,043	140.7
Onions	66	131		0	2,451	1,961	4.6
Grass peas	550	926		87	6,375	5,100	13.0
Common beans	3,730	5,667		556	4,629	3,703	18.3
Safflower	547	2,124		2,672	2,183	1,747	9.3
Dryland barley	97,499	179,290		177,145	328,375	262,700	1,045.0
Olives	14,586	17,931	18,635	3,883	567,000	453,600	1,79.5
Grapes	55,704	66,677	5,600	2,240	0		130.2
Pasture							
Fallow	420,668			336,535			757.2
Dehesa pasture	331,684			398,021			729.7
Floodable pasture	980,000			784,000			1764.0
Forestry							
Poplars	23,923		34,771	11,993	13,760	11,008	95.5
Total	3,143,656	2,931,506	62,890	4,104,006	1,741,433	1,393,146	13,376.6

given a similar value to weeds. In this way, we obtain a dry root biomass figure for weeds of 47,300 kg of dry matter.

The sum of the total dry biomass of the crop would come to 565.7 Mg of dry matter.

For cereals, we have used the conversion factors for old varieties given in Appendix I, which refers to harvest indices typical before 1940, which are surely more similar to those used in 1752, than those used today.

The gross energy of the aerial biomass of the crop is obtained by multiplying the fresh biomass of the crop and of residues by the corresponding gross energy value. Specifically, in the case of broad beans, the harvested biomass (124,568 kg of fresh matter) is multiplied by 15.59 MJ/kg fresh matters, while the biomass of the residues (194,637 kg fresh matter) is multiplied by 15.57 MJ/kg fresh matter (beans talks).

The gross energy of the roots of the crop and of the weeds (aerial part and roots) have been calculated by multiplying their respective dry biomass values by 17.57 MJ/kg dry matter, which, as explained in the text, is an approximate value for biomass composed fundamentally of carbohydrates. The gross energy of the biomass generated in the municipality of Santa Fe in 1752 is shown in Table 3.4.

For pastures, the aerial biomass of the *dehesa* (905.4 kg dm/ha) was calculated by applying an algorithm adapted to the growing conditions in Santa Fe (Passera

Table 3.4   Net Primary Productivity (Gross Energy) of the Santa Fe Agroecosystem in 1752

	Crops				Weeds		
	Aerial Part			Root (MJ)	Aerial Part (MJ)	Root (MJ)	
	Harvest (MJ)	Residues (MJ)	Accumulated Perennial Structures (MJ)				Total GJ
Broad beans	1,942,015	3,030,405		3,037,227	1,038,992	831,193	9,880
Hemp	108,540	27,135		24,918	313,079	250,463	724
Wheat	14,258,264	39,670,874		35,571,815	8,666,450	6,933,160	105,101
Flax	983,230	245,808		221,886	3,066,330	2,453,064	6,970
Corn	207,393	271,796		118,946	96,686	77,349	772
Irrigated barley	1,433,967	2,624,413		2,593,013	801,113	640,891	8,093
Chick peas	9,457	16,006		1,535	59,853	47,883	135
Millet	669,984	856,323		374,735	308,475	246,780	2,456
Onions	1,069	2,300		0	43,078	34,462	81
Grass peas	11,064	16,268		1,533	112,033	89,626	231
Common beans	52,319	99,593		9,762	81,339	65,071	308
Safflower	10,544	37,333		46,949	38,367	30,694	164
Dryland barley	1,721,485	3,150,622		3,112,926	5,770,471	4,616,377	18,372
Olives	215,843	338,529	425.142	68,236	9,963,778	7,971,022	18,983
Grapes	543,434	1,285,978	108.005	39,363	0	0	1,977
Pasture							
Fallow	7,392,320			5,913,856			13,306
Dehesa pasture	5,828,621			6,994,345			12,823
Flood-able pasture	17,221,344			13,777,075			30,998
Forestry							
Poplars	442,567		653.319	210,752	241,802	193,441	1,742
Total	53,053,460	51,673,383	1,186,466	72,118,872	30,601,846	24,481,476	233.116

Photograph 3.1   Cover crops in olive groves increased net primary productivity of the agro-ecosystem when it transitioned from industrial to organic management.

(a)   (c)

Photograph 3.2   Polycultures allow for an increase in the socialized biomass of the agroecosystem, strengthening the internal energy loops. (a) Maize and beans, (b) lettuce and leeks, and (c) broad beans and spinach.

Sassi et al., 2001). However, the productivity of floodable pasture (1400 kg dm/ha) and fallows (356 kg dm/ha) is based on studies with similar agroclimatic and management conditions (Campos and Naredo, 1980; San Miguel Ayanz, 2009). The root:shoot ratio of grass is 0.8, except for the *dehesa*, which has been considered 50% of Mediterranean scrub (ratio: 1.6) and 50% herbaceous grass (ratio: 0.8) (Table AI.2, Appendix I). Gross energy of pasture is 17.57 MJ/kg dry matter.

From this, again taking as a basis the historical sources and information, we can determine the biomass socialized by the human population that was used to maintain the livestock and that which was available for the remaining heterotrophic organisms. This part, together with that consumed by livestock, is the recirculating biomass of the agroecosystem.

The high livestock population at the time meant that straw and stubble were all consumed, and so we suppose that they were not burnt during that period. Likewise, we have supposed that the pruning and sucker waste was not burnt in the fields, since the firewood demand by the local population for cooking and heating far exceeded availability in the municipality.

We can also estimate the biomass that was accumulated annually in perennial vegetation (trees and shrubs) both in the root system and the aerial part. The annual biomass accumulated in the olive groves has been estimated on the basis of Almagro et al. (2010). These authors estimated the accumulated dry biomass in the aerial part to be 17,298 kg dry matter/hectare and 3604 kg dry matter/hectare in the roots, in 100-year-old dry-farmed olive groves with trees planted in a 10×10 m^2 pattern. Such olive groves are similar to those in the Santa Fe case. This would mean an accumulation of 2.1 kg of dry material annually per tree (1.7 in the aerial part and the rest in the roots). In our case, there were 57 trees per hectare and 189 ha of olive groves. Therefore, the annual accumulation would be 18,635 kg of aerial dry matter and 3,883 kg dry matter in the roots in the olive groves in the municipality. This is a simplification, since the process is not linear. To calculate the amount of aerial biomass accumulated annually in poplars, we have divided the total amount of wood obtained after felling by the number of years of growth until the felling (15 years). The dry root biomass accumulated annually has been calculated taking into account the root:shoot ratio of the poplar. For grape-vine, we have considered 30-year-old vine and so the total biomass accumulated in the plant is divided by the total number of years of the plantation.

In our case, the direct appropriation of biomass (socialized vegetable biomass) by the population represented 7% of the dry matter, that used for animal feed was 30%, that available for other heterotrophic species came to 62.5%, although most of this (66%) recirculated in the soil (Table 3.5). Very little biomass was accumulated annually in perennial plants (0.5%) due to the small crop area devoted to perennial crops or forestry (Table 3.5). The *dehesa* pastureland was without trees and had an herbaceous and shrub cover, according to descriptions from the time.

The agroecosystem in Santa Fe in 1752 provided the flows of biomass necessary to maintain the human population and livestock, which in turn guaranteed the supply of the flows of energy and nutrients necessary to sustain agricultural production, achieving very high levels of sustainability (González de Molina and Guzmán, 2006).

Table 3.5   Distribution of the Vegetable Biomass Produced Annually by the Santa Fe Agroecosystem in 1752

		Biomass (kg Dry Matter)	Gross Energy (MJ)	%
Socialized vegetable biomass	Foodstuffs	868,349	13,321,555	
	Fiber	62,676	1,102,313	
	Wood and firewood	89,455	1,707,377	
Subtotal		1,020,480	16,131,245	7.0
Reused biomass[a]		3,968,050	69,048,225	30.0
Unharvested biomass[b]	Aboveground	2,828,064	50,149,219	21.5
Unharvested biomass[b]	Belowground	5,497,152	96,600,348	41.0
Accumulated biomass		62,890	1,186,466	0.5
	Total	13,320,035	233,115,503	100

[a] For animal feed and bedding, seed, and so on.
[b] Available for other species.

About 37% of the nonaccumulated aboveground biomass would have been available for nondomesticated species, allowing the maintenance of wild heterotrophic organisms in the municipality. At the other extreme, there is the biomass available for edaphic heterotrophic organisms since, to the enormous amount of belowground biomass that was directly recirculated (5,497 Mg of dry matter), it must be added to the biomass of the manure, which became incorporated into the soil and which amounted to 2,831 Mg (González de Molina and Guzmán, 2006). Such a high recirculation of biomass in the soil guaranteed the adequate condition of the resource, as well as edaphic biodiversity, which was not damaged by the use of biocides, which were unknown at the time.

CHAPTER 4

The Input Side
Calculating the Embodied Energy of Agricultural Inputs

Eduardo Aguilera, Gloria I. Guzmán, Juan Infante, David Soto, and Manuel González de Molina

CONTENTS

4.1 INTRODUCTION

During sociometabolic transitions from traditional to industrial societies, the role of agriculture as the major source of energy and materials in preindustrial societies gave place to fossil fuels and minerals in industrial societies (Krausmann and Haberl, 2002; Fischer-Kowalski and Haberl, 2007; Infante Amate et al., 2015). In the specific case of agriculture, metabolic transitions are characterized by large quantitative and qualitative changes in agrarian inputs that were usually linked to increases in outputs (increased land productivity) and decreases in human labor (increased labor productivity) (Boserup, 1981; Giampietro et al., 1999) Typically, solar-based local, organic inputs produced on farm such as manure and animal draft power were substituted by high amounts of fossil fuel-based external inorganic inputs such as synthetic fertilizers and pesticides, machinery, fuel, and electricity (Guzmán and González de Molina, 2009).

Despite the growing body of research addressing energy balances of agricultural systems from a historical perspective (e.g., Bayliss-Smith, 1982; Cleveland, 1995; Krausmann, 2004; Cussó et al., 2006; Carpintero and Naredo, 2006; Guzmán and González de Molina, 2009, 2015; Infante-Amate et al., 2014; Tello et al., 2016), to our knowledge, the changes in the energy efficiency of the production of inputs have scarcely been taken into account. Only studies based on monetary data instead of on a process analysis systematically consider these changes because their calculations are based on year specific energy efficiencies (e.g., Cleveland, 1995; Cao et al., 2010, see Section 4.11). Another interesting study (Pelletier et al., 2014) on egg production in the United States in 1960 and 2010, accounts for temporal changes in the energy efficiency of agricultural inputs from a life-cycle assessment (LCA) perspective.

Today there is still a scarce, although growing, body of information on the changes that have occurred in the production of most agricultural inputs. In terms of energy, the changes in inputs have not only been driven by the changes in their quantities and qualities, but also in the energy required to produce them. Technology improvements are responsible for a general trend in the twentieth century toward increased energy efficiency in the production of most agricultural inputs, such as nitrogen and phosphate fertilizers or steel for machinery production (Smil, 1999, 2013b; Jenssen and Kongshaug, 2003; Ramírez and Worrell, 2006; Dahmus, 2014). In some periods, such as the energy crisis of the 1980s, this trend was intensified due to increased energy prices and concerns about the security of energy supply (Bhat et al., 1994).

There are some agricultural inputs, however, which required relatively low energy use in the early stages of their industrial developments, because their production energy is mainly used in mining activities, and easy to extract, high-grade ores were exploited first. The progressive depletion of these resources means increasing energy consumption to extract and refine the materials (Meadows et al., 1972), as lower-grade ores typically demand more energy to extract the resource (Gutowski et al., 2013). Therefore, despite technological improvements, the energy efficiency of the production of raw materials may ultimately decline. For example, this is the case of oil and gas production in the United States (Hall et al., 2009, 2014), and also in other countries and in the world as a whole (Gagnon et al., 2009; Hall et al., 2014), whose energy return on investment (EROI) is already declining. As another example, the energy efficiency of potash fertilizer production in the United States did not increase during 1979–1987, despite high energy prices that boosted energy efficiency improvements in N and P fertilizers (Bhat et al., 1994).

In this chapter, we aim to provide a comprehensive compilation of embodied energy coefficients for the major agricultural inputs with a historical perspective. We exemplify how to use and choose the coefficients with a practical case study, but we also aim to provide a framework where researchers can situate their own choices. With these objectives, we review the history of the agricultural use and production processes of agricultural inputs, and construct reasonable estimates, as disaggregated as possible, of the energy employed in the different phases of these production processes. Our main focus is on industrial inputs at the world level, for which we have aimed to construct a coherent, self-referenced database (fully shown in Appendix II) starting from the production of fuels and other energy carriers, raw materials and finally manufactured goods delivered to the farm. In the case of nonindustrial inputs such as different types of biomass, animal work, human labor, or nonmaterial services, we just describe the most usual approaches for the estimation of their embodied energy. In Section 4.13, we develop an example of the application of these embodied energy coefficients to one case study, Spanish agriculture in 2008. In this example, we choose specific factors and methods. The applicability to these procedures to other case studies would depend on specific data availability and study system boundaries and goals.

4.2 THEORETICAL AND METHODOLOGICAL CONSIDERATIONS

In this chapter, the embodied energy of a given input refers to the sum of the higher heating value (HHV, gross energy [GE]) of the input plus the energy requirements for the production and delivery of the input. Thus, in most cases this metrics would be equivalent to the "cumulative energy demand" concept used in life-cycle assessments, and also to the "energy intensity" concept used in some energy studies. All components of the embodied energy are expressed in terms of higher heating value or gross energy.

Energy requirements refer to the energy employed in the production of a given input. They are divided in direct and indirect energy requirements. Direct energy requirements refer to the gross energy of the fuels directly used in the production process. Indirect energy requirements include all remaining processes needed for the production of the input and its use at the farm, including fuel production and transport, raw materials production and transport, energy embedded in buildings and equipment, and transport of finished products up to the farm. It has to be clarified that only physical processes are included.

We follow the definition of energy carriers stated by Murphy and Hall (2011): "a primary energy source is an energy source that exists in nature and can be used to generate energy carriers (e.g., solar radiation, fossil fuels, or waterfalls). An energy carrier is a vector derived from a primary energy source (e.g., electricity, gasoline, or steam)." The EROI would represent the relationship between the energy carriers produced in an energy production process and the energy carriers employed in the process.

In this paper, nonrenewable energy (NRE) includes fossil fuels, nuclear and, when the data are available (primarily when the data are gathered from ecoinvent), nonrenewable biomass, which always represent a very small portion. Renewable energy is represented by hydro, renewable biomass, geothermal, wind, and solar. The distinction between renewable and nonrenewable energy sources is essential for the assessment of agroecosystem sustainability. Therefore, we provide data on NRE use for all items considered, as described in Section 2.4.

The energy content of fuels and biomass products can be measured as the lower heating value (LHV) or the higher heating value (HHV), also called net energy (NE) and gross energy (GE) values, respectively. As fuels usually have trace amounts of water, the LHV or NE considers only the energy that can be obtained from fuel combustion without recovering the energy in the evaporated water, while the HHV or GE considers all fuel energy (enthalpy) without correcting for water evaporation. The NE typically represents about 95% of the GE of liquid fossil fuels, and about 90% in the case of natural gas (International Energy Agency [IEA], 2004) (see Section 4.4). Despite a consensus is far from being reached (Kim et al., 2014), we employed the HHV or GE, as in many other energy analyses of cropping systems (e.g., Patzek, 2004; Pimentel, 2003) and in life-cycle impact assessment (LCIA) methods implemented in ecoinvent such as cumulative energy demand (Frischknecht et al., 2007a). This choice is coherent with the use of GE values for agricultural energy outputs (Chapter 3). On the other hand, we did not apply any quality correction factor to the heat value of the different fuels.

In this chapter, we define the energy embodied in agricultural inputs as the higher heating value of the primary fuel plus any other energy contribution, including: (1) harvesting (or extracting), (2) refining, (3) manufacturing (in the case of manufactured products such as fertilizers or machinery), (4) transport to the farm, and (5) maintenance (in the case of capital goods).

Which specific agricultural inputs are to be studied and what amount of energy is estimated to be embodied in these inputs depend on system boundaries, which in turn depend on study objective. We review some usual approaches for accounting each of the studied inputs in the corresponding sections of this chapter, and a deeper discussion can be found in Aguilera et al. (2015c).

The geographical representativity of the data is a particularly important point given the significant differences in energy efficiency between world regions that can be observed for many processes. When possible, we provide dynamic, world averaged coefficients. This was not always possible, and in those cases the estimations are based on a single country or region accounting for a significant share of world production (usually United States or Europe). Likewise, the estimation of dynamic factors was not possible in some cases, so that fixed factors had to be used instead. In some cases, we provide information of differences in energy efficiency between world regions for a single recent time point or for various time points.

We estimated the relative share of NRE in world primary energy consumption. The reconstruction of long-term series of world primary energy consumption by source has been attempted in few occasions, usually including very gross assumptions particularly for the estimation of biomass energy. We took the data from Koppelaar (2012), who compiled some of the available series (e.g., Fernandes et al., 2007; Krausmann et al., 2009; Smil, 2010; BP, 2011), and constructed a unique long-term series of world primary energy consumption by source. The resulting estimation of the relative share of NRE in global primary energy production is given in Table 4.1. In the same table, we also included an estimation of the relative share of NRE in world electricity production (Table 4.1; see Section 4.4).

Table 4.1   Relative Share of Nonrenewable Energy in World Primary Energy Production
and World Electricity Production, 1900–2010 (%)

	1900	1910	1920	1930	1940	1950	1960	1970	1980	1990	2000	2010
Primary energy (%)	47	56	61	62	64	70	77	83	85	86	86	86
Electricity (%)				99	98	95	91	88	89	91	92	91

Source: Author data.

4.3 HUMAN LABOR

The assessment of the energy embodied in human labor is highly controversial and varies widely depending on system boundaries and researchers' criteria. Herein we describe, largely following and continuing the review by Fluck (1992), some human labor accounting methods in a hierarchical way, from narrower (being the narrowest exclusion of this input) to broader system limits.

Many studies exclude human labor of agricultural energy assessments, particularly in industrialized systems. Some authors employed the muscular power output of human labor (e.g., Rappaport, 1971; Bayliss-Smith, 1982), which would represent the "direct energy input" in our terminology, and has also been termed "applied power" (Giampietro and Pimentel, 1990). This energy was estimated to represent 0.3–1.3 MJ/h in a range of agricultural tasks in a tribe of New Guinea, and 0.8 MJ/h for the average agricultural worker in a variety of examples of agricultural systems around the world (Bayliss-Smith, 1982). An accepted average value is 0.27 MJ/h (75 W) (Pimentel and Pimentel, 1979; Giampietro and Pimentel, 1990). Probably the majority of agricultural energy analyses estimate the energy in human labor as the dietary energy consumption, that is, the metabolic requirements or the energy content of the food consumed by the workers. Direct dietary energy consumption may range between 0.35 and 0.61 MJ/h, for diets of 2000 and 3500 kcal/day, respectively. Fluck (1992) identified three methods for assessing the dietary energy of a working hour: as the partial energy consumed from metabolized food during work, excluding basal energy consumption; as the total food energy metabolized during work (e.g., Tello et al., 2016); or as the total dietary energy consumed by workers (during working days or the whole week). A value of 2.2 MJ/h, based on the data offered by Fluck (1992), has been widely used in the literature (e.g., Kaltsas et al., 2007, Guzmán and Alonso, 2008; Alonso and Guzmán, 2010). A further step, which would still fit within our definition of "embodied energy," is to take into account the energy required to produce the food consumed by the labor. This indirect energy input of the human diet would depend on the energy efficiency of the food production system, which has been estimated to be 7.3 energy units consumed per dietary unit energy in the modern U.S. agrifood system (Heller and Keoleian, 2003). The marginal substitution ratio (Fluck, 1992), also called marginal energy requirement of employment (Jones, 1989) represents the additional energy produced by the agricultural system per each hour of added labor at a given yield and technological level, and is calcu-lated using iso-yield functions. This approach has rarely been followed in the lit-erature. Finally, the widest system boundary to consider is the energy required for supporting the lifestyle pattern of the worker, and in some cases also his family or the people who depend upon him (male workers are usually assumed). Giampietro and Pimentel (1990) estimated this energy by extending labor energy to the whole per capita energy use in society, obtaining an energy cost of 151–250 MJ/h. Another approach that maintains the philosophy of the lifestyle support energy methods but also tries to avoid double counting is net energy analysis. Fluck (1981) provided a value of 74.3 MJ/h for agricultural labor in the United States in 1973. The reviewed methods for the estimation of the embodied energy of human labor are visually compared in Figure 4.1.

In conclusion, there is a wide disparity of criteria to account for human labor, which yield values that might differ in two orders of magnitude. Thus, human labor represents the clear example of the importance of the definition of system boundaries in line with the study objective. For our purposes, we consider most appropriate to employ the total dietary energy of human labor (2.2 MJ/h), which partially includes embodied energy of labor but avoids a possible problem of cir-cular reference if the embodied energy of the food consumed by the labor was included.

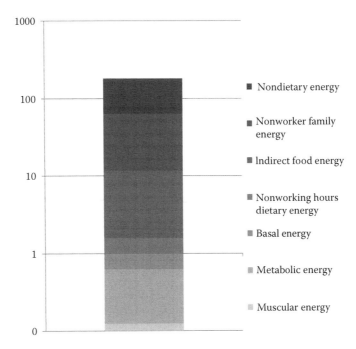

Figure 4.1 An ideal composition of the energy expenditure of human labor (MJ/h). (Author data.)

4.4 ENERGY CARRIERS (FUELS AND ELECTRICITY)

Fossil fuels are widely used as the main direct energy source in mechanized agriculture, and they are also employed in the production of all other industrial inputs. As energy carriers, the embodied energy of fuels and electricity include their direct and production energy. From a historical perspective, there have been large changes in the EROIs of fossil fuels and in the efficiency of electricity power generation, resulting in changes in the energy requirements and embodied energy of these energy carriers.

Direct energy use in fuel and electricity consumption refers to the fairly constant and well-defined gross energy (GE) content of fuels and to the consumption of electricity. We also provide typical density and net energy values (Tables 4.2 and AII.1.1), gathered from the sources detailed in Appendix II.

Indirect energy use in fuel production refers to the energy invested in extracting and transporting the resource, transforming it into a commercial fuel (refining) and distributing the fuel. We estimated the evolution of the EROI of world oil, gas, and coal production based on the data by Hall et al. (2014) and Guilford et al. (2011), as described in Appendix II. Refining oil to commercial fuels such as gasoline and diesel has an energy efficiency of 83%–94% in the United States (Wang, 2008). There exist opposite historical trends in refining energy requirements: on the one side, efficiency gains due to technological improvements; on the other side, the fact that much of the new oil production is heavy oil with a higher sulfur content and therefore requires more energy to refine (Bredeson et al., 2010; Karras, 2010). In addition, environmental and health regulations are imposing higher refining costs (Guseo, 2011). The procedures for the estimation of the evolution of refining energy costs and transport distances and embodied energy are described in Appendix II.

Table 4.2   Density and Gross Energy (Higher Heating Value) of Fossil Fuels Selected in This Work

	Density g/L	Higher Heating Value	
		MJ Gross Energy/kg	MJ Gross Energy/L
Fuel oil, kerosene	802.6	46.2	37.1
Gasoline	740.7	47.1	34.9
Diesel	843.9	45.7	38.5
Naphta	690.6	47.7	33.0
Distillates	823.3	45.9	37.8
LPG	522.2	50.1	26.2
Natural gas (m³)	799.6	50.4	40.0
Average liquids	795.7	46.3	36.9
Coal		22.4	

Sources: IEA (ed.), *Energy Statistics Manual*, OECD Publishing, Paris, 2004; Frischknecht, R. et al., *Ecoinvent Report No. 1*, Dübendorf, Swiss Centre for Life Cycle Inventories, 2007b; Audsley, E. et al., *Final Report Concerted Action AIR3-CT94-2028*, European Commission DG VI Agriculture (ed.), 2003.

The resulting total energy requirements values, including resource extraction, raw resource transport, refinery or processing energy and distribution of oil products to the farm are shown in Table 4.3.

The energy embodied in electricity refers to the amount of energy consumed to produce and deliver electricity, including fuel energy and the energy required to produce the fuels and the facilities employed in electricity production, as well as the grid losses and the maintenance of the grid infrastructure until the electricity reaches the final consumer. We estimated total energy requirements of electricity production for each energy source (Table 4.4). Using these embodied energy coefficients and the energy mix described in Appendix II, we reconstructed the world average embodied energy of electricity production from 1930 to 2010, also including the series in Table 4.4. In addition, we included the embodied energy of electricity at the point of use (after grid losses). On the other hand, grid construction and maintenance may

Table 4.3 **Historical Evolution of the Total Energy Requirements for the Production, Refining, and Transport of the Major Fossil Fuels (MJ/MJ Direct), 1900–2010**

	1900	1910	1920	1930	1940	1950	1960	1970	1980	1990	2000	2010
Fuel oil	0.16	0.15	0.15	0.15	0.15	0.15	0.16	0.17	0.16	0.16	0.17	0.19
Gasoline	0.24	0.24	0.23	0.23	0.23	0.23	0.24	0.25	0.24	0.24	0.26	0.28
Diesel	0.21	0.21	0.20	0.20	0.20	0.20	0.21	0.22	0.21	0.21	0.22	0.24
Oil fuels	0.20	0.20	0.20	0.19	0.19	0.19	0.20	0.21	0.20	0.20	0.22	0.24
Coal	0.10	0.10	0.10	0.10	0.10	0.10	0.06	0.06	0.05	0.05	0.05	0.05
Natural gas	0.04	0.04	0.04	0.04	0.03	0.03	0.03	0.03	0.04	0.05	0.07	0.08

Source: Author data.

Table 4.4 **Historical Evolution of Total Embodied Energy of Electricity Production with Different Energy Sources and in the Global Mix, at Power Plant Gate, 1930–2010 (MJ/MJ Electricity)**

	1930	1940	1950	1960	1970	1980	1990	2000	2010
Coal	4.95	4.18	3.68	2.63	2.67	2.92	2.93	3.00	3.14
Oil	5.74	4.78	4.23	3.43	3.38	3.19	3.16	3.28	3.32
Natural gas	6.22	5.47	4.67	3.58	3.45	3.49	3.31	3.15	2.66
Nuclear				3.26	3.26	3.26	3.26	3.26	3.26
Hydro	1.05	1.05	1.05	1.05	1.05	1.05	1.05	1.05	1.05
Solar					1.17	1.17	1.17	1.17	1.17
Wind					1.05	1.05	1.05	1.05	1.05
Mix, plant gate	4.88	3.89	3.22	2.37	2.36	2.58	2.64	2.71	2.62
Mix, farm gate	5.29	4.21	3.48	2.57	2.55	2.79	2.86	2.93	2.84

Source: Author data.
Note: The global mix is also expressed at the consumer point (farm gate).

also consume significant amounts of energy, but we did not have enough data to model their contribution to electricity embodied energy.

4.5 RAW MATERIALS

Steel and other iron-based materials are the basic components of machinery, and their production is responsible for the majority of machinery production energy requirements. These materials are also a major component of irrigation systems, greenhouse infrastructures, and buildings. The energy efficiency of iron smelting has drastically increased in the last 250 years (Smil, 1999; IEA, 2007; Dahmus, 2014). From 1760 to 1990, the direct energy required to smelt pig iron decreased from 270 to 16 MJ/kg (Smil, 1999). Aluminum production has also experienced significant efficiency gains (Dahmus, 2014; IEA, 2007). As electricity is the main energy source of aluminum production, the changes are affected by electricity efficiency gains. The estimated evolution of the embodied energy, including direct and indirect energy requirements, of all metallic materials is shown in Table 4.5.

A wide range of nonmetallic materials are used in agricultural systems. Plastics are probably the most important ones from an energy point of view. They are widely used for the manufacture of pipes and greenhouse covers, among other uses. An examination of sources used in the literature reveals important differences between estimated energy requirements in the early 1970s (Batty and Keller, 1980) and those in the early 2000s (Ambrose et al., 2002; Piratla et al., 2012; Du et al., 2013). Concrete is commonly used for the foundations of greenhouses and in the construction of ditches and other irrigation infrastructure. We offer estimations of the evolution of the embodied energy of cement, concrete, and reinforced concrete. Glass is used in greenhouses, buildings, and machinery. Table 4.6 shows the estimated energy requirements of all nonmetallic materials studied.

Table 4.5 Historical Evolution of Total Embodied Energy of Metallic Materials Used in Agricultural Systems, 1910–2010 (MJ/kg)

	1910	1920	1930	1940	1950	1960	1970	1980	1990	2000	2010
Pig iron	69.3	59.2	51.7	46.9	44.8	33.4	28.9	27.4	23.1	23.1	23.1
Steel (machin.)	70.9	60.5	52.9	48.0	45.8	34.1	29.5	28.1	23.6	23.6	23.6
Steel (irrig.)			73.3	65.6	61.4	50.1	42.4	39.4	33.9	32.2	32.2
Chromium steel			189.5	165.4	150.2	140.7	115.6	103.5	92.1	80.7	80.7
Lead		42.3	36.1	31.5	28.6	26.8	22.0	19.7	17.5	15.4	15.4
Aluminum			540.0	390.1	297.2	196.9	180.4	162.7	150.8	146.2	142.2
Other metals		102.2	87.3	76.2	69.2	64.9	53.3	47.7	42.4	37.2	37.2

Source: Author data.

Table 4.6 Historical Evolution of Total Embodied Energy of Nonmetallic Materials Used in Agricultural Systems, 1950–2010 (MJ/kg)

	1930	1940	1950	1960	1970	1980	1990	2000	2010
Plastics									
Polyethylene (PE)			264.7	205.8	160.0	124.4	96.7	75.2	58.5
PVC		192.2	164.2	140.4	120.0	102.5	87.6	74.9	64.0
PVC-O							102.8	87.9	75.1
Plexiglass			314.6	260.5	215.8	178.7	148.0	122.6	101.5
Construction									
Cement	13.8	12.2	10.7	9.5	8.3	7.4	6.5	5.7	5.0
Concrete	1.8	1.6	1.4	1.2	1.1	1.0	0.9	0.8	0.7
Reinforced concrete	5.0	4.3	3.8	3.4	3.1	2.5	2.2	2.0	1.7
Other									
Glass			26.0	21.0	17.9	15.6	13.5	11.8	10.3

Source: Author data.

4.6 TRACTION POWER

Agricultural operations in traditional systems were made with renewable local materials and powered by animal and human power sustained mainly on on-farm production. At the beginning of industrialization, steam power engines provided motive power for some particular tasks, but most of the power was provided by animals. The invention of internal combustion engines and their application to farm machinery resulted in the first tractors in the early twentieth century. The mechanization process took place in various periods during the twentieth century in different parts of the world, and it is still ongoing in some areas.

A common approach to estimate the energy embodied in animal power is to allocate all gross energy of feed to animal work, as many draught animals do not have any other significant purpose in the agroecosystem. Therefore, their replacement and maintenance costs can be attributed solely to work. González de Molina and Guzmán (2006) offered values of 938 MJ/working day for a team of 2 equids (mules or horses) (469 MJ/working day per animal head) and 1060 MJ/working day for a team of 2 oxen (530 MJ/working day per ox). The number of days worked by the animals was 188 days. In the case of double-purpose animals (production of meat and/or milk and work), some authors have segregated the gross energy employed by the animal in food production from that employed in work (Zerbini and Shapiro, 1997).

The energy consumption of machinery is attributable to four factors: production of raw materials, manufacture, repair and maintenance, and fuel consumption. For the first three factors, related to the embodied energy of the machinery itself, we largely follow the approach developed by Doering (1980), based on raw materials embodied energy, manufacture energy, and the energy employed in repairs and maintenance expressed as a proportion of original equipment energy costs.

Different works have estimated the energy consumption in the production of farm machinery, and Stout and McKiernan (1992) outlined some changes in the energy requirements that have taken place during the technological development of farm machinery.

Machinery design has greatly changed during the history of mechanized agriculture. The first step was the use of metals in farm implements. The first decades of the nineteenth century witnessed the invention of cast iron and steel ploughs and other tillage and farm machines such as threshing machines. Steam traction engines were already available in the last decades of the nineteenth century, but they were heavy, dangerous machines and their yearly installed capacity never grew above that of horses. The invention of the tractor in the turn of the century was followed by important improvements in tractor design (White, 2008). The average power of tractors has increased (Aguilera et al., 2015c) and their specific weight has decreased (Appendix II, Table AII.4.1) significantly during their history.

We described the evolution of the embodied energy of the raw materials employed in machinery manufacture in Section 4.5. The next step to calculate machinery energy requirements is to know the relative share of each material in machinery composition. Steel is the major component of machinery both in terms of weight and raw materials energy requirements. Lighter and more efficient engines imply lower material consumption in machinery production and lower fuel consumption in machinery, but more energy is demanded for the use of scarce metals in alloys or more complex production processes (Stout and McKiernan, 1992). The use of more energy-intensive materials in machinery construction has increased in the last two decades with the increasing use of electronics. We estimated the historical changes in the composition of the machinery (Appendix II, Tables AII.4.2 through AII.4.4). Multiplying raw materials embodied energy by their relative share in the composition of the machinery, we estimated the evolution of the energy embodied in raw materials per kilogram of machinery (Tables AII.4.6 through AII.4.8) and per kilowatt engine rated power (Table AII.4.9).

We also estimated the energy embodied in manufacture (Tables AII.4.10 and AII.4.13 through AII.4.15) and maintenance and repairs (Table AII.4.10) of farm machinery, as described in Appendix II. The sum of raw materials production, manufacture, and maintenance and repairs results in the total embodied energy values shown in Table AII.4.16, and the disaggregated values in Table AII.4.17.

Once we have all energy inputs related to machinery production and maintenance, we have to know the average useful life to estimate an hourly machinery energy use. The published estimations suggest that the average useful life of self-propelled farm machinery has changed over time (see Appendix II and Table AII.4.11). We multiplied the embodied energy per kilogram of machinery by the specific weight and by an example rated power (50 kW) and divided by the useful life to obtain hourly embodied energy values for self-propelled machinery use along the studied period (Figure 4.2 for tractors, and Table AII.4.18 for all machinery). We also estimated the evolution of the energy intensity of 1 hour of use of each implement (Table AII.4.19), multiplying the embodied energy per kilogram of machinery by the specific weight of each implement, and dividing by the useful life.

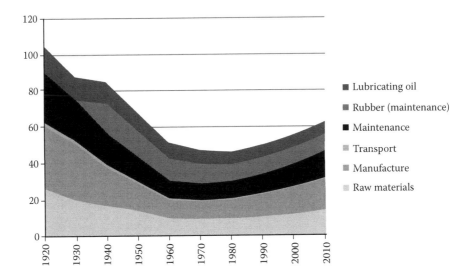

Figure 4.2   Historical evolution of the embodied energy of the hourly use of a 50 kW tractor, 1920–2010 (MJ/h). (Author data.)

When there is a lack of data on fuel consumption, it is necessary to estimate fuel consumption based on the available management information. Typical values of hourly fuel consumption would depend on the efficiency of the engine, which has changed over time (Appendix II, Table AII.4.21). In Table AII.4.22, we show our estimation of the evolution of the specific fuel consumption, taking into account field conditions. The values in Table AII.4.22 represent parameter c in the following equation, which can be used to estimate fuel consumption of a tractor of a given rated power:

$$FC = c \times P \times R$$

where FC is fuel consumption (L/h), c is the specific fuel consumption under field conditions (L/kWh), P is the rated power of the machinery (kW), and R is the ratio of the equivalent power to the rated power (the percentage of the full load that is being used). We show reference values of R for typical tasks in Table AII.4.23. In Table AII.4.24, we show our estimation of direct and indirect fuel energy and machinery production and maintenance energy use per hour of tillage work and kilowatt of rated tractor power, applying our own embodied energy coefficients and assuming that the task is performed with a 50 kW tractor at full load.

If data on hourly tractor use are not available, we can use literature values of time spent in each task for a given level of tractor power (Table AII.4.25). We used these values to construct series of typical fuel consumption and total energy requirements for each agricultural task (Figure 4.3, Tables AII.4.26 and AII.4.27).

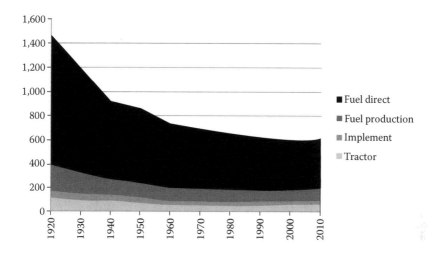

Figure 4.3 Historical evolution of total embodied energy per hectare for a tillage (cultivator) operation with a 50 kWh tractor, 1920–2010 (MJ/ha). (Author data.)

4.7 SYNTHETIC FERTILIZERS AND PESTICIDES

The industrial production of mineral fertilizers started in the mid-to-late nineteenth century and grew rapidly during the twentieth century, while the use of synthetic fertilizers in combination with new crop varieties was associated to major yield increases (Isherwood, 2003). Nutrients in synthetic fertilizers were more concentrated (and thus easier to handle) and easily assimilated by plants than nutrients in organic fertilizers. Table AII.5.1 shows nutrient content of some common mineral fertilizers.

The artificial fixation of *nitrogen* (N) and its industrial development has been one of the major events in agricultural history. The first external nitrogen sources for agriculture were guano and Chilean Nitrate ($NaNO_3^-$), but they were physically and geographically limited. Ammonia gas recovered from coke ovens represented a significant fraction of total world supply of mineral nitrogen in the beginning of the twentieth century, although it was limited by the low quantity of nitrogen contained in coal (1%–1.6%) and by the inefficiency of the process. An obvious alternative to these limited nitrogen sources was to exploit the enormous stock of this element contained in the atmosphere, but this task proved to be technically challenging. After, initial attempts with cyanamide and electric arch, the breakthrough discovery for ammonia synthesis was known as the Haber–Bosch process, which drastically reduced the energy need for ammonia production and became the first global source of mineral N in the early 1930s. After first developments with coal, natural gas soon became the main source of H and energy for the process. Heavy fuel oil and coal are still common today, although more energy intensive (Rafiqul et al., 2005). The energy efficiency of NH_3 production increased rapidly from more than

100 GJ/Mg N–NH$_3$ after the invention of the Haber–Bosch process to 30 and 44 GJ/Mg N–NH$_3$ in best and average modern plants, respectively (Smil, 2001b; Jenssen and Kongshaug, 2003), and about 63 GJ/Mg N when upstream energy consumption is also accounted for (Kool et al., 2012). Ammonia undergoes further chemical and physical processes until obtaining commercial fertilizers. The energy required for these processes has also experienced significant reductions in the last decades, in some cases resulting in net energy exports (Ramírez and Worrell, 2006; Jenssen and Kongshaug, 2003). However, the composition of the mix of fertilizer types employed has also changed, and now more energy-intensive fertilizers such as urea are more common (Ramírez and Worrell, 2006). Total embodied energy series of some selected fertilizers are shown in Figure 4.4 and Tables AII.5.11 through AII.5.13, while the calculation procedure is described in Appendix II, and the disaggregated data for all N fertilizers are shown in Tables AII.5.2 through AII.5.10.

We can identify a trend toward increased energy efficiency in most of the studied fertilizers. However, the weighted average trend during the first decades of the twentieth century suggest an increasing energy consumption due to the transition from mining and subproduct sources of N to artificially fixed sources, which were still very inefficient at this time. We must acknowledge, however, the high uncertainty of our estimations during this early period, particularly regarding to transport distances and efficiency assumptions. On the other hand, the rate of efficiency gain of Haber–Bosch ammonia is very high during the first half of the studied period but it is greatly reduced from around 1970, as some of the efficiency gains in ammonia production are offset by increases in feedstock production energy and the shift to more energy-intensive production countries (China).

Agricultural *phosphorus* (P) sources were of organic origin up to the mid-nineteenth century, being recycled mainly from crop residues and animal manure. Some sources such as guano and slag from iron ore were developed during the

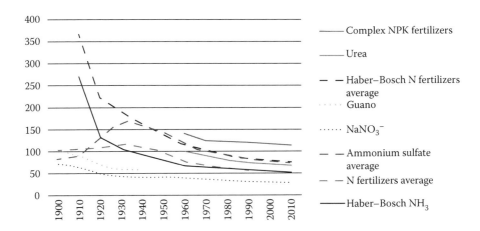

Figure 4.4 Historical evolution of total embodied energy of selected N fertilizers and NH$_3$, 1900–2010 (GJ/Mg N). (Author data.)

nineteenth century, but they never represented a large share of global phosphorus use. Superphosphate production by acidifying mineral phosphates with sulfuric acid allowed phosphates to be easily released into the soil and absorbed by plants. Thus superphosphate fertilizers became the main external source of phosphorus to agricultural systems before the end of the nineteenth century, more than half of total inputs by 1955, and about 85% since 1975 (Cordell, 2009). Energy requirements involve mining and beneficiation of phosphate ore, sulfur production at crude oil refinery, phosphate rock and sulfur transport, sulfuric acid production, superphosphate manufacturing, and granulation of the final product. The process involving the exothermic reaction of rock phosphate and sulfuric acid generates useful energy (steam) in modern plants and consumes it in old ones (Jenssen and Kongshaug, 2003). The literature shows a relatively high variability in the energy requirements of phosphate fertilizers (Aguilera et al., 2015c). Table 4.7 shows our estimation of the evolution of total embodied energy of the main phosphate fertilizers, calculated as explained in Appendix II, while Tables AII.6.1 through AII.6.7 show disaggregated values.

Potassium (K) is an essential nutrient of plants. Potash fertilizers include many K-bearing minerals, of which the most important is potassium chloride (KCl), also known as muriate of potash (MOP). The first mines of potash were opened in Germany in 1861, but the use of potassium as fertilizer really took off in the 1960s with the development of Canadian mines (Khan et al., 2014). Energy use in potash production includes mining and processing of the ores, as well as packaging and transport of the final products. Published values of potash embodied energy range between 4 and 14 MJ/kg K_2O for recent periods (Aguilera et al., 2015c). Our own estimations are shown in Table 4.8, described in Appendix II and disaggregated in Tables AII.7.1 through AII.7.7.

Pesticide use in agriculture has been recorded since ancient times (Taylor et al., 2007), including compounds based on sulfur, arsenic salt, or plant extracts. New plant extracts such as rotenone or tobacco appeared in the seventeenth century, and

Table 4.7 Historical Evolution of Total Embodied Energy of Phosphate Fertilizer Production, 1950–2010 (MJ/kg P_2O_5)

	1950	1960	1970	1980	1990	2000	2010
PK 22–22	107.1	77.5	61.9	49.2	41.3	36.2	32.1
AP	45.6	33.6	27.4	22.4	19.3	17.1	15.4
TSP	56.6	41.6	33.5	27.4	23.4	21.0	19.1
SSP	59.4	42.3	35.9	30.4	27.8	26.3	25.2
MAP	50.5	37.5	30.5	24.8	21.2	18.7	16.6
DAP	40.8	29.7	24.3	19.9	17.4	15.6	14.2
NPK	71.5	52.3	41.3	34.1	28.2	23.8	20.3
Slag	55.8	34.6	31.8	28.1	29.2	30.5	31.7
Ground rock	20.6	14.0	13.1	12.0	12.3	12.7	13.1
P fertilizers average	56.8	40.0	32.9	27.0	23.5	20.4	18.5

Source: Author data.

Table 4.8   Historical Evolution of Total Embodied Energy of Potassium Fertilizer Production, 1950–2010 (MJ/kg K₂O)

	1900	1910	1920	1930	1940	1950	1960	1970	1980	1990	2000	2010
KCl	21.3	20.8	20.4	19.9	19.4	19.4	15.3	14.4	13.3	13.0	12.7	12.4
NPK						24.0	19.0	18.2	18.0	18.8	19.4	19.4
K fertilizers average	21.3	20.8	20.4	19.9	19.4	19.4	15.7	15.3	14.6	14.9	14.6	14.4

Source: Author data.

in the nineteenth century other pesticides such as pyrethrum, derris, copper sulfate compounds or mixtures, copper–arsenic mixtures, and petroleum oils. Stronger pesticides, such as those based in lead–arsenate or organic mercury compounds, expanded in the late nineteenth and early twentieth centuries, triggered by the development of spraying methods. The production of modern pesticides started in the 1930s with the first synthetic organic chemicals, and remarkably with the discovery and expansion of dichlorodiphenyltrichloroethane (DDT) use as insecticide. New regulations in the 1960s and 1970s responded to environmental and health concerns about early pesticides. New pesticides usually required more energy to be produced and were used in larger quantities per hectare, although after ca. 1980 the recommended application doses of most pesticides decreased (Audsley et al., 2009). Two series are shown in Table 4.9: one of the total energy requirements of new pesticides released in each period, including active ingredient production, formulation, packaging and transport energy, and another one of the estimated total energy requirements of the pesticides actually used in each period, also including all embodied energy components. All disaggregated data are shown in Table AII.8.1. We also provide a table compiling all openly published values of individual pesticides (Table AII.8.2).

We propose that the embodied energy of a given pesticide in a certain time point could be calculated summing the active matter embodied energy values in Table AII.8.2 with the formulation, packaging, and transport embodied energy of that specific year provided in Table AII.8.1. If the active matter of the pesticide under study is not included in Table AII.8.2, its energy could be estimated based on its release date, following the equation in Audsley et al., (2009) ("New pesticides" series in Table AII.8.1). If this information is not available, the average production values for used pesticides in each period could be used ("Average used pesticides" series in Table AII.8.1).

4.8 IRRIGATION

By removing water limitation, irrigation is associated to productivity increases in water deficit areas, and now contributes significantly to the overall primary productivity of global croplands (Ozdogan, 2011). The energy embodied in irrigation involves the energy required to extract the water, store it, deliver it to the farm, and distribute it within the field. Modern irrigation systems such as drip irrigation or sprinkle systems lower the amount of water used for irrigation but usually show

Table 4.9   Historical Evolution of Total Embodied Energy of Synthetic Pesticides, 1940–2010 (MJ/kg Active Ingredient)

	1940	1950	1960	1970	1980	1990	2000	2010
New pesticides	76	184	293	390	497	603	712	820
Average used pesticides	76	130	185	228	281	333	388	442

Source: Author data.

increased energy demand per cubic meter of water used due to pressurizing requirements and the use of more energy-intensive water sources (Daccache et al., 2014).

Energy is directly used in irrigation by electric or diesel pumps. Increased pressurizing needs make trickle irrigation less energy efficient when water energy cost is low, but decreased water consumption in this type of irrigation increases the overall efficiency when water energy cost is high. This relationship can be observed in the data provided by Batty and Heller (1980), who estimated energy requirements for various types of irrigation systems taking into account the efficiency in water delivery of each system. In Appendix II (see Tables AII.9.1 and AII.9.2), we include the irrigation efficiency and head pressures data of Batty and Heller (1980) and we use them to calculate direct electricity energy requirements per 500 mm net irrigation per hectare (Table AII.9.3).

Indirect energy for the production of electricity or fuel should preferentially be calculated using specific information about the energy mix of electricity power generation used by the system. If this information is not available at a local level, our estimations of the global average energy efficiency of electric power generation could also be used (Section 4.4) to estimate total embodied energy use in pumping (Tables AII.9.4 through AII.9.6).

In addition to direct energy consumption, the energy embodied in irrigation infrastructure is the other major component of irrigation energy requirements. Main types of irrigation systems are surface irrigation (with or without runoff return system, IRRS), sprinkler irrigation (solid-set, permanent, hand-moved, sider-roll, center pivot, and traveler), and trickle irrigation (Batty and Keller, 1980). The differences observed in the estimations of energy requirements of irrigation systems by Batty and Keller (1980) are mainly due to differences in their material requirements. In turn, these material requirements depend on their useful lives (see Appendix II). The materials used for irrigation pipelines and equipment have changed over time from metal to plastic (Melby, 1995). Plastic pipes can have relatively thin walls and thus low mass per meter pipe (Piratla et al., 2012). Besides the dynamic factors of metallic and nonmetallic materials calculated in Section 4.5, in Table 4.10 we show the energy required for manufacturing metallic components and those of grading and ditching.

We classified irrigation systems into four categories: surface with or without IRRS, sprinkler, and drip irrigation. The estimations of infrastructure energy are described in Appendix II and the coefficients are shown in Tables AII.9.7 through

Table 4.10   Historical Evolution of Total Embodied Energy of Some Irrigation System Processes, 1930–2010 (MJ/unit)

	Unit	1930	1940	1950	1960	1970	1980	1990	2000	2010
Manufacture metallic	kg	39	31	26	19	19	21	21	21	21
Grading (m³)	m³	15	15	15	15	15	15	15	15	15
Ditching (m)	m	57	54	52	50	48	46	45	43	42

Source: Author data.

AII.9.19. Surface irrigation systems without IRRS typically require very little infrastructure energy, mainly for earth movements and concrete ditches. Sprinkler irrigation systems show a wide variability of material requirements, but usually PVC tubes and metal components are the major contributors to energy requirements. Trickle irrigation systems are usually very energy demanding due to the high amount of polyethylene used and its relatively short lifetime. Our results suggest that the embodied energy of the different irrigation infrastructure technologies has decreased considerably in the studied period due to the use of lower amounts of materials and increased energy efficiency of materials production.

Total energy use in irrigation results from the sum of direct energy use, indirect energy for energy carrier production, and embodied energy of irrigation system materials. Wider system boundaries, including the water distribution network, could also be applied. We provide an example of total irrigation energy requirements for 500 mm net irrigation using water from 0 to 100 m depth wells with the four types of irrigation systems studied, assuming that the energy used is electricity, which is produced with the world average efficiency calculated in Section 4.4. The results are shown in Figure 4.5 and Tables AII.9.20 through AII.9.26.

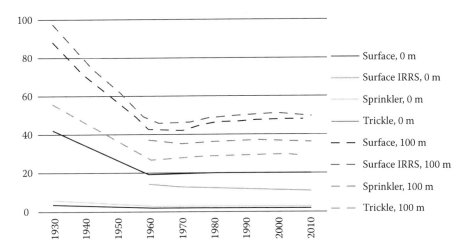

Figure 4.5   Historical evolution of total embodied energy for the net application of 500 mm water in one hectare with different irrigation systems using water from 0 (solid lines) to 100 m (dashed lines) wells, 1930–2010 (GJ/ha). (Author data.)

4.9 OTHER INFRASTRUCTURE

The relative importance of *buildings* in the overall energy balance of agricultural systems is generally very low. For example, they represented 0.1% or less of total energy consumed in a set of apple cropping systems in the United States (Funt, 1980). We did not estimate the historical evolution of farm buildings energy costs, given the lack of available information and the relative low contribution of this input to total energy use. We suggest applying the value of residential buildings from Doering (1980) to farm machinery buildings, whereas the value of industrial buildings from Audsley et al. (2003) could be applied to buildings for intensive livestock production.

Greenhouses are structures that allow trapping solar heat, thus overcoming temperature limitations of certain crops in cold areas or during cold months. There is a high variety of greenhouse types, covering more or less permanent structures with more or less heat trapping capacity. Glass greenhouses typically require a very high energy investment for their construction, whereas plastic greenhouses typically require much lower initial energy investment. Plastic covers have a very limited useful life, of 1.5–3 years. In Appendix II we provide some examples of the typical life-cycle inventory and the historical evolution of the estimated energy requirements of some greenhouse types from the literature, including Almeria "Parral" type, glass greenhouse in Austria, tunnel greenhouse in Austria, and multitunnel in Spain (Alonso and Guzmán, 2010; Theurl et al., 2013) (see Appendix II). The resulting embodied energy values are shown in Table 4.11. The complete set of coefficients are provided in Tables AII.10.1 through AII.10.8.

4.10 TRANSPORT

Transport is a required process in many stages of the production chain of agricultural inputs, from distribution of fuels and raw materials to manufacturing plants to the final distribution of manufactured products to regional stores and finally to the farms. Freight energy efficiency depends on transport mode, the efficiency of the given transport mode in the selected place and time and the efficiency of the

Table 4.11   Historical Evolution of Total Embodied Energy of Greenhouse Infrastructures, 1950–2010 (GJ/ha yr)

	1950	1960	1970	1980	1990	2000	2010
Almeria vineyard	388	304	243	196	159	132	111
Glass, Austria	1623	1231	1088	996	904	857	817
Tunnel, Austria	194	151	125	109	93	83	76
Multitunnel, Spain	1222	952	786	679	578	515	468

Source: Author data.

production of the materials and energy carriers used in transport. Moreover, the energy embodied in transport of farm inputs to the farm also depends on the distance traveled in each transport mode.

The direct energy efficiency of each transport mode usually increased along history, although there are many exceptions in certain modes, time periods or countries (e.g., Dahmus, 2014; Kamakaté and Schipper, 2009; Ruzzenenti and Basosi, 2009). We estimated the evolution in the energy efficiency of the following transport modes: rail freight transport, road freight transport, and maritime freight transport (Appendix II, Table AII.11.1). The energy efficiency of rail transport has experienced important historical changes, resulting from a combination of changes between technologies and improvements of the technologies. Between 1950 and 1970 the energy efficiency of rail freight in the United States improved due to the substitution of coal-burning steam engines by diesel engines (Hirst, 1973). In parallel, important efficiency gains were achieved in diesel-fueled rail freight transport in the United States (Dahmus, 2014). The changes in road transport are less evident than those in rail or sea transport. Ruzzenenti and Basosi (2009) studied changes in road transport efficiencies in selected EU countries between 1970 and 1998, and in the majority of cases they did not show clear downward trends along the period. The highest transport energy efficiency is achieved by water transport, ranging from 0.1 to more than 1 MJ/Mg-km in the present (Hirst, 1973; Weber and Matthews, 2008; Spielmann et al., 2007; Kamakaté and Schipper, 2009). Sail transport dominated until the nineteenth century. This technology does not require direct external energy inputs, only the embodied energy of ship building and maintenance. By the end of the nineteenth century, however, coal powered steamers had already substituted sail boats by a large extent due to their capacity to achieve higher speeds. The energy efficiency of steamers greatly improved during their history (Geels, 2005). By 1910, internal combustion engines powered by oil fuel started to substitute steamers. The data offered by Stopford (2009) suggest that the increase in energy efficiency was invested in increasing the average speed of the boats. Other transport modes are pipelines and air freight. Energy consumption by pipeline transport was estimated by Hirst (1973) to be 0.73 MJ/Mg-km. Air freight energy consumption data published in the literature shows a large variability, from 10 MJ/Mg-km (Weber and Matthews, 2008) to 37–71 MJ/Mg-km (Hirst, 1973).

Indirect energy in transport is consumed in the production of fuels and electricity, the production and maintenance of vehicle and the construction and maintenance of infrastructure such as ports, roads, and railways. Our estimations are described in Appendix II and the results are shown in Tables AII.11.2 and AII.11.3. Embodied energy coefficients of transport modes result from the sum of direct and indirect energy consumption (Tables 4.12 and AII.11.4).

We estimated distance traveled by farm inputs, in each transport mode (Appendix II and Table AII.11.5). We considered this series a very rough estimation, valid for all industrial inputs as a gross approach. We multiplied the embodied energy coefficients of transport modes by the distance traveled to estimate the energy embodied in transporting industrial inputs to the farm (Tables 4.13 and AII.11.6).

Table 4.12 Historical Evolution of Total Embodied Energy of Transport Modes, 1930–2010 (MJ/Mg-km)

	1930	1940	1950	1960	1970	1980	1990	2000	2010
Truck	5.93	5.92	5.93	5.98	5.77	4.61	4.44	4.31	4.21
Rail	5.86	5.84	6.37	2.12	1.61	0.88	0.75	0.64	0.53
Water (container and bulk)	0.55	0.49	0.44	0.40	0.36	0.32	0.28	0.26	0.24
Water (tanker)	0.27	0.24	0.22	0.20	0.18	0.16	0.14	0.13	0.12

Source: Author data.

Table 4.13 Historical Evolution of Total Embodied Energy of Transporting Industrial Inputs to the Farm, 1900–2010 (MJ/kg)

	1900	1910	1920	1930	1940	1950	1960	1970	1980	1990	2000	2010
Farm inputs				4.11	4.10	4.37	2.26	1.96	1.59	1.71	1.83	1.95
Oil products				1.19	1.18	1.19	1.20	1.16	0.93	0.90	0.87	0.85
Guano, NaNO$_3$	8.89	7.95	5.52	4.54	4.46	4.31	4.02	3.54	3.16	2.92	2.70	2.49

Source: Author data.

4.11 NONINDUSTRIAL INPUTS

Seeds energy includes inherent energy of seeds and the energy required to produce the seeds. The inherent energy of the seeds of grains and pulses can be equaled to the energy content of the corresponding agricultural products, which were reviewed in Appendix I. The energy used in the production of seeds varies widely depending on the energy profile of the seed production system. In any case, seed production in modern agriculture is usually a very sophisticated process. Graboski (2002) estimated that hybrid corn seed required 4.7 times the energy required for commercial corn production. Heichel (1980) classified the methods to account for the fossil energy embodied in seeds. The first method estimates them as a multiple of the enthalpy or the digestible energy content of the seed. The second method assumes that the energy cost of producing the propagation seed is similar to the energy cost of producing the commercial product, and thus subtracts the amount of seed from the total yield of the crop. This method could only be applied when the commercial product and the propagation material are similar. The third method is based on the economic costs of propagation materials. The fourth method reviewed is based on a specific process analysis of the energetic costs of producing the propagation material, using a detailed inventory of its production process. The calculations of Heichel (1980) show that the third method (economic based) is the one that yields the energy values that are most approximated to the ones obtained with the fourth (process analysis) method.

According to Beccaro et al. (2014), a *nursery* is a primary system of crop production, providing materials (seedlings and young plants, in general) for use in secondary systems such as horticulture, orchards, and forestry. The nursery stage of the life cycle of these crops has been usually neglected or overlooked in energy analyses and LCA studies, probably due to the lack of available information on these processes. However, nursery production is an energy-intensive, complex process that has been shown to represent a significant fraction of the ecological footprint of crop production systems (Beccaro et al., 2014).

Replacement of livestock is a frequent input in many agroecosystems. One option to account for this input is to consider that a fraction of the herd has to be replaced every year (González de Molina and Guzmán, 2006). The difference between the replacement fraction and the livestock raised in the agroecosystem is the amount that had to be imported for replacement. In energy terms, the cost of these imported animals would be the reproduction and feeding costs up to their entrance in the agroecosystem. For simplification, the fraction of energy represented by these costs is considered to be equivalent to the same costs within the agroecosystem.

Manure, as other organic materials, is a renewable, energy-rich material, and also nutrient-rich and carbon-rich, which performs numerous ecological functions in the soils. Two methods for estimating this energy are considering the gross energy of manure or the energy value of its major nutrients (González de Molina and Guzmán, 2006). The gross energy of manure is mainly dependent of its dry matter content. Gross energy content (HHV) of manure dry matter ranged 11.9–19.4 MJ/kg in a set of manures and manure mixtures of various species (Choi et al., 2014). Another method for calculating the gross energy of the manure is based on the energy balance partitioning of livestock animals. Starting from gross and metabolizable energies in feed, we can estimate the amount of energy that is rejected as feces (the nonmetabolizable fraction of the gross energy) and methane, and the energy that is metabolized into retained energy, heat, and urine. The application of this method involves the risk of double counting. Fresh manure energy would correspond to the sum of feces and urine. If this manure is collected, it is usually subjected to different types of management that affect its energy content. On the one hand, straw or other bedding materials are usually mixed with the manure in solid manure management systems, adding to the energy of urine and feces. Energy contents of crop residues were reviewed in Appendix I. In addition, different storage methods result in unavoidable losses of organic matter due to mineralization processes. These losses may account for 25%–53% of the carbon, mainly as CO_2 but also as CH_4, and 17%–45% of the nitrogen, mainly as NH_3 but also as N_2O (Pardo et al., 2014). The most common management method, simple storage, is associated to average carbon losses of 42%. Carbon losses can be taken as a proxy for dry matter losses.

Most *organic inputs* to cropland soils are produced within the cropping system in the form of unharvested aboveground and belowground crop residues and weeds. These organic materials reused in the system are very important in energetic terms, and in many occasions their magnitude is much higher than that of the embodied energy of external inputs. As in the case of manure, they provide nutrients but also have other important ecological roles in the system. Therefore, it is necessary to account for them in full energy balances, and they can be used for constructing certain indicators. The estimation of the energy in crop residues usually requires the reconstruction of net primary production (NPP) from crop production data (Chapter 3). Organic inputs may also include external organic residues such as agroindustry waste, municipal solid waste, sewage sludge, or others. These materials are residues and therefore the energy credit for their production is usually not allocated to them but to the main process responsible for their production. Only specific processes addressed to the transformation of the residue for its land application are usually included in their embodied energy, as well as the transport energy from the production source to the field. Some of these processes are drying, composting or unmanaged storing. However, it is necessary to take into account that residues have to be managed in any case. Hence, some residue management energy might be allocated to the main product.

Feed production represents the majority of modern livestock production energy requirements for most animal species (Smith et al., 2015). Feed energy includes the inherent energy content of the ingredients, most of which can be found in Appendix I and may include the energy required to produce the raw agricultural commodities, transport them to the feed production facility, process them, and distribute them to the farm. Pelletier et al. (2014) found that the energy used in the production of feed products in the United States increased from 1960 to 2010 in most cases but the energy efficiency of egg production still increased in the studied period, due to the increase in feed conversion efficiencies of layers.

Auxiliary services are fundamental for the functioning of modern cropping systems. They include financial services to make investments in capital inputs, insurance services to assure that fixed costs are paid during years of harvest failure, administrative services to provide support including research and extension services, agricultural subsidies, or market regulations. However, these services are usually excluded from agricultural energy analyses, which are usually process-based analyses focused only on physical inputs. This gap is covered by studies in which the embodied energy is estimated with input–output models based on the energy intensity of economic sectors (Cleveland, 1995). Hybrid energy analyses (e.g., Suh et al., 2004; Crawford, 2009; Prieto and Hall, 2013) combine process-based analysis with input–output data. This way, the precision of process-based analysis is complemented by the exhaustiveness of input–output analysis.

4.12 SOME CONCLUSIONS

The energy requirements for the production of agricultural inputs have experienced some opposite trends during the historical evolution of agricultural technology. A clear, usually dominating trend toward increased energy efficiency can be identified during the majority of the studied period in most industrial processes involved in inputs production, such as electricity power generation, ammonia production, fertilizer manufacturing or iron smelting. Other technological changes have reduced the material and energy requirements at the farm, such as lighter and more fuel efficient farm machinery and more efficient fertilizers and pesticides.

In spite of these improvements, our results show that efficiency gains are slowing down in recent times. First, the energy efficiency in the production of many materials is approaching the thermodynamic limit (Gutowski et al., 2013). Second, the decreases in the EROI of primary energy sources, particularly of fossil fuels as they approach their production peaks, and the depletion of highly concentrated metal ores, have imposed an additional thermodynamic constraint to the advances in the energy efficiency of industrial processes in the last decades. Third, the changes toward better performing inputs have pushed the demand for more energy-intensive raw materials. Fourth, some production has been delocalized to countries such as China, where industrial energy efficiency is generally low. Fifth, human labor has experienced a spectacular decrease in terms of units used per hectare or unit product, but its embodied energy may have also sharply increased with the rise in societal energy use.

Our estimations unveil the magnitude of the changes that have taken place, underlining the need to account for them in the analysis of agricultural systems and to intensify the research on the changes in the energy efficiency of agricultural inputs. Important knowledge gaps need to be filled in order to be able to make precise energy analyses of the temporal changes in agricultural energy use, especially during sociometabolic transitions and during the development of industrial agricultures. We have aimed to provide approximate values that could be used meanwhile information gaps are filled with specific studies.

4.13 WORKING EXAMPLE: AGRICULTURAL INPUTS IN SPAIN IN 2008

As a working example for the calculation of the embodied energy of the inputs employed in an agricultural system, we have chosen Spanish agriculture in 2008. This system represents a mature modern agricultural system that includes all major inputs described in this chapter. We worked on an average of 5 years around 2008 (2006–2010) to buffer annual variability in the amounts of inputs employed. The total embodied energy of agricultural inputs is basically calculated as the product of the amount of inputs employed, in our case obtained mainly from official statistics, by their embodied energy coefficients, which can be found in Sections 4.3 through 4.11 of this chapter and in Appendix II. As year 2008, to which this example refers, is not included in the series, we estimated 2008 coefficients by linear interpolation between 2000 and 2010 values.

The amount of energy carriers, including electricity and fuels, is provided by official statistics from the Spanish ministry of industry (Ministerio de Energía Industria y Turismo [MINETUR], 2015). We converted these official data, which are expressed as net energy values, to gross energy values using the coefficients in Table AII.1.1. The resulting amounts were multiplied by their embodied energy coefficients to obtain total embodied energy of energy carriers in Spanish agriculture (Table 4.14).

The embodied energy coefficients of fuels were estimated by the sum of direct energy (1, as the primary data are expressed as energy) and energy requirements,

Table 4.14   Embodied Energy of Electricity Consumed in Spain in 2008

	Production in Spain		Energy Requirements for Each Type of Technology			Embodied Energy of the Spanish Mix		
			Direct	Fuel Production	Total at Plant Gate	Total at Plant Gate	Grid Losses	Total at Farm Gate
	TWh	%	MJ/MJ Electricity	MJ/MJ Electricity	MJ/MJ Electricity	MJ/MJ Electricity	MJ/MJ Electricity	MJ/MJ Electricity
Hydro	32.2	11.8	1	0.05	1.05	0.12	0.01	0.13
Coal	46.5	17.1	3.11	0.17	3.27	0.56	0.05	0.61
Oil	3.9	1.4	2.73	0.64	3.37	0.05	0.00	0.05
Gas	94.0	34.6	2.05	0.17	2.22	0.77	0.06	0.83
Nuclear	55.3	20.3	3.06	0.20	3.26	0.66	0.06	0.72
Biomass	4.4	1.6	3.11	0.17	3.27	0.05	0.00	0.06
Solar	3.3	1.2	1	0.17	1.17	0.01	0.00	0.02
Wind	32.3	11.9	1	0.05	1.05	0.12	0.01	0.14
Total	271.9					2.35	0.20	2.55

Sources: MINETUR, *Balances de energía final (1990–2013)*, MINETUR, Madrid, 2015; MITYC (Ministerio de Industria, Turismo y Comercio), *Estadística de la industria de energía eléctrica 2008*, Secretaría General Técnica, Madrid, 2009; REE (Red Eléctrica de España), *Balances de energía eléctrica 1990–2014*, Red Eléctrica Española (ed.), Madrid, 2015; and Appendix II (this volume).

calculated by interpolation of 2000 and 2010 global averages shown in Table AII.1.10. In the case of electricity, we estimated its embodied energy using specific data for Spain, complemented with global data from Tables AII.2 when Spanish data were absent. We used specific Spanish data of energy efficiency of thermal electricity production for each type of technology: coal, oil, and gas (MITYC, 2009). Nuclear was taken from Table AII.2.1, and biomass was equated to coal. Fuel production energy requirements, obtained from Table AII.1.10, were summed with direct energy requirements in electricity production to obtain total embodied energy of electricity production at the farm gate for each type of technology (Table 4.14). Then, we calculated the embodied energy of the Spanish electricity mix taking into account the relative contribution of each source to the mix (REE, 2015; see Table 4.14). The resulting value, representing embodied energy of electricity at farm gate, was summed with grid losses (Section 4.4) to obtain the embodied energy of electricity at the point of consumption (Table 4.14).

In our study, the calculation of machinery emissions is the most complex of all agricultural inputs. The calculation of the annual embodied energy of a given machine can be done multiplying its rated power by the specific weight and embodied energy coefficients of that specific year (Tables AII.4.1 and AII.4.17), and dividing by its useful life (Tables AII.4.11 and AII.4.12). However, the machines employed in the country are a mix of different ages, each of which has different characteristics such as specific weight and embodied energy coefficients. Thus, previously to the multiplication of activity data by energy coefficients, it is necessary to adjust both parameters to the actual machinery mix. We estimated the ages of this mix based on the annual registrations and removals from 1900 to 2008. Every year, we obtained the number of machines and their power form official statistics (Aguilera et al., forthcoming) and the characteristics of the new machines added to the census from the global data in Tables AII.4.1 through AII.4.17 (calculating intermediate years through interpolation). We assumed that each year removals from the census corresponded to the oldest machines. The age of these removed machines was assumed to be the estimated useful life of the machines in a given year. This way, we could estimate the evolution of the total weight and embodied energy of the machines composing the census up to our studied year, 2008. The procedure and the references employed are explained in detail in Aguilera et al. (forthcoming). Some of the parameters employed are shown in Table 4.15.

Once we have prepared our activity data and coefficients, we proceed to multiplying the annualized weight of machinery by the average embodied energy coefficients to obtain the annual embodied energy of the machinery in 2008 (Table 4.16).

We gathered fertilizer and pesticides amounts used in Spain from Spanish official statistics (MAGRAMA, 2013), averaging the data from 2006 to 2010. In the case of sulfate and copper pesticides, we assumed that the amount had remained constant since 1970, the last year with available data. The embodied energy coefficients were estimated through interpolation of the data in Tables AII.5.12, AII.6.5, AII.7.5, and AII.8.1. In the case of N fertilizers, we chose the "World Excluding China" category.

Table 4.15   Parameters for the Calculation of the Embodied Energy of Agricultural Machinery in Spain in 2008

	Unit Power	Total Power	Specific Weight	Total Weight	Useful Life	Annualized Weight	Energy Coefficients	
	Average		New				New	Average
	Units	HP/Unit	MW	kg/u[a]	Tg	Years	Tg/year	MJ/kg	MJ/kg
Tillage implements	1,138,990			985	1,122	25	45	72	83
Other implements	2,138,546			801	1,685	20	84	62	66
Tractors, 2 axes	1,027,076	64	48,130	68	3,439	45	77	158	176
Tractors, 1 axis	281,481	13	2,639	68	204	33	6	62	77
Harvesters	59,267	107	4,678	68	378	42	9	102	113
Other motors	59,433	76	3,320	68	234	25	9	62	60
Irrigation motors			9,042	68		25			
Total machinery			67,808		4,256				

Sources: Aguilera, E. et al., *Industrial Inputs in Spanish Agriculture, 1900–2008*, Agroecosystems History Laboratory Working Paper, forthcoming; and Appendix II (this volume).
[a] Specific weight: In the case of implements, it is referred to units; in the case of self-powered machines, it is referred to kilowatts.

Table 4.16   Annual Embodied Energy of Agricultural Machinery Employed in Spanish
             Agriculture in 2008

	Unit	Amount (units)	Coefficient (GJ/unit)	Total Embodied (TJ)
Tillage implements	Mg (annualized)	44,879	83	3,718
Other implements	Mg (annualized)	84,263	66	5,576
Tractors, 2 axes	Mg (annualized)	77,272	176	13,634
Tractors, 1 axis	Mg (annualized)	6,250	77	479
Harvesters	Mg (annualized)	9,020	113	1,019
Other motors	Mg (annualized)	9,378	60	563
Total machinery		231,062		24,990

Sources: Aguilera, E. et al., *Industrial Inputs in Spanish Agriculture, 1900–2008*, Agroecosystems
History Laboratory Working Paper, forthcoming; Author data.

In the case of pesticides, we chose "Average used pesticides" category. In the case of copper and sulfur pesticides, for which there is no embodied energy coefficient available in Table AII.8.2, we assumed the same coefficient as potassium. The amounts, coefficients, and resulting embodied energy values of fertilizers and pesticides are shown in Table 4.17.

Next, we estimated the embodied energy of some of the infrastructure elements present in Spanish agriculture, particularly those associated to irrigation (on-farm only) and crop protection (Table 4.18). We did not include buildings because of the lack of data. The published official statistics include the surface areas of different types of irrigation (surface, sprinkle, and drip) (MAGRAMA, 2015) and crop protection systems (greenhouses, tunnels, plastic mulches, and sand mulches) (MAGRAMA, 2013). We multiplied these values by the corresponding embodied energy coefficients in Tables AII.9.19 and AII.10.8, interpolating between 2000 and 2010. We took into account that 3% of the greenhouses in Spain were highly technified whereas 97% were low or medium technified. We took "Glass, Austria" category for highly technified and "Almeria, vineyard" for medium and low technified greenhouses.

The number of hours of human labor was estimated based on the number of employed workers in the agriculture sector from the official yearly report that the National Statistics Institute (INE) publishes since 1960 ("Encuesta de Población Activa" report, see www.ine.es). As these data are absent in 1900, we estimated them by subtracting the estimated agriculture sector unemployment rate (15%) from the total number of workers in the agriculture sector. We estimated the energy in human labor as the dietary energy consumption (2.2 MJ/h), based in the data offered by Fluck (1992). This has been widely used in the literature (see a discussion in Section 4.3). The resulting values are shown in Table 4.19.

Imported biomass (feed and seeds) data for 2008 were obtained from DATACOMEX database of Spanish overseas trade (Ministerio de Economía y Competitividad, 2015). For 1960, we used the Food and Agriculture Organization

Table 4.17 **Embodied Energy of Fertilizers and Pesticides Used in Spanish Agriculture in 2008**

	Unit	Amount (units)	Coefficient (GJ/unit)	Total Embodied (TJ)
Saltpeter	Mg N	2,925	29	86
Ammonium sulfate	Mg N	75,698	43	3,288
Ammonium nitrate	Mg N	41,283	56	2,331
Calcium ammonium nitrate	Mg N	199,809	65	12,922
Urea	Mg N	244,492	62	15,205
NPK	Mg N	202,753	108	21,967
Other	Mg N	116,531	68	7,882
Total N	Mg N	883,490	72	63,681
Slag	Mg P_2O_5	1,247	31	39
Superphosphate	Mg P_2O_5	27,552	25	700
NPK	Mg P_2O_5	324,347	21	6,820
Other	Mg P_2O_5	22,943	19	433
Total P	Mg P_2O_5	376,089	21	7,992
Potassium chloride	Mg K_2O	96,736	14	1,396
Potassium sulfate	Mg K_2O	17,547	14	253
NPK	Mg K_2O	221,193	19	4,295
Other	Mg K_2O	91	14	1
Total K	Mg K_2O	335,567	18	5,945
Total nutrients	Mg nutrient	1,595,145	49	77,618
Copper and sulfur	Mg active matter	23,869	14	344
Synthetic pesticides	Mg active matter	38,803	447	17,334
Total pesticides	Mg active matter	62,672	115	17,678

Sources: MAGRAMA, *Anuario de Estadística Agraria 2013*, MAGRAMA, Madrid, 2013; Appendix II (this volume).

of the United Nations (FAO) database (FAOSTAT, 2015). Finally, for 1900, we used overseas trade statistics for Spain.* The energy in imported biomass is expressed as the higher heating value (gross energy) of the different products, which was estimated using the corresponding coefficients in Appendix I. We also included transport energy requirements (based on the data in Section 4.10), but not the energy required to produce the biomass to avoid double counting, as its cost should be allocated to the

---

* The original Trade Yearbooks are available online at the Ministerio de Economía y Competitividad. http://www4.mityc.es/BibliotecaCOM/abwebp.exe. Accessed on April 3, 2015.

Table 4.18   Embodied Energy of Infrastructure Elements of Spanish Agriculture in 2008

	Unit	Amount (Units)	Coefficient (GJ/unit)	Total Embodied (TJ)
Almeria vineyard	Ha	47,971	115	5,538
Glass, Austria	Ha	1,484	829	1,231
Greenhouses, total	Ha	49,454	137	6,769
Tunnels	Ha	12,916	479	6,187
Plastic mulches	Ha	49,603	10	501
Sand mulches	Ha	11,556	0	0
Total greenhouses and mulches	Ha	123,529	626	13,457
Surface	Ha	1,093,610	0.5	537
Sprinkler	Ha	736,119	1.8	1,317
Drip	Ha	1,538,987	2.0	3,124
Total irrigated surface	Ha	3,368,716		4,977

Sources: MAGRAMA, *Anuario de Estadística Agraria 2013*, MAGRAMA, Madrid, 2013; MAGRAMA, *Encuesta sobre superficies y rendimientos de cultivo. Informe sobre regadíos en España*, MAGRAMA, Madrid, 2015; Appendix II (this volume).

Table 4.19   Embodied Energy of Nonindustrial Inputs

	Unit	Amount (Units)	Coefficient (GJ/unit)	Total Embodied (TJ)
Feed	Gg dry matter	10,109	19	187,842
Seeds	Gg dry matter	308	17	5,335
Human labor	Mhour	1,481	2	3,257

Authors' elaboration from agricultural statistics and Appendices I and II.

exporting agroecosystems. The total amounts, coefficients and resulting embodied energy values of seeds and feeds are shown in Table 4.19.

Now we can overview the total embodied energy of external inputs in Spanish agriculture and the relative contribution of each of them and their different functions such as traction, fertilization, irrigation, and crop protection (Figure 4.6). Total embodied energy of industrial inputs in Spanish agriculture in 2008 represented 308 PJ, rising to 505 PJ if nonindustrial inputs are also included. Feed imports were the main energy input to Spanish agriculture, making livestock the component with the largest contribution to total embodied energy (40.3%). Livestock is followed by traction, which represented 21.5%, and fertilization and irrigation (15.4% and 14.9%, respectively), while all other components represented 7.9%, including 3.5% pesticides, 2.7% greenhouses, or just 0.6% human labor. Spanish agriculture in 2008 benefited from relatively low embodied energy coefficients of fertilizers and electricity, which were at their historical minimums (Section 4.4) due to the improvements in the energy efficiency of industrial processes and to the increasing share of renewable

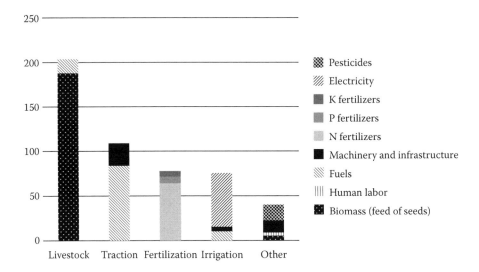

Figure 4.6 Embodied energy of the main industrial inputs of Spanish agriculture in 2008, grouped by function (PJ). (Author data.)

energy in electricity production. On the other hand, the effect of the 2008 crisis on fertilizer consumption is reflected in our 5 years average. Our analysis shows the relative importance of each agricultural input in the overall energy budget, and allows, together with the estimation of the output side (Chapter 3), the calculation of the different EROI indicators.

PART II

Case Studies

CHAPTER 5

Diachronic Analysis at a Local Scale
Santa Fe, Spain

CONTENTS

5.1 INTRODUCTION

There have been few studies of the energy efficiency of the different farming methods over time and in the same agroecosystem. These analyses have usually been made between different agroecosystems or between different farming methods, but synchronously. A long-term analysis in the same system makes it possible to discover the main changes that have occurred since the time when organic farming methods were of low intensity and until the present day. It should not be forgotten, in this regard, that historical analysis is a valuable tool to contrast the different uses made of energy in agrarian systems and to compare their efficiency.

To evaluate its usefulness, we have applied this proposal to an agroecosystem representative of the Mediterranean agroenvironmental conditions whose evolution over the last 250 years has been studied in depth: the Santa Fe agroecosystem in South Spain. A detailed analysis, as we developed (González de Molina and Guzmán, 2006), can only be carried out at a local level. Its soil and climate conditions are representative of the Dry Spain, with crops that are also typical of the Mediterranean (cereals, olives, horticultural products, etc.), where there is both dry and irrigation farming; among all of the different types of agricultural landholding, there are large, medium, and small farms; a place that from very early times has been connected to the markets; a place where problems today are similar to those facing European agriculture as a whole. The changes that have occurred in the ecosystem under study are, then, paradigms of Mediterranean agriculture. The evolution of the agroecosystem from the middle of the eighteenth century has been described in some detail in previous publications (Guzmán Casado and González de Molina, 2007, 2009, 2015), but some information of interest is offered in the following text.

5.2 DATA COLLECTION AND DESCRIPTION OF THE STUDY SITE

The study area is located in the municipality of Santa Fe, in the center of the region of La Vega de Granada, some 12 km west of the city of Granada (Figure 5.1) in the southeast of the Iberian Peninsula. This area has great agricultural potential, being an agroecosystem with a high response to green revolution technologies. Some 85.6% of the municipality's land has a slope of less than 3% (AMA, 1991), deep soils with potentially high production. These mainly fall into the category of Xerofluvents, except for those in the extreme south in sloped areas, which belong to the Great Group Xerochrepts. However, this potential is only fulfilled when access to water is guaranteed through irrigation; this is the key given that annual rainfall is only around 390 mm. Months with excess of water are nonexistent, causing a strong deficit during the summer and early autumn (González de Molina and Guzmán Casado, 2006).

Different degrees of energy efficiency were studied by means of the selection of large synchronic sections to characterize the evolution of the different farming methods that have been applied in Mediterranean agriculture since preindustrial times. In each chronological section, we have tried to characterize the structure and functioning of the agroecosystem, which is a reflection of the methods used by farmers. The different moments of analysis were chosen to combine the availability of detailed

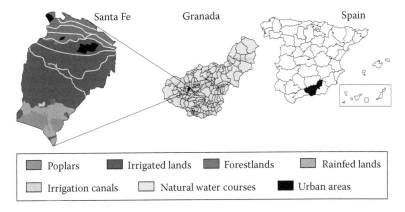

Figure 5.1 Location of Santa Fe case study and land uses at the end of twentieth century.

sources of information with key moments, which accurately capture the transformations in the agroecosystem. As a starting point, we took the mid-eighteenth century, specifically 1750–1752, the date of the Cadaster of the Marquis of Ensenada. There were two reasons for this choice. At that time, the technological changes that would later transform European agriculture had still not begun. Furthermore, the Cadaster of the Marquis of Ensenada is the most detailed source of information available for the moments immediately prior to the introduction of the market and of modern private property. The different documents that make up the Cadaster also provide the essential information needed to accurately reconstruct the structure and operation of the agrarian system before the aforementioned changes took place. The next moment studied is 1904, to capture the consequences of the so-called *Fin de Siècle* Crisis at the turn of the century, where historians pinpoint the beginnings of the "modernization" of Spanish agriculture and the appearance of the first chemical fertilizers. For this period, as well as other sources, we have a detailed report made by the Town Council on the state of agriculture in Santa Fe. The next milestone is 1934, the year in which the decline of beet and the expansion of new crops on the La Vega area of Granada began, a time when organic agriculture had practically peaked and neither mechanization nor the green revolution had yet appeared. The last point in time studied is 1997, the most recent date for which we have homogenous data. The information for 1997 was obtained from secondary sources and complemented with primary information obtained from interviews.

A painstaking process of reviewing historic files was required to obtain the necessary information. This information has allowed us to determine the organization of the agricultural ecosystem in Santa Fe, the structure of the landscape, livestock numbers, distribution of crops, production, goods consumed, human labor required for the tasks, and so on, in 1752, 1904, and 1934. The main sources for these three periods are as follows: the information for 1752 was mostly taken from "Respuestas Particulares del Catastro del Marqués de la Ensenada" available at the Municipal Archive of Santa Fe (A.M.S.), the "Apeos de Marjales de 1754" from A.M.S. box 380, doc. 1, and postmortem inventories (notarized documentation deposited in the Protocol archives in Granada). The information for 1904 is from "Ayuntamiento de Santa Fe, Contestación al cuestionario remitido por el Exc. Sr. Gobernador Civil de la Provincia" (A.M.S. box 391, doc. 1), of the "Estadística de Ganados" (A.M.S. box 49), of the "Junta Consultiva Agronómica para 1919" (1921) on the use of fertilizers, and of the "Cartilla Evaluatoria de 1896" containing the reports for the neighboring town of Pinos Puente, with similar edaphological features, crops, and handling.

The land uses and the crop distribution for 1934 have been calculated from data provided by three different sources: the "Padrón de labradores de Santa Fe, 1929" (A.M.S. box 391, doc. 16); the "Estadística Agrícola. Resumen de las declaraciones presentadas por los agricultores del término municipal de Santa Fe, 1929" (A.M.S. box 391, doc.13); and a summary report corresponding to the 1937–1938 campaign for knowing the distribution of the rain feed surface (A.M.S. box 387, doc. 60). The yields per unit area and other biophysical data have been obtained from the agronomic studies carried out in this area by the Spanish Institute for Agrarian Reform (IRA) in 1932–1936. The basic information for 1997 is found in the following documents: "Superficies

Photograph 5.1   Current landscape in Santa Fe, where poplar groves coexist with traditional
crops such as vegetables and wheat.

ocupadas por los cultivos agrícolas" drawn up by the Local Chamber of Agriculture;
Agricultural Census for 1999 (INE, 1999); Santa Fe Livestock Census, provided by the
District Office for Agriculture; and from López Pérez (1998), which describes the technical itineraries of the crops. Information on the handling of livestock was obtained
by interviewing local livestock farmers. More detailed information can be found in
González de Molina and Guzmán (2006). Véanse las características básicas de la agricultura de Santa Fe en las Tablas 1 y 2.

5.3 AGROECOSYSTEM IN TERRITORIAL EQUILIBRIUM

In the mid-eighteenth century, the agroecosystem was largely geared toward
crop production. A certain level of specialization could already be seen and some
crops were destined for outside markets. Alongside wheat for human consumption
and barley for animal feed, flax and (to a lesser extent) hemp for the making sails and
rigging for the Royal Navy constituted the major part of irrigated cultivation.

The main source of energy was solar, which meant that through the management of
biological converters, all the fuel, foodstuffs, and fibers, along with the feed needed to
maintain draft animals and revenue livestock, had to be obtained from the land. In general, the best land was given over to the production of quality foodstuffs for humans,
while livestock or forestry production took place on less suitable land. In this way,
there was less competition between the different land uses and virtually all the land
could be put to good use. However, the edaphic and climatic conditions of the area and

more generally in the South-Eastern Iberian Peninsula (González de Molina, 2002), did not make multiple uses of the same plot of land feasible, obliging people instead to dedicate extensive areas to the production of timber and firewood, pasture for livestock or crops for human foodstuffs. Nevertheless, the inhabitants of Santa Fe had adapted well to these limitations, striving to make the best possible use of the agroecosystem.

Their organization tended toward a balance between the diverse agricultural uses of the land: each portion of territory was devoted to a particular use and could satisfy the needs generated by the others (Guzmán Casado and González de Molina, 2009). The inhabitants of Santa Fe had only appropriated the part of the territory where they were able to cultivate with their available workforce and level of technology. Poplar and (to a far lesser extent) ash, were planted along the river crossing through the territory, mainly to defend the cultivated land from frequent floods, but also to provide timber and firewood. Offcuts from the pruning of olive trees and firewood from some mulberry and walnut trees also added to these resources, but this was not enough to meet the fuel requirements of the inhabitants. Firewood and timber had to be imported from neighboring wooded areas.

Fodder and forage requirements for cattle were provided by land, which remained unfarmed: thickets and meadows in the flood zone, whose production was short-lived, but above all with the land situated in the highest part of the municipality dominated by thyme (*Thymus* sp.) and esparto grasses (*Stipa tenacissima*), known as *dehesa boyal* (lit. ox meadow) (366 ha, see Figure 5.2). The scarcity of pasture produced on open Mediterranean hillside also forced inhabitants to devote a substantial part of farming land to the production of grains to complement livestock feed. This meant that cereal production for livestock was competing with cereal production for human consumption. Nevertheless, an equilibrium had been reached based on the use of draft animals that satisfied both traction and fertilization needs.

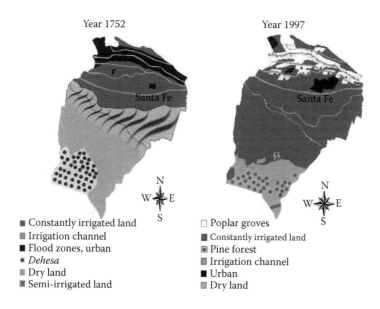

Figure 5.2 Evolution of land use in Santa Fe from 1752 to 1997.

This relative equilibrium in the mid-eighteenth century can be demonstrated with a simple appraisal of the agroecosystem's capacity to replace nutrients exported from each harvest. Annual manure production from livestock in 1752 was sufficient to satisfy the demands of the lands under constant irrigation (these were regularly fertilized). On lands that were watered on an occasional basis (referred to henceforth as *semi-irrigated*), olives and grapevines were grown in association, and above all, wheat and barley, on a 1-year sown/1-year fallow regime (these were not fertilized) (see Table 5.1).

It is worth considering, however, whether it would have been possible to increase the numbers of livestock, especially the draft animals, which had a greater capacity for producing organic fertilizer. In fact, the number of heads per hectare was very low in comparison with the north of the Iberian Peninsula and with other European

Table 5.1   Land Uses, Population, Productivity, and Farm Size in Santa Fe, 1752–1997

Data	1752	1904	1934	1997
Cropland (A)				
Constantly irrigated lands (ha)	288	1,333	1,153	2,134
Semi-irrigated lands (ha)	1,281	466	496	0
Rain feed lands (ha)	1,128	1,239	1,379	785
Forestland (B)				
Poplars groves (ha)	3	53	223	440
Pinar (ha)	0	0	0	210
Pastureland (C)				
Dehesa/pastures (ha)	366	93	93	0
Wet pastureland (ha)	700	600	0	0
Total (A+B+C)	3,763	3,784	3,344	3,569
Population (no. of inhabitants)	2,384	7,228	9,344	12,387
Farms (no.)	314	633	768	405[a]
Active Agrarian Population (no.)	550	1,675	2,123	600
Population density (inhab per km²)	61.7	187.2	242	338.2
Average size of farms (arable ha)	8.6	4.9	4.2	8.8
Land available per inhabitant (ha)	1.28	0.53	0.41	0.29
Arable land available per inhabitant (ha)	1.13	0.43	0.35	0.27
Productivity in Pesetas in 1904				
Final Agricultural Production (pesetas of 1904)	7,00,018	16,67,166	22,91,288	5,771,681
FAP per active worker (pesetas of 1904)	1,276	1,296	2,231	12,749
FAP ha^{-1} (pesetas of 1904)	228	524	685	1,617
FAP per inhabitant (pesetas of 1904)	294	231	245	466

Source: Author data.
[a] Agrarian Census 1999 (INE, 1999).

countries (Wrigley, 1993, pp. 55–56). The type of plants grown required a lot of manual labor and relatively little in terms of traction, but even so, the available live-stock barely covered needs in the months of September, October, and November during harvest, soil preparation, and sowing. Livestock could have been brought in from elsewhere, but there was an iron law of transport in this era where communication networks were severely restricted: it was not worth investing more energy in the haulage of a product than the energetic content of the product itself (Sieferle, 2001). Under these circumstances, an increase in livestock would have forced inhabitants to devote more land to the production of animal feed and forage, reducing the area dedicated to the production of human foodstuffs and raw materials. Table 5.2 shows the precarious equilibrium reached between the nutritional requirements of the live-stock and the production of hay and most importantly, grains, the mainstay of their diet. The maintenance of draft livestock obliged farmers to devote most of their nonirrigated lands to barley, and even to sow it on irrigated land and include forage

Table 5.2   Physical Data for Agricultural Production in Santa Fe, 1754–1997

Concept	1754	1904	1934	1997
Net agricultural production (Mg)	1,737	26,524	16,374	32,444
Net livestock production (Mg)	346	164	567	2,881
Human foodstuffs (Mg)	1,253	10,654	7,427	17,505
Industrial crops (Mg)	68	14,106[b]	4,716	1,200
Animal feed (Mg)[a]	385	1,291	2,263	4,828
Forestry production (Mg)[c]	30	472	1,968	8,910
Heads of cattle	2,609	3,050	3,591	4,220
Traction requirements (heads)	77	113	108	4,156[d]
Draft livestock (heads)	122	225	158	8,780[e]
Feed demand (Mg)[a]	388	770	1,147	–
Production of useful manure (Mg)[f]	2,593	3,552	5,162	20,935
Net demand (Mg of manure)[g]	1,667	22,751	16,949	79,991
Deficit	–	19,199	11,787	59,056[h]
Labour demand (UALs—unit of agricultural labor)	548	1,286	1,027	453
Active agrarian population (UALs)	550	1,675	2,123	600
Water (irrigation) (hm^3 year^{-1})	0.48	3.82	4.53	13.03

Source: Author data.

a   Only grain for draft livestock has been taken into consideration; the total cereal by-products produced by the agroecosystem is enough to satisfy the requirements of this type of livestock.

b   Includes sugar beet (crop destined for sugar factories).

c   Refers to the cultivation of poplars for timber.

d   Represents the equivalent horsepower required for tasks involving tractors, cultivators or combine harvesters.

e   Power installed in equivalent horsepower.

f   Excluding, except for 1997, manure from sheep and goats due to difficulties in collecting this from animals not kept in stables.

g   The calculation for demand was performed via a balance of nutrients (crop extractions—inputs) to standardize the calculations, and to avoid the variations and approximations of historical sources.

h   Manure is no longer used for fertilization, meaning that the real deficit is equal to the demand.

plants or cereals fodder such as broad beans, millet, and maize in their rotations. The area dedicated to producing wheat for bread-making or plants for use in industry, all of which were more marketable and profitable, was restricted.

In the aftermath of the epidemic crises and the price crashes of the first third of the nineteenth century, a new expansion in agricultural activity began that would permanently upset the established equilibrium. The institutional reforms of liberalism stimulated a significant expansion of cultivated land. Population and market growth in the city of Granada meant growing demand for wheat, wine grapes, oil and to a lesser degree pulses and vegetables. The agroecosystem specialized even more intensively in the production of cereals. The cultivated land increased including some sections recovered from the river Genil and some resulting from the drying out of part of the marshlands. A large part of the *dehesa* was also plowed. The biggest crop growth seen in the central decades of the nineteenth century was that of the potato. Between 1851 and 1867 its production quadrupled, driven by the population increase in Granada and its surrounding towns and villages. As could be foreseen, the demand for labor grew significantly and the population grew along with it. The population almost doubled in just over a 100 years from 1752 to 1856, mainly due to immigration.

More and better irrigated farming land multiplied traction and transport requirements, forcing farmers to increase their draft herd by approximately 50%. This was done at the cost of livestock for meat, milk, and wool, whose numbers fell to coincide with the loss of grazing land in the *dehesa* and floodplains. Bovine stock, most commonly used for drawing vehicles in the past, diminished in favor of horses. Competition between the production of human foodstuffs and animal feed continued, limiting the size of the draft herd and fertilization capacity. Draft livestock was sufficient to meet traction demands, but not fertilizer requirements, generating a high deficit in manure, which had to be brought in from surrounding villages. The progression of cultivated land and the destruction of the agrosilvopastoral equilibrium that had previously characterized agricultural production in the mid-eighteenth century, revealed by an almost complete predominance of the agricultural landscape, can only be explained by the importation of large quantities of nutrients from nearby agroecosystems, transferring to these the territorial footprint, in this case hidden, of their agricultural metabolism, whether via the maintenance of abundant grazing land or via the preferential devotion of large expanses of land barley and hay production, as occurred, for example, in the neighboring populations of Colomera, Deifontes, Moclín, or Iznalloz (Calderón Espinosa, 2002).

5.4 DISEQUILIBRIUM OF AN "ADVANCED ORGANIC ECONOMY"

The fin de siècle crisis paved the way for the introduction of sugar beet during the 1880s. Despite its peaks and troughs, it would be the predominant crop for the next 40 years. Within a new rotation that included wheat and, in some cases, potatoes, its cultivation spread until it occupied no less than 86% of the irrigated area, giving rise to a heretofore unseen level of productive specialization. Traditional crops for personal consumption went into considerable and almost permanent decline. As a consequence, the inhabitants of Santa Fe increasingly had to import foodstuffs for themselves and feed for their animals from other areas.

It would seem that production of feed for livestock had almost doubled. This increase, however, was due to the broad beans crop, which formed part of the sugar beet rotation. The drop in barley prices made its cultivation decline significantly to practically half the previous levels. As is common knowledge, broad beans are no good as the main source of food for livestock, particularly in the case of horses, which are highly dependent on barley. In real terms, the capacity of the agroecosystem to feed the livestock diminished as the farming became more intensive.

The sugar beet crop represented an increase in nutrient requirements. Three-year rotation (broad beans—sugar beet—wheat or potato) shortened the previous rotation by a half and moreover increased the doses of manure used. The average annual dose would have been between 13 and 15,600 kg/ha/year. To cover these requirements solely with manure, supplies would need to be doubled, rendering it impossible to take on the consequent increase in costs. In fact this expense was already 24% of total costs; if the dose was doubled then the expense would also double, making it half of the total costs of cultivation and reducing profits from 21% to 1.7%.

All in all, specializing in sugar beet increased the nutrient deficit (see Table 5.2). From 1750 to 1885, the price of wheat became 2.5 times higher, whereas the price of manure became five times higher and its transport costs doubled. Under these circumstances, substitution with chemical fertilizers was becoming advisable not only from an agricultural perspective, but also from a financial one. In fact, the spread of sugar beet was made possible by this nonorganic input, allowing farmers to reduce "territorial costs" in the agroecosystem. From the outset, its introduction was accompanied by a new intensification of agricultural use of the land. The constantly irrigated area increased by 10% at the cost of semi-irrigated land. In this way, intensive production had now come to represent more than a third of the total municipal territory.

Since the agricultural intensification broke the balance between different land uses, such a notable increase in production and yield would not have been possible without recourse to chemical fertilizers, whose active ingredients came from outside Granada and even outside the country (Algeria and the United States, according to the 1921 report of the Junta Consultiva Agronómica [lit. Agronomic Consultancy Board]). The deepening disequilibrium of the advanced organic economy, still limited by the availability of land, made the search for soil or nutrient substitutes ever more pressing. The spread of plants for industry and wheat led to soil importation in the form of manure and barley in the first instance and later of nutrients and fossil fuels from relatively distant ecosystems. Santa Fe's agroecosystem was now unable to maintain this intensity of biomass production without the support of nonlocal input.

5.5 1934: THE BEET CRISIS AND THE DIVERSIFICATION OF CROPS

The data for 1934 allows us to describe the functioning of the agroecosystem and how the agrarian metabolism was structured in the years prior to the application of the technologies of the green revolution, which led to the complete industrialization of agriculture in Santa Fe. In the late 1920s, the crop area devoted to beet had fallen considerably in Santa Fe. The crop was becoming less profitable, yield fell, probably due to plant health problems, and farmers began to seek alternatives. They once again took up the traditional crops: potatoes, vegetables, alfalfa, and so on. Then, given its agronomic qualities, tobacco appeared as an alternative to beet. The useful crop area again rose (5%) at the expense of apparently unused land. However, the distribution of land use showed significant extensification of production in comparison with the early years of the century. *Per capita* agricultural production fell from 11.78 GJ/person per year in 1904 to 9.14 GJ/person per year in 1934. The combined effect of the beet crisis, rising wages, and the Great Depression of 1929 explains the phenomenon of extensification.

Nevertheless, the market orientation of production had become consolidated by this stage of the twentieth century, conditioning the organization of the agroecosystem. This can be seen from the livestock farming component that, until 1904, formed part of the system as a whole, supplying basic materials for human and animal feed and for the fertilization of the land, as well as a labor force to undertake the heaviest work in the fields. In line with the extensification seen in agricultural crops, the number of working animals fell. The crop area of cereals no longer bore any relation to livestock, greatly exceeding its needs. The abundance of food explains the proliferation of pig rearing, in which numbers multiplied by five with respect to 1904, and the appearance for the first time of a considerable number of head of cattle, which were not used as draft animals. In the same way, the expansion of poplar trees would explain the relative substitution of sheep by goats that is recorded in the sources.

The 33% increase in the live weight of livestock was paralleled by an increase in the availability of manure. This increase produced only a relative reduction in the chronic nutrient deficit in the agroecosystem. While in 1907, the livestock provided one-sixth of the nutrient needs of the crops that were fertilized, in 1934 this had risen to one-third, but the deficit persisted (Table 5.2). This was no longer so important. The use of chemical fertilizers, often combined with organic fertilizers, had become generalized. According to our calculations, almost 694 Gg of chemical fertilizer of all types were used during that year. Net marketable production *per capita* fell by half, agricultural employment fell by 260 persons with respect to 1904, causing an unemployment rate of over 50% among the active agrarian population or, in some cases, a substantial reduction in the number of day's labor worked per year.

The system required a further increase in external energy inputs. To function, rising to 11,940.1 GJ, and for the first time incorporating animal labor from neighboring towns in the month of November, when demand for draft animals peaked. However, the acquisition of these inputs from outside the municipality increased the loss of autonomy in the functioning of the agroecosystem. The incorporation of fossil energy was still restricted to chemical fertilizers, which by now came to 6291 GJ. Human labor was provided mainly by the population of Santa Fe, with only 18.8 GJ coming from outside the municipality. The restocking of livestock (4222 GJ) completed the imports.

In short, the agrarian metabolism in 1934 had suffered a progressive loss in the autonomy of the system, evident in the significant increase in energy imports with respect to the energy extracted from the system. In the same way, energy exports were consolidated, despite the fall in sugar beet, its place being taken by the poplar. The internal destructuring of energy flows and the dependence of the agroecosystem on external energy sources from the markets, where the exported energy also had its uncertain destination, had received its definitive impulse.

5.6 AGRICULTURAL PRODUCTION PARTIALLY DISSOCIATED FROM ITS TERRITORY

The industrialization of Spanish agriculture began to take root in the 1950s and even more so in the 1960s (see Chapter 6). Cultivated land increased by 8.7% from 1904 to 1997. But the most significant transformation was the disappearance of lands classified as semi-irrigated and the increase in constantly irrigated land to 60%. Water shortfalls in the summer were compensated by water drawn from a reservoir (Los Bermejales) and by wells drilled by private individuals and irrigation communities. The expansion of nonirrigated land, which had not stopped since the eighteenth century, aided by population growth and the demands of local livestock and the national cereal market, came to a permanent halt, reducing its surface area to 63% of that of 1904. The part of this surface at lowest altitude was reconverted into irrigated land and the rest, now freed of any pecuniary or nutritional obligations, was repopulated with pine forest. The mechanization of many agricultural tasks and an end to restrictions on the use of chemical fertilizers allowed farmers to devote the remaining part of dry land to alternative uses. The self-sufficiency that agriculture in Santa Fe had within its own territory and which was characteristic of agriculture based on organic energy had disintegrated once and for all.

Mechanization and the mass spread of fertilizers therefore made the segregation of land uses possible and paved the way for the almost complete predominance of agricultural use over the rest. The structure of production radically changed direction, now focusing on commercial crops that could be sown on irrigated land. Cereals, tobacco, garlic, and onions, among the other vegetables and fruit trees, were crops that played a leading role in the industrialization of agriculture in Santa Fe. Crops for local consumption such as wine grapes disappeared, or completely changed their commercial focus, as is the case with olive groves over recent decades. There was no longer a place for legumes either, which by 1997 had stopped being grown in combination with cereals, an irrefutable sign of the abandonment of traditional rotation techniques. The number of draft animals continued to dwindle and stabled revenue livestock began to consume compound feeds supplied in part from outside the agroecosystem. The depletion of livestock, particularly draft animals, favored in turn the abandonment of manure and the exclusive use of chemical fertilizers, a decisive phenomenon in the dynamics of the agroecosystem.

From a strictly ecological perspective, the complete commercialization of the production process represented the disintegration of local energy and nutrient cycles. Reusing waste products and by-products (everything from cereal stubble to poplar firewood, including of course manure) was no longer common practice. Santa Fe's agriculture became integrated into a considerably wider nutrient cycle. The agroeco-system's self-sufficiency was now circumscribed only to water supply and the terri-tory itself, which contributed less and less to agricultural activity.

But with time, the farmers of Santa Fe had to face up to the progressive loss of profitability as a consequence of the sustained increase in intermediate inputs and the progressive fall in perceived prices. Loss of income was a constant problem that has become more acute in recent decades, attenuated only by substantial sub-sidies from the *Common Agricultural Policy* (CAP). This has led to two important changes: the slow but steady increase in building land and the devotion of a con-siderable portion of the best farming land to poplar plantations. The construction of housing and commercial premises on flat terrain surrounding the metropolitan area of Granada and land demands for industry and services have offered far more substantial profits in the short term than agricultural activities. The loss of prestige in farming, and farmers growing old or changing profession, have converted the sale of their land as building plots into a type of "compensation" for the cessation of their activities. What is more, the pressure for building land has now eased off, forcing the price of farming land higher due to market speculation and greatly hindering the incorporation of new farmers into the system.

5.7 BIOMASS PRODUCTION AND EXTERNAL INPUTS

5.7.1 Net Primary Productivity

In Chapter 3, we used the case study of Santa Fe as an example of the application of our methodology for calculating the net primary productivity of agroecosystems (output side), so we will not repeat it here (see Section 3.4).

The actual net primary productivity (NPP_{act}) ha^{-1} of the agroecosystem (Figure 5.3) has multiplied by 4.1 over the period studied, rising from 62 GJ ha^{-1} in 1752 to 251 GJ ha^{-1} in 1997. However, the evolution of the productivity of the territory, depend-ing on the use made of the land, has been variable and, up to a point, contradictory.

The NPP_{act} ha^{-1} of cropland increased spectacularly between 1752 and 1904. Over this period, it multiplied by 1.55 as a result of the expansion of irrigation and the use of chemical fertilizers, which allowed the intensification of rotation in irriga-tion farming. During this period, the intensification did not affect rainfed cropland.

Between 1904 and 1934, the NPP_{act} ha^{-1} of cropland increased by 16%. However, progress in this period was due to the intensification of dry farming. The addition of fertilizers with mineral phosphorus made it possible to move from a three-field rotation system to the more intensive cereal-legume rotation. On irrigated land, it fell slightly due to the partial substitution of beet (a crop, which generated a large amount of biomass) by tobacco.

However, from 1934 to 1997, the NPP_{act} ha^{-1} of cropland fell to levels even slightly lower than in 1904. This reduction was seen on both irrigated and dry farm-ing land (Figure 5.3). It occurred despite the dramatic increase in chemical fertilizers

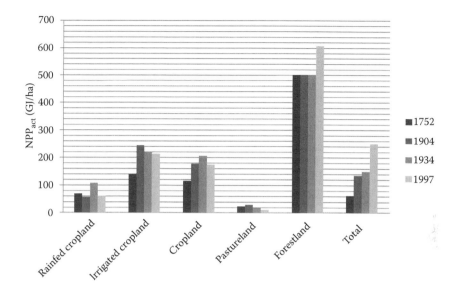

Figure 5.3   Evolution of Santa Fe NPP$_{act}$ (GJ ha^{-1}).

and pesticides and the dosage of irrigation water, which rose from 287 to 615 mm. On irrigated farmland, this fall can be partially explained that the crops that generate abundant biomass (sugar beet and wheat) were totally or partially substituted by horticultural products (garlic, asparagus, etc.) whose production of dry matter is lower. The cultivation of sugar beet disappeared, while the crop area of wheat was reduced by half. As a result, although the NPP$_{act}$ increased for the same crop (e.g., for wheat, it rose from 212 GJ ha^{-1} in 1934 to 335 GJ ha^{-1} in 1997), overall, it fell from 222 to 216 GJ ha^{-1}. Second, the use of herbicides reduced weed biomass from 30.7 to 15.4 GJ ha^{-1}. Third, the reductions arouse the suspicion that there could be serious degradation processes affecting the fund elements of the agroecosystem, which became more visible in dry farming areas (NPP$_{act}$ fell from 108 to 60 GJ ha^{-1}). On this land, the lack of water made it impossible to disguise the environmental deterioration by the addition of growing amounts of inputs of fossil origin.

On pastureland, the fall in productivity in 1997 with respect to previous years was due to the disappearance of the most productive pastureland, flooded riverbank pastureland that was wet for most of the year. Furthermore, *dehesa* was converted to pine forest, where production of pasture is insignificant. The reduction in these spaces was more than compensated for by the increase in forest land, due to the poplar coming to be considered a crop in the second half of the twentieth century. The old cultivars were replaced by hybrids, and fertilizers and herbicides were used.

Figure 5.4 breaks down the NPP$_{act}$ by its use in relative terms. The *socialized vegetable biomass* (*SVB*) rose from 1752 to 1904, paralleled by a fall in *reused biomass* (*RuB*), which reached a minimum in 1904. This is the result of the expansion of agricultural usage (agriculturalization) of the territory, which progressively

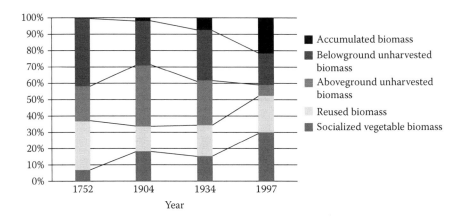

Figure 5.4 Evolution of total NPP_{act} (MJ) by its use in relative terms.

reduced livestock farming, due to the privatization of common land promoted by the liberal laws, thereby hindering the feeding of the livestock, especially for small farmers (González de Molina and Guzmán, 2006). This allowed an increase in the area devoted to crops for human use (wheat and sugar beet), and at the same time increased the average yield of crops thanks to increased irrigation and the use of chemical fertilizers. The expulsion of livestock farming considerably increased the aboveground unharvested biomass ($AUhB$), which became available for other heterotrophic species. In 1904, however, there was a fall in relative terms in belowground unharvested biomass ($BUhB$) brought about by the cultivation of sugar beet and potato, whose harvesting involved the extraction of the root from the soil. The accumulated biomass (AB) grew slightly between 1752 and 1904 due to the increased crop area of poplar, although its cultivation was no more intensive.

Between 1904 and 1934, the changes in the destinations of the biomass were less dramatic and were due to the substitution of some crops by others of different characteristics. It was due, fundamentally, to the increase in the crop area of poplar, which multiplied by 4, although its cultivation was no more intensive, increasing the percentage of accumulated biomass in the agroecosystem from 2% to 8% (Figure 5.4). It was also due to the partial substitution of beets by tobacco, two very different crops with regard to the harvested part. This made it possible to increase the belowground unharvested biomass, but reduced that which was left on the surface.

Between 1934 and 1997, SVB multiplied by 3.1 in absolute terms (Table 5.3) and almost doubled in relative terms (Figure 5.4). The case of wheat is paradigmatic. For this crop, to the increase in NPP_{act} were added the highest harvest index of the modern varieties and the chemical control of weeds. In all, this meant that the population went from socializing 13.4% of the NPP_{act} of a hectare of wheat to socializing 31.3%. However, the crop that most contributed to the increase in SVB was the poplar. In 1997, 61.5% of the SVB was timber and firewood, compared with 32.3% in 1934.

The RuB was multiplied by 1.9 in absolute terms and by 1.2 in relative terms. This increase was not due to an increase in livestock farming with respect to 1934. In

Table 5.3   Biomass Production (GJ) and External Inputs (GJ) in Santa Fe (1752–1997); Basic Data

	1752	1904	1934	1997
NPP$_{act}$ (a+c+d+e)	233,116	521,624	555,846	896,508
Socialized vegetable biomass (SVB) (a)	16,131	94,230	85,265	265,437
Socialized animal biomass (SAB) (b)	6,400	7,227	7,836	11,187
Socialized biomass (SB) (a+b)	22,531	101,457	93,101	276,624
Reused biomass (RuB) (c)	69,048	81,505	104,699	201,401
Unharvested biomass (UhB) (d)	146,750	335,494	323,552	234,484
Aboveground unharvested biomass (AUhB)	50,149	193,227	150,992	59,519
Belowground unharvested biomass (BUhB)	96,600	142,266	172,560	174,964
Recycling biomass (RcB) (c+d)	215,798	416,999	428,251	435,885
Accumulated biomass (AB) (e)	1,186	10,395	42,329	195,187
External input (EI) (f)	2,391	9,215	11,940	126,914
Total inputs consumed (TIC) (c+d+f)	218,188	426,214	440,192	562,799

Source: Author calculation.

fact, the percentage of *RuB* devoted to animal feed fell from 93.8% to 57.6%. In 1997, animal feed for livestock (mainly intensive dairy farming) was partially imported, representing 13.5% of external input (*EI*). This increase in the *RuB* was due to the generalization of the burning of waste (straw, thin pruning waste, etc.) in the fields.

Finally, the *AB* grew spectacularly thanks to the poplar. Overall, the relative increase in these uses of NPP$_{act}$ in 1997 was possible partly as a result of the sharp fall in *UhB*, which became available to sustain wild food chains in the soil and in the air. Wildlife has been practically expelled from the municipality.

5.7.2 External Inputs

Figure 5.5 shows the evolution of EI. In 1752, the imported energy was 100% renewable and consisted of human work and restocking of draft animals, which were bred in neighboring, more mountainous villages. In 1904, synthetic fertilizers were already present. Superphosphates (204.5 t), potassium chloride (222.7 t), and ammonium sulfate (192.3 t) were the fertilizers most used. For the first time, nonrenewable energy was incorporated to the agroecosystem. Overall, EI increased almost fourfold from 1752, and approximately 50% of this energy was nonrenewable. In 1934, fertilizer consumption keeps growing, driven by the expansion of fertilization to rainfed areas.

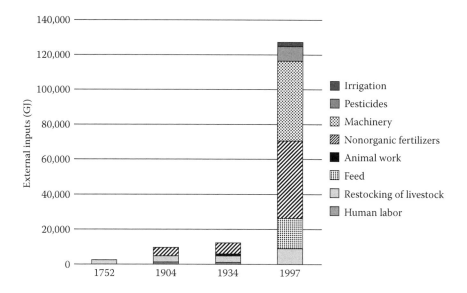

Figure 5.5   Evolution of external inputs (GJ).

A new input also appears, "imported animal work," showing the loss of the capacity of Santa Fe agroecosystem to generate its own energy and material fluxes. This loss of autonomy is already completed in 1997, when the modernization process of Spanish agriculture is fully consolidated. Mechanization and synthetic fertilizers represent the two most important inputs in terms of embodied energy. At the end of the twentieth century, only 20.7% of the imported energy is in the form of biomass (feed, restocking of livestock, and human labor), and approximately 75% of total EI are nonrenewable.

5.8 ENERGY RETURN ON INVESTMENTS FROM AN ECONOMIC POINT OF VIEW

The agroecosystem of Santa Fe as a whole has undergone a process of simplification of the landscape or agriculturalization since the mid-eighteenth century. The territory has ceased to provide the nutrient and energy flows necessary for the functioning of the agroecosystem, and they have been replaced by chemical fertilizers and machinery powered by fossil fuels. It is only partially able to feed revenue livestock. In parallel, the agroecosystem has been reforested. The area of timber species rose from 3.4 ha (poplar) in 1752 to 53 ha in 1904 (poplar), 223 ha in 1934 (poplar), and 650 ha in 1997 (440 ha poplar—210 pines). The existence of these processes and their consequences for the sustainability of the agroecosystem should be reflected in the raft of economic and agroecological energy return on investments (EROIs) we propose.

As we have seen in Chapter 2, the economic EROIs tell us about the return on the energy investment made by society in agroecosystems. *Socialized biomass* (*SB*) is a net supply of energy carriers able to be consumed by the local population or for use in other socioeconomic systems. As we also have seen in Chapter 2, the final EROI (FEROI) can be broken down into two different EROIs: external final EROI (EFEROI) and internal final EROI (IFEROI) (Tello et al., 2015). The external final EROI relates *EI* to the final output crossing the agroecosystem boundaries (Carpintero and Naredo, 2006; Pracha and Volk, 2011). In the academic literature, it has frequently been called "net efficiency," and it is one of the indicators most commonly used to evaluate agriculture from the energy perspective (Guzmán and Alonso, 2008). This ratio links the agrarian sector with the rest of the energy system of a society—and thus assesses to what extent the agroecosystem analyzed becomes a net provider or rather a net consumer of energy. On the other hand, the internal final EROI explains the efficiency with which the biomass that is intentionally returned to the agroecosystem is transformed into a product that is useful to society. This indicator has not habitually been used but its usefulness is growing since this biomass can have alternative uses (e.g., biofuels), since poor management can cause environmental problems (e.g., pig slurry pollution) or due to the ecosystem services it can provide (e.g., soil carbon sequestration).

The strategy followed by industrialized agriculture in the twentieth century has been to reduce the inputs invested per unit of production, relying on external inputs and savings of internal inputs. The results obtained in 1904 appear to show that this strategy is appropriate in the beginning of the process in zones with a high potential response, such as Santa Fe. The import of chemical fertilizers appeared to very successfully replace the high investment made by society in livestock-based biomass to replenish the fertility of the soil. In other words, the introduction of superphosphate and other chemical fertilizers made a process of agriculturalization possible, which produced an increase in energy return. The synergy established between both types of input was possibly based on the high quality of the fund elements of the agroecosystem, such as the soil, water, and biodiversity in the early twentieth century. In effect, organic agriculture in Santa Fe had previously provided large quantities of energy in the form of *RuB* and *BUhB* (Figure 5.4) for the functioning of edaphic food chains and the maintenance of organic material in the soil. Furthermore, the territorial equilibrium necessary to generate sufficient *RuB*, while also providing AUhB and the absence of biocides, maintained high levels of aerial biodiversity (Guzmán and González de Molina, 2009). In this context, the addition of phosphorus, an essential nutrient for the symbiotic fixing of nitrogen by legumes included in the rotation, could have had a synergic effect. This synergy does not appear to exist today when external inputs are partially replaced by internal inputs in the conversion to organic farming of very intensive agroecosystems. And so, Ponisio et al. (2015) reported an average fall in yield of 19.2% in organic farming, based on a meta-analysis of 115 studies. This is almost certainly because the situation is the opposite. The degradation of fund elements requires an energy investment over years for its improvement before it can once again see increasing returns in *SB*. The distance between the pathways of degradation and restoration is known as the hysteresis of land rehabilitation (Tittonell et al., 2012).

However, the increasing application of this strategy throughout the twentieth century has revealed its limitations. Between 1904 and 1934, there was a fall of 29% in these EROIs, which was due above all to the partial substitution of beet by tobacco, which brought about a fall in the socialized biomass (see Table 5.3), and at the same time caused an increase in the use of EI, as synthetic fertilization expanded to dry farming land and RuB increased due to the increase in livestock farming mentioned in Sections 5.5 and 5.7. The fall in Final EROI in this period reflects the inability of inorganic inputs to replace biomass flows in the maintenance of fund elements. Proof of this was the early appearance of edaphic phyto-pathological problems in sugar beet. On irrigated land in Santa Fe, the use of chemical fertilizers and the removal of roots reduced *RcB* to a minimum. The reduction in the energy available to sustain complex food chains encouraged plant health problems which caused significant falls in productivity in this crop in the 1920s and, later, its substitution by other crops.

Viewed in perspective, during the twentieth century the process of agriculturalization continued, though less vigorously. In fact, the consumption of biomass by livestock with respect to NPP_{act} fell from 14.5% (1904) to 12.9% (1997). Nevertheless, final EROI fell by 25%, due to the low return on external inputs invested, whose EROI fell by 80% (see Table 5.4). The massive substitution of *RuB* flows by *EI* hardly managed to raise internal final EROI in 1997.

5.9 EROIs FROM AN AGROECOLOGICAL POINT OF VIEW

The agroecological EROIs inform us of the real productivity of the agroecosystem, taking into account the ability to reproduce the fund elements, that is, in the structure of the agroecosystem (e.g., biodiversity, spatial heterogeneity, and the complexity of agroforest landscapes or soil quality) and the provision of the basic ecosystem services. As we have seen in Chapter 2, we have proposed four EROIs: the NPP_{act} *EROI* (= NPP_{act}/total inputs consumed, being TIC = RcB + EI) tells us the real productive capacity of the agroecosystem, whatever the origin of the energy it receives (solar for the biomass or fossil for an important portion of the external inputs). This indicator measures the degradation of productive capacity of agroecosystems, affecting natural resources such as soil salinization or erosion, genetic erosion, and so on. To compensate this loss, an increasing amount of energy should be devoted to palliate the loss of productive capacity.

In energy terms, the return of TIC, considering total productivity of the agroecosystem (NPP_{act} EROI), grew by 49% between 1752 and 1997 and only by 30% during

Table 5.4   EROIs from an Economic Point of View (Santa Fe, 1752–1997)

	1752	1904	1934	1997
Final EROI (FEROI)	0.32	1.12	0.80	0.84
External final EROI (EFEROI)	9.42	11.01	7.80	2.18
Internal final EROI (IFEROI)	0.33	1.24	0.89	1.37

Source: Author calculation.

Table 5.5 EROIs from an Agroecological Point of View (Santa Fe, 1752–1997)

	1752	1904	1934	1997
NPP$_{act}$ EROI	1.07	1.22	1.26	1.59
Agroecological final EROI	0.10	0.24	0.21	0.49
Biodiversity EROI	0.67	0.79	0.74	0.42
Woodening EROI	0.01	0.02	0.10	0.35

Souce: Author calculation.

the twentieth century (Table 5.5). This growth accompanied the expansion of irrigation which, in semiarid climates such as the Mediterranean, is essential in order to achieve a response to the use of *EI*, such as chemical fertilizers (Lacasta and Meco, 2011). It is, in truth, difficult to know to what extent the exponential growth in the amount of water consumed (Table 5.2) masks problems of the degradation of fund elements, which could be deteriorating productive capacity. However, we have seen that the modernization of the agroecosystem was not uniform, but occurred in waves. The first, in 1904, was focused on irrigated cropland, which expanded progressively. The second, in 1934, centered on rainfed cropland, later becoming consolidated. Finally, the third wave affected the poplar, which went from being a riverbank tree to being just another crop, in 1997. Figure 5.3 shows similar waves in the fall of NPP$_{act}$ per hectare, in parallel to the progressive modernization of the agroecosystem. Thus, in 1997, the NPP$_{act}$ EROI was sustained by the poplar, a crop for which the NPP$_{act}$ EROI reached 4.6, compensating the fall to 1.1 in the rest of the territory.

On the other hand, this agroecosystem is paradigmatic due to its high response to the technology of the green revolution in Mediterranean areas with access to irrigation. To this are added changes in soil use based on crops with high biomass production such as sugar beet and poplar. From this perspective, the increase in NPP$_{act}$ EROI can be considered the maximum that can be achieved with this technological package.

The *agroecological final EROI* (= *SB/TIC*) shows the real amount of energy investment required to obtain the biomass intended to society. From an ecological point of view, the return that society receives from agroecosystems is not only the result of the energy expressly invested by society, but also that which is really recycled without human intervention. This indicator grew more than NPP$_{act}$ EROI, multiplying by 4.8 in the period 1752–1997 as a result of the sharp increase in the portion of biomass socialized with respect to that produced.

In Chapter 2, we have drawn attention to the relationship between agroecological final EROI and final EROI (Biodiversity EROI = 1–AE-FEROI/FEROI = *UhB/TIC*). A sustainable management of agroecosystems should leave biomass available for other species allowing the generation of complex food chains in order to guarantee ecosystem services. But doing this at the expense of *RuB* might also entail reducing the need for an integrated land use management. Thus getting rid of *RuB* per unit of TIC might lead to a decrease in the spatial heterogeneity and complexity of agroforest landscapes, and a reduction in the species richness they can shelter (Gliessmann,

1997; Guzmán and González de Molina, 2009; Perfecto and Vandermeer, 2010). Furthermore, a drastic reduction in *RuB* would lead to an increase in the use of *EI* for the functioning of the agroecosystems.

The industrialization of agriculture has been accompanied by structural and functional changes that are evidenced by this agroecological EROI (see Table 5.5). The premise that the industrialization of agriculture has led to land sparing appears to apply, in this case, 210 ha of pine forest, which has been protected from agricultural activities to the benefit of wild biodiversity. However, the agroecological EROIs question this benefit. The biodiversity EROI falls to 0.42 in 1997 and this alerts us to the fact that society has massively appropriated the flows of biomass generated. The *UhB* collapsed, mainly the aboveground *UhB*, expelling many wild heterotrophic species from the agroecosystem. From this perspective, land sparing is not so important as the liberation of biomass, whose flows are, in the final analysis, what sustain the food chains. This liberation of biomass for use by wildlife does not necessarily need to be linked to land sparing. Our case study shows that traditional agriculture combined productive activity with conservation in the same territory (land sharing), since it was able to free biomass for heterotrophic species at the same time as it generated a complex, heterogenous spatial matrix (Guzmán and González de Molina, 2009), which facilitated the movement of organisms between fragments of natural habitats. This balance is reflected in the Biodiversity EROI of 0.67 in 1752. In 1904 and 1934, the use of chemical fertilizers brought down the *RuB* per unit of TIC, leading to a slight reduction in the heterogeneity and complexity of the Santa Fe landscape (Guzmán and González de Molina, 2009) and a higher proportion of *UhB* that was reflected in a slight rise in this ratio. The effect that this phenomenon might have on biodiversity is low, with the exception of the effect mentioned above on land devoted to rotation with sugar beet.

The woodening EROI (=*AB/TIC*) shows that in the period studied the transformation into accumulated biomass of the energy invested strongly increased, in line with the process of specialization in poplar. For this reason, this indicator has risen from 0.01 to 0.35. In 1997, more energy was accumulated annually in the form of biomass than that input in the form of EI, transforming fossil energy (75% of EI) into bioenergy.

The maintenance of this EROI values, however, is linked to a high consumption of irrigation water (880 mm). However, Mediterranean region is one of the so-called climate change hot spots. Climate models have forecast a rise in temperature and a fall in precipitation that will affect the quality, quantity, and management of water resources (García-Ruiz et al., 2011). Therefore, it is possible that, even if this conversion is positive from an energetic point of view, it may not be so from a hydric point of view.

5.10 CONCLUSIONS

The application of the proposed economic and agroecological EROIs to the case study has informed us, first, of the return on energy investments for society and, second, of processes that affect the fund elements of agroecosystems and their capacity to generate flows of ecosystem services. From a social point of view, the return was highest in the early twentieth century. Both the increase in *SVB* induced by the expansion of irrigation and the introduction of chemical fertilizers, as well as the process of agriculturalization that this allowed, contributed to this. Although, throughout the twentieth century, the process of agriculturalization continued, the return enjoyed by society fell, possibly as a result of the degradation of the fund elements. Specifically, cropland soil suffered a drastic reduction in the flows of the biomass necessary to maintain its quality. This situation was aggravated by the fact that an important portion of *RuB* was burned in the fields during the second half of the twentieth century. However, the NPP_{act} EROI grew and this does not allow us to state that generalized degradation processes affecting fund elements are taking place, though these may be masked by the significant expansion of irrigation and the poplars, an intensive "crop" in industrial inputs. In fact, the agroecological EROIs tell us of other processes, which undermine the sustainability of the agroecosystem. The Biodiversity EROI alerts us to the low return to nature in the form of *UhB* available to aerial and underground wildlife. This low return is not compensated by the isolation of a protected "nonagrarian" space, questioning the hypothesis of *land sparing* for the purpose of sustaining biodiversity, rather than *land sharing*. As a counterpoint, the process of agriculturalization has allowed the reforestation of the agroecosystem, and this must be valued positively in the context of a deforested area such as Santa Fe. Although the high water consumption in which it is based casts doubt on its sustainability.

The results obtained confirm the usefulness of agroecological EROIs for a more profound comprehension and evaluation of the energy functions in agroecosystems from the point of view of sustainability.

CHAPTER 6

Diachronic Analysis at a National Scale
Spanish Agriculture, 1900–2008

Gloria I. Guzmán, Manuel González de Molina, David Soto Fernández,
Juan Infante-Amate, and Eduardo Aguilera Fernández

CONTENTS

6.1 INTRODUCTION

In this chapter, we will study the energy efficiency to a more aggregate level than used in Chapter 5. We want to know if the processes described in the local case study have happened in the same way at national level or whether new properties have emerged that make energy throughput different. In particular, we want to know whether changes in energy efficiency observed in the case of Santa Fe can be moved to a larger context. For this purpose, we will analyze the historical evolution of Spanish agriculture since the early twentieth century, when the main industrial transformations began with the introduction of chemical fertilizers.

Although this process of industrialization was not as vigorous as in other developed countries and followed their own chronology, the evolution experienced by the Spanish agriculture since the early twentieth century is representative of the industrialization process of the European agriculture. Therefore, this chapter, as the previous, has a strong historical component that allows us to compare different levels of efficiency among different metabolic arrangements during the transition to industrial agriculture and provide explanations for each of them. The advantage of applying this methodology to the transition process from traditional organic agriculture over to industrialized agriculture is that it provides diachronic scenarios with different land use intensities. This issue has been put forward by various authors to evaluate the state of fund elements and the ecosystem services of agroecosystems (Berlin and Uhlin, 2004; Tuomisto et al., 2012). History allows comparisons to be made without having to resort to constructed scenarios in which the definition of variables is always arbitrary. Other authors such as Tittonell (2014) and Wehrden et al. (2014) have also highlighted the importance of history when understanding the configuration and functioning of today's agroecosystems.

6.2 DESCRIPTION OF THE STUDY SITE AND DATA COLLECTION

The varied geography of Spain, along with its location in South latitudes of the temperate area of the Northern Hemisphere (Figure 6.1), is responsible for a notable climate diversity. Spain can be divided into three different regions: humid, semiarid, and arid. The humid region has a marked influence of the Atlantic Ocean, with mild temperatures along the year, and an average precipitation of 1100 mm. It comprises eight provinces in the North-West. The semiarid region has an average precipitation of 500 mm and hot, dry summers that constitute the dry season (according to Gaussen index). It comprises 33 provinces. The arid region has an average precipitation of 300 mm and comprises 9 provinces. The dry season in these regions extends from summer (in the northern provinces) to the whole year (in the Canary Islands and the South-East extreme of the Iberian Peninsula). Despite these regional differences, however, the climate in most of the territory (except the humid region that is mostly temperate, and the Canary Islands that are subtropical) can be classified as Mediterranean.

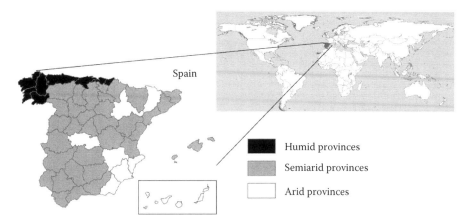

Figure 6.1   Spain in the world.

To study the evolution of Spanish agriculture, copious quantitative information must be compiled and processed. Annual crop production series—cereals, legumes, grapes, and olives—are available beginning in the late nineteenth century (GEHR, 1991) and until the 1930s, but data about total agricultural production are only available since 1922, and disaggregated per year since 1929. We also have the most significant information for 1902 products (MAICOP, undated, 1905; Ministerio de Fomento, 1892). The missing information was reconstructed from complementary sources, namely the annual memoirs published about assorted topics by the Junta Consultiva Agronómica (Ministerio de Fomento, 1912, 1913, 1914a,b, 1915). Thus, the Spanish agricultural production has been estimated for 1900. Since 1960, annual series of agricultural production have been published in the Anuarios Estadísticos de las Producciones Agrícolas (Ministerio de Agricultura, 1959, 1960a, 1961, 1962, 1963) and 2008 in the Anuarios de Estadística Agraria (MAGRAMA, 2006–2010).* We considered 5-year averages with centers in the years 1960 and 2008. This study also takes into account agricultural residues, calculating the production using converters (Appendix I). The amount of straw of cereals and legumes with an economic use is provided by the Spanish Agrarian Yearbooks.* This straw is mainly used for livestock. Based on land uses reconstructed using the sources cited previously, the production of pastureland and fallowland was calculated. The production of timber and wood was also estimated (Infante-Amate et al., 2014). Also using 5-year averages, the exports and imports of biomass were calculated from foreign trade sources. For 1900, these figures were taken from the Directorate General of Customs and Excise (1899, 1900, 1903a, 1903b, 1903c). For 1960, they were taken from the Food and Agriculture Organization of the United Nations (FAO) (FAO, 2015), and 2008 from the DATACOMEX database (Ministerio de Economía y Competitividad, 2015). Production figures from the livestock subsector were obtained from censuses (Ministerio de Fomento, 1892; GEHR, 1991; Ministerio de Agricultura, 1960b; MAGRAMA, 2006–2010). To calculate the feed consumed by livestock, an animal feed balance model was applied, similar to the model used by Krausmann et al. (2008). This model takes into account the different food requirements of livestock on the basis of numerous variables (production, mobility, age, breed, etc.).

6.3 INPUTS/OUTPUTS CALCULATION

6.3.1 Outputs Calculation

As discussed in Chapter 3, actual net primary productivity (NPP_{act}) is the amount of energy really incorporated into plant tissues and is the result of the opposed processes of photosynthesis and respiration. NPP is expressed in terms of energy accumulated (e.g., joules/hectare/year) or in terms of the organic material synthesized (e.g., kg/hectare/year) over a given period of time. As we also show in Chapter 3, we have taken root biomass into account, due to its relevant role in the maintenance

---

* The Spanish Agrarian Yearbooks are available online: http://www.magrama.gob.es/es/estadistica /temas/publicaciones/anuario-de-estadistica/.

of complex food chains and in the accumulation of organic material in the soil (Paustian et al., 2016).

The NPP_{act} of historical and modern agroecosystems does not appear in agricultural statistics, which usually focus on the harvested portion of the NPP_{act}. It therefore has to be estimated, for which there are different approaches. For example, experimentally recreating past conditions and carrying out direct measurements that can be extrapolated (*experimental history*), or using algorithms that take into account variations in vegetation and soil and climatic conditions, or using conversion factors that allow the NPP_{act} to be estimated from the harvested biomass, taking into account the changes that have occurred over the period studied, or extrapolating from studies based on different agroclimatic and management conditions. The three latter options have been preferred in this case study. As stated in Chapter 3, the NPP_{act} of the Spanish agroecosystems has been broken down into different portions according to the use made by society. They are summarized as follows:

1. *Socialized vegetable biomass* (*SVB*): This is the phytomass that is directly appropriated by human society, considered as it is extracted from the agroecosystem.
2. *Reused biomass* (*RuB*): This is the portion of phytomass that is intentionally returned to the agroecosystem by human work. The reincorporation into the agroecosystem of this vegetable biomass has a agronomic purpose that is recognized by farmer. The product of livestock farming that is available to society as a result of this reuse (live weight of meat at the farm-gate, milk, wool, etc.) is called *socialized animal biomass* (*SAB*).
3. *Unharvested biomass* (*UhB*): This is the phytomass that is returned to the agroecosystem by abandonment, not in the pursuit of any specific aim, and can be divided into *aboveground unharvested biomass* (*AUhB*) and *belowground unharvested biomass* (*BUhB*). The sum of *RuB* and *UhB* comprises the *recycling biomass* (*RcB*) that is the phytomass intentionally or unintentionally reincorporated into the agroecosystem.
4. *Accumulated biomass* (AB): This is the portion of phytomass that accumulates annually in the aerial structure (trunk and crown) and in the roots of perennial species. The sum of all these categories includes the NNPact as follows:

$$NPP_{act} = SVB + RuB + UhB + AB$$

We have also taken into account other categories essential for the analysis of biomass flows. We have used some basic economy-wide material flow accounts (EW-MFA) categories (see Chapter 2) such as domestic extraction (DE) and the impact on apparent consumption of the results of overseas trade (imports and exports) calculating the Physical Trade Balance (PTB) and domestic material consumption (DMC). In this way

$$DE = SVB + RuB$$

$$PTB = Imports - Exports$$

$$DMC = DE + PTB$$

We have calculated the *cropland NPP$_{act}$* as the biomass of the crops and also of the associated weeds. It has been obtained from the conversion factors (see Chapter 3 and Table AI.1) that allow it to be estimated from crop production (*SVB*), which is the most commonly available data, especially in historical sources. The conversion factors allow the user to calculate the total biomass (aboveground + belowground biomass), to convert the fresh biomass into dry biomass and vice versa, and finally to convert the biomass into gross energy. To estimate the aboveground biomass of weeds in traditional agricultural systems, we have used data from contemporary organic farming trials and for more recent periods, data from conventional agriculture (Table AI.4). For most coefficients, we do not expect large variations over time. For some, like the harvest index, which changes over time, we have also provided information for preindustrial time periods in some crops.

The *grassland NPP$_{act}$* has been collected from studies performed in Spain and based on different grassland types and climatic conditions (CIFA, 2007; Correal et al., 2007; Hernández Díaz-Ambrona et al., 2008; López-Díaz et al., 2013; Robles et al., 2001; Robles, 2008; San Miguel, 2009). The productivity of root biomass was calculated using conversion factors (see Chapter 3 and Table AI.2).

The *woodland NPP$_{act}$* (fuel wood and timber for society + aboveground wood accumulated on trees) can be seen in Infante-Amate et al. (2014). By applying partition coefficients, the annual production of leaf biomass and reproductive structures (flowers and fruit) was calculated, along with the root biomass. The proportion of root biomass that is accumulated and recycled every year in the soil was calculated taking into account the root:shoot ratio of the adult holm oak (0.84) and pine (0.3). Basic data regarding these transformations can be seen in Almoguera Millán (2007) and CMAOT (2014). The conversion factors for biomass into gross energy can be found in Table AI.5.

To calculate the domestic material consumption (DMC), we have considered all of the exports and imports of primary and transformed biomass. We have used 5-year means. In the case of overseas trade, we have continuous series for the entire period, although the methodology used and the categorization available has varied significantly over the years. Between 2000 and 2008, we used the DATACOMEX database of Spanish overseas trade (Ministerio de Economía y Competitividad, 2015). For 1960 and 1990, we used the FAO database (FAOSTAT, 2015). Last, for the period from 1900 to 1950, we used overseas trade statistics for Spain.* We distinguished between five categories: food, feed, seeds, wood and fuel wood, and other raw materials.

6.3.2 Calculation of External Inputs

The *external inputs* (EI) include industrial inputs (chemical fertilizers, machinery, etc.) and nonindustrial inputs (biomass, human labor, etc.). The allocation of energy to each type of input is summarized as follows.

---

* The original publications are available online at the Ministry of Industry website: http://www4.mityc .es/BibliotecaCOM/abwebp.exe.

Table 6.1   Embodied Energy (TJ) of Industrial Inputs Invested in the Spanish Agriculture

		1900	1960	2008
Machinery	Implements	443	922	9,290
	Motorized	24	2,088	15,266
Traction Fuels	Fuel production	6	1,728	18,680
	Fuel use	56	8,027	86,430
Irrigation	Infrastructure	799	1,081	4,977
	Fuel production	15	930	1,966
	Fuel use	149	4,552	8,335
	Electricity	32	3,206	60,129
Fertilizers	N Fertilizers	784	24,715	63,681
	P Fertilizers	1,047	12,600	7,992
	K Fertilizers	54	1,454	5,945
Crop Protection	Pesticides	26	872	17,678
	Greenhouses and mulches	0	12	13,457
Total		3,434	62,185	313,826

Source: Author data.

Industrial inputs: In this study, the energy allocated to industrial inputs is embodied energy. In other words, it is the sum of the gross energy of the input and the energy requirements for producing and delivering them. The embodied energy of industrial inputs evolved over time, as the energy efficiency of the production and delivery of the inputs changed as shown in Chapter 4. Table 6.1 provides the total embodied energy of industrial inputs, using specific information from Spanish historical sources and conversion factor compiled in Appendix II.

Nonindustrial inputs: We estimate the energy in human labor as the dietary energy consumption (2.2 MJ/h), based on the data offered by Fluck (1992) (Table 6.2). This has been widely used in the literature (see Section 4.3 in Chapter 4). This method for accounting for energy in human labor does not include the energy required to produce the food consumed by the labor (embodied energy). This avoids a problem of circular referencing or double counting, which can stem from this method, since the product (food) is used as an (important) input of the system.

Net imported biomass energy is the gross energy of the different products, calculated using conversion factors (see Table AI.5). The cost of transport was added to this (Section 4.10 in Chapter 4, and Appendix II). The energy required to produce the biomass was not considered, to avoid problems of double counting, since this cost should be attributed to the agroecosystems of origin.

6.4 INDUSTRIALIZATION OF SPANISH AGRICULTURE

Agricultural production was multiplied by 3.33 between 1900 and 2008, reaching its peak in the first years of the twenty-first century with over 104 million tons, almost four times more than in 1900. This gradual evolution can be divided into

Table 6.2 Energy of Nonindustrial Inputs

	Nonindustrial Inputs (Unit)			Total Energy (TJ)		
	Unit	1900	1960	2008	1900	1960	2008
Feed	Gg d.m.	72.7	461.0	10,109.1	1,270	10,289	187,842
Seeds	Gg d.m.	18.0	94.8	307.7	315	1,785	5,335
Human labor	M hour	8,502	8,378	1,481	18,705	18,431	3,257
Total energy					20,289	30,505	196,434

Sources: Dirección General de Aduanas, 1899, 1900, 1903a, 1903b, 1903c, *Estadística general del comercio exterior de España en 1898, 1899, 1900, 1901, 1902*, Dirección General de Aduanas, Madrid; FAO 2015; Ministerio de Economía y Competitividad, DATACOMEX-Estadísticas del comercio exterior español, Ministerio de Economía y Competitividad, Madrid, http://datacomex.comercio.es/principal_comex_es.aspx, accessed April 2, 2015.

four different periods. The first corresponds to the period from 1900 to 1933, when production grew by 52% (Table 6.3). The "modernization" (GEHR, 1991; Simpson, 1997; Pujol et al., 2001) became apparent in the decrease in the relative importance of cereals and legumes that were replaced by more commercial crops, some of which, such as olives, fruit, vegetables, potatoes, industrial crops, and fodder plants, were destined to the international market. This growth in agrarian production improved the quantity and quality of the Spanish diet (González de Molina et al., 2014).

In this regard, the Civil War (1936–1939) and dictatorship of Franco was also a veritable tragedy. The diet of the Spanish people suffered an even greater setback than that of the late nineteenth century agrarian crisis, both in terms of daily calories per capita and of composition. International isolation and the ill-advised economic policy of the regime led to a reduction in agrarian production, which would only be resolved after the 1960s, once the "green revolution" had begun (González de Molina et al., 2014). In fact, the final phase of agricultural industrialization in Spain began in the late 1950s, coinciding with the slight economic liberalization of the Franco regime, and was to last a couple of decades. Performance per surface unit multiplied thanks to the use of the complete package of the green revolution: improved seeds, synthetic chemical fertilizers, phytosanitary products, and irrigation. Production recovered in a short period of time; in 1960 it was doubled that of 1900, by 1980 it had tripled, and by the end of the century it had almost quadrupled. Cereal production, which had been decreasing up until then as a clear result of the abandonment of traditional agriculture, recovered swiftly due to a growing demand for cereals for fodder. This abandonment of traditional agriculture translated into a continued decline in legume production.

Since the 1970s, per capita consumption of meat and dairy has increased considerably, whereas the intake of carbohydrates has fallen below the recommendations of the World Health Organization. The Spanish are gradually abandoning the Mediterranean diet and moving to a more meat-based diet that requires high livestock production (Schmidhuber, 2006; González de Molina et al., 2013). It should be noted that there has been an increase in fodder crops since the 1960s, accompanied by an increase in livestock and the numerous intensive farms that use combined fodder and forage, abandoning pastures. Paradoxically, the main production specialization of Spanish

Table 6.3   Evolution of Agricultural Production (1900–2008) in Million Tons of Fresh
            Matter

	1900	1933	1960	1990	2008
Cereals	6.24	8.96	8.62	18.83	23.00
Leguminous	0.50	0.72	0.90	0.24	0.23
Tubers and vegetables	5.37	9.02	9.77	16.77	14.99
Fruits	1.85	3.10	3.66	8.59	10.20
Vineyards	3.78	3.29	2.63	5.33	5.40
Olives	1.09	1.82	1.95	2.93	6.26
Industrial crops	1.27	2.47	4.43	9.39	6.73
Fodder crops	8.40	13.93	25.52	35.61	28.33
Total	28.50	43.31	57.48	97.70	95.14
In % over the total	**1900**	**1933**	**1960**	**1990**	**2008**
Cereals	21.9	20.7	15.0	19.3	24.2
Leguminous	1.8	1.7	1.6	0.2	0.2
Tubers and vegetables	18.8	20.8	17.0	17.2	15.8
Fruits	6.5	7.2	6.4	8.8	10.7
Vineyards	13.3	7.6	4.6	5.5	5.7
Olives	3.8	4.2	3.4	3.0	6.6
Industrial crops	4.5	5.7	7.7	9.6	7.1
Fodder crops	29.5	32.2	44.4	36.4	29.8
Total	100.0	100.0	100.0	100.0	100.0

Source: Author data.

agriculture was consolidated in this period: fruit, vegetables, and olive oil, products, which are mainly aimed at central European markets (Pinilla, 2001; Clar et al., 2014).

At least since the year 2000, a final phase has become distinct in agrarian statistics, although some of its trends started taking shape earlier, particularly since Spain's incorporation into the European Economic Community (1986). This phase is characterized by a notable reduction in agrarian production compared with the year 2000. The only crops that have increased are fruit trees—minimally—cereals, and olive products. The rest (legumes, vegetables, vines, industrial, and fodder crops) have decreased noticeably. This does not mean that the domestic consumption of vegetable products has fallen, but rather the opposite, it has continued to grow. These products have been replaced by imports from other European Union and Latin American countries (Witzke and Noleppa, 2010; Infante and González de Molina, 2013). This phenomenon is the logical consequence of the abandonment of land for crops and the underutilization of the pastures mentioned earlier. The comparatively low prices of these "commodities," particularly cereals and legumes for fodder, explain abandonment among other things. We do not know the extent to which replacement through imports and preferential agreements between the EU and third countries can account for the fall in vegetable production between 2000 and 2008. In any case, the economic crisis has introduced major distortions in the evolution of all these indicators, so that it is still too soon to fully confirm what appears to be an obvious and visible tendency, as we will see later: the progressive integration of the Spanish agrarian sector into global agrarian markets and the growing importance of biomass imports for the operation of a Spanish agrifood system, increasingly in need of raw materials.

The reason for this is not the disproportionate growth of the population, since this is not the case in Spain, but rather the disproportionate rise in consumption, specifically, the change in food consumption patterns where meat and dairy products are becoming increasingly important. Table 6.4 shows that livestock production, mostly consumed in Spain, increased in the first third of the twentieth century at the same rate as agricultural production. This continued to be the case until the 1970s, and coinciding with the progressive abandonment of the Mediterranean diet, livestock production grew almost exponentially. At present it is eight times greater than in 1900, while agricultural production has tripled and its growth rate was less than half of this. As it can be observed, the main protagonist of the increase in livestock production was the demand for meat, dairy products, and eggs, which had been consumed in rather small proportions until the 1970s.

This impressive growth of agrarian production has been equally seen in monetary terms. The value of the final agricultural production (FAP) at constant 1995 prices was multiplied by nearly 5.3 between 1953 and 2009, peaking in the year 2003 when the value of FAP was nearly 6 times that of 1953. In 1953, Spain was still suffering the atrocious agricultural policies of the Franco dictatorship; hence the volume of agricultural production was lower than it was 20 years earlier during the second Spanish Republic. Despite that, the figures show a spectacular growth attributable to the application of the green revolution technology, that is, to the economic effects of the industrialization of agriculture. According to the analysis, Prados (2003) made of the GDP in Spanish agriculture, its value multiplied by 3.6 between 1900 and 1990.

Table 6.4   Evolution of Livestock Production (1900–2008) in Million Tons of Fresh Matter

	1900	1933	1960	1990	2008
Meat	0.52	0.92	0.97	5.20	8.01
Milk	1.36	1.91	3.25	6.74	7.31
Eggs	0.05	0.08	0.20	0.56	0.66
Wool	0.03	0.03	0.03	0.03	0.03
Honey and wax	0.01	0.01	0.01	0.02	0.03
Total	1.96	2.95	4.46	12.55	16.03
In % over the total	**1900**	**1933**	**1960**	**1990**	**2008**
Meat	100	178	188	1007	1550
Milk	100	141	240	497	539
Eggs	100	156	366	1034	1217
Wool	100	116	117	109	106
Honey and wax	100	100	130	377	517
Total	100	151	228	641	818

Source: Author data.

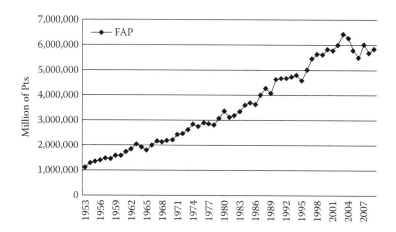

Figure 6.2 Evolution of final agricultural production in Spain (in pesetas of 1995). 166,386 pesetas = 1€. (Author data.)

Figure 6.2 details the always increasing evolution of the value of final production until the late twentieth century, when a decrease in value began related to the phenomenon mentioned earlier, that of the replacement of domestic production with imports and the abandonment of agrarian activity by many farmers, the relative extensification of agroecosystems and the abandonment of land for crops and underutilization of pastures. This can be observed with a simple comparison of the Agrarian Censuses of 1999 and 2009 (INE, 1999, 2009). In the 2009 census, 54.27% of the farms (818,560) existing in 1999 had disappeared as the crop surface had been reduced by 9.13% and that for permanent pastures by 10.84%. Between both dates, the utilized agrarian surface in Spain had decreased by almost 10% (9.74%).

6.5 SOCIAL METABOLISM OF SPANISH AGRICULTURE

From a biophysical perspective, the evolution of Spanish agriculture offers a positive, yet much more nuanced, view and offers more realistic explanations of the growth of agrarian production in developed countries. NPP_{act} (dry matter) grew moderately between 1900 and 2008 (28%), although its components showed varied behavior. There was a greater increase in cultivated cropland areas (57%) and forests (42%) but an increase of only 8% in pastureland (Figure 6.3b). This means that human pressure increased, especially on cultivated cropland, increasing the production of biomass. It should be borne in mind that the cultivated cropland area hardly grew between 1900 and 2008 (Figure 6.3a). On forestland, on the other hand, the increase was due to more productive use of the land, the increase in the area of forestland and the abandonment of many traditional modes of forest exploitation. The comparatively small rise in the NPP of pastureland is explained, first, by the reduction of the area devoted to this use and, second, and contradictorily, by the abandonment and underuse of pastureland in Spain.

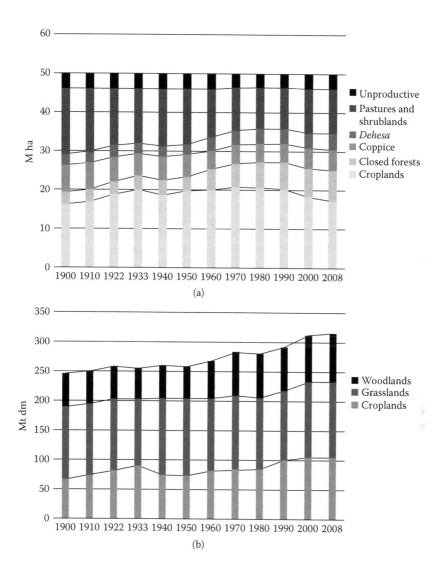

Figure 6.3 Spain NPP. (a) Land uses (million ha). (b) Origin of NPP$_{act}$ in Mt dm. (Author data.)

Figure 6.4a shows the evolution of the amount of vegetable biomass extracted from Spanish agroecosystems between 1900 and 2008, measured in megatons of dry matter. This is the most suitable way to measure the true scope of the changes, as this avoids the effect on the entire production of the current crops, with their characteristic higher water content, as well as the change toward crops with higher hydric demands. This method is particularly suited to Mediterranean agricultures such as the Spanish one where the expansion of irrigation was a key element for explaining agrarian growth. Irrigation has made it possible to multiply performances

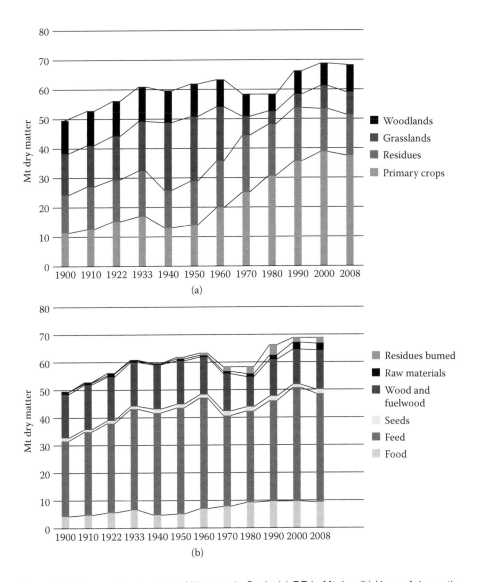

Figure 6.4 Domestic extraction of biomass in Spain (a) DE in Mt dm. (b) Uses of domestic extraction in Mt dm. (From Soto, D. et al., *Ecol. Econ.*, 128, 130–138, 2016, http://dx.doi.org/10.1016/j.ecolecon.2016.04.017.)

per surface unit and even allowed crops that are impossible to grow in rainfed conditions, such as fruit and vegetables, to become the main specialization in Spanish agriculture. The data show in physical terms that the domestic extraction of biomass (i.e., plant biomass extracted from the Spanish agroecosystems, not only the portion of commercial value) grew at a moderate rate throughout the twentieth century, contrasting with the monetary growth. While in monetary terms the agricultural production increased nearly sixfold, in physical terms it grew only by 38%, showing the contrast between monetary and biophysical measurements.

The breakdown of the analysis also shows, in this case, significant changes. The growth in domestic extraction was concentrated in primary crops, rising by 236% with respect to 1900, as against a small 8% growth in residues. Nevertheless, pasture was reduced by 46% and forests by 17% (Table 6.5).

This evolution shows that there has been a very significant change in the different uses of the biomass over the period studied, which has favored primary crops above other uses. In this regard, it can be said that the industrialization of agriculture has led to a significant increase in the biomass produced but, above all, to a concentration of the biomass extracted from primary crops. Total domestic extraction has risen from 33% of the NPP of crops to 50% in 2008. Cereals, olives, industrial crops, and artificial meadowland and forage are the crops that have grown most over the period studied. This is the biophysical translation of the specialization seen in the Spanish agrarian sector in livestock, fruit and vegetable, and olive oil production.

Table 6.5   Basic Data of Biomass Production in Spanish Agriculture, 1900–2008 (Megatons of Dry Matter)

	1900	1933	1950	1970	1990	2008
Vegetable Biomass						
NPP_{act}	245	256	258	283	293	314
DE	50	61	62	58	66	69
Primary crops	11	17	14	25	36	38
Crop residues	13	16	15	19	18	14
Grazed biomass	14	17	23	6	5	8
Woodlands	11	11	11	8	8	9
Imports	1	1	1	5	14	32
Exports	0	0	0	1	5	13
DMC Vegetable	50	61	62	62	75	87
DMC Vegetable Biomass						
Food	4.4	6.4	5.1	7.7	10.9	11.7
Feed	27.8	36.7	39.0	36.0	40.7	50.8
Seeds	0.6	1.0	0.7	1.1	1.2	1.7
Wood and fuelwood	16.3	16.2	16.0	14.5	16.9	18.1
Raw materials	0.7	0.8	0.7	0.9	1.8	3.1
Residues burned	0.0	0.0	0.2	1.5	3.6	1.3
DMC Animal Biomass						
Food	0.33	0.54	0.52	1.54	3.15	4.05
Raw materials	0.03	0.05	0.06	0.15	0.24	0.36

Source: Author data.
Notes: DE, domestic extraction; DMC, domestic material consumption; NPP, net primary productivity.

In fact, the evolution of the Spanish agriculture during the last century can be considered as a process of growing commoditization of production and of the factors that make it possible. The technological and productive efforts were oriented toward maximizing the portion of biomass having higher market value and mobility, reducing the multifunctionality of crops. In other words, the growth of agricultural production did not correspond to an equal increase in total biomass extracted from Spanish agroecosystems, but only of the portion having high market values.

At the same time, since the 1960s the livestock production had a considerable growth while it changed its composition to meet the increasing demand of meat and dairy products of the Spanish population. Working livestock has practically disappeared. Pigs and poultry today represent 53% of total livestock compared with 9% in 1900. Bovine had the same weighting in 2008 as in 1900 (32%), but this data hide a significant change in the breeds of cattle, which are today much more specialized in milk and meat (Figure 6.5). This specialization process has also fostered a change in the functionality of cattle. The mixed use races, providing both labor and manure as meat and milk, have gradually become marginal. Likewise, the number of monogastric animals has increased significantly. These animals depend on high-quality processed feed, unlike species able to feed off pastureland and residues (sheep, goats, cattle on extensive farms, etc.). The transition from an organic livestock farming to an industrial livestock breeding has meant an increase in feed conversion ratio (% animal product output:input feed) from 1.3 to 7.9 between 1900 and 2008. This ratio stood at the average in Western Europe in 2000 (7.8), which, with the ratio of Eastern Europe, is the highest in the world (Krausmann et al., 2008a).

However, such expansion was not sustained by Spanish agroecosystems, instead, since 1970 the abandonment of croplands and grasslands has acquired worrisome proportions. This abandonment process was related to the low profitability of extensive livestock production and the fall of fodder prices in the international markets. Thus the growth of Spanish livestock production—mostly intensive and practiced on small land surfaces—depended on massive imports of cereals and legumes (or

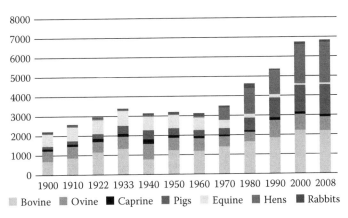

Figure 6.5 Evolution of livestock. (From Soto, D. et al., *Ecol. Econ.*, 128, 130–138, 2016, http://dx.doi.org/10.1016/j.ecolecon.2016.04.017.)

even processed animal foodstuff) for feeding livestock. This fact explains why the domestic extraction and domestic consumption of biomass (domestic extraction plus imports minus exports) have displayed a divergent evolution and a decoupling trend. As seen in Table 6.5, until 1970 the domestic extraction and consumption remained closely linked, but since that year the role of imports became increasingly important, such that by 2008 the net balance of foreign trade of biomass amounted to 23% of the domestic consumption (imports reaching up to 37%).

From a monetary standpoint, Spain was traditionally a net exporter of agricultural products (Pinilla and Gallego, 1996). From a biophysical point of view, however, it never was. In fact, foreign trade played a rather modest role in the evolution of the Spanish agricultural sector until the late 1970s. But, as a consequence of the globalization process, foreign trade has become a key factor of the Spanish food and agriculture sector: on the one side, it enables the specialization of Spanish agricultural products (oil and horticultural and fruit products) to enter the international markets, particularly European; and on the other side, it allows for sustaining the growing consumption of meats and dairy products of Spaniards, supplying an important percentage of animal foodstuffs (Figure 6.6). In fact, the current availability in Spain of animal foodstuffs exceeds the demand, given that an appreciably amount of biomass from grasslands and crop residues that is potentially useful for feeding animals remain unharvested.

As we have seen, the main vector of socioenvironmental change in agriculture was the increase in commercial biomass. We have verified that this part of the total biomass grew much more quickly and intensely. Part of this growth can be explained by the technology associated with the "green revolution" where genetic change has played a major role, leading to a type of crops with less "residue" weight. This process can be observed clearly in the evolution of the relationship between grain and straw of cereals and legumes (Figure 6.7). For example, agrarian industrialization was accompanied by the replacement of traditional cereal and legume varieties, where straw was

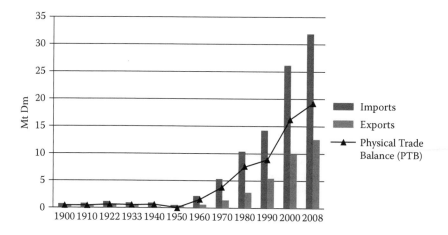

Figure 6.6 Trade of plant biomass in Spain. (Author data.)

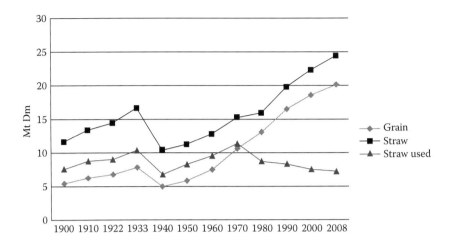

Figure 6.7 Evolution of the relationship between grain and straw of cereals and legumes. (Author data.)

an essential part of animal feed, with high stalk varieties and less grain, in favor of varieties that aimed to increase the amount of grain to the detriment of straw. At the same time, increasing amounts of straw have either been burned or abandoned on the land, in parallel with the drop in feed prices, the greater protagonism of monogastric livestock and the almost complete disappearance of labor livestock.

Yields per unit area increased thanks to the addition of higher amounts of chemical fertilizers, especially in irrigated lands and in the territories that used new seed varieties, both hybrid and improved. Table 6.1 shows how the use of fertilizers grew by 100% between 1960 and 2008 although this growth had become more moderate due to the results of the economic and financial crisis. The most immediate effect of the application of this land-saving technology was not only the possibility of using these varieties, far more productive than traditional ones in optimum nutrient and water supply conditions, but it also put an end to the rotations imposed by traditional farming to adapt to the scarcity of both factors. This is how monoculture progressed, with the alternation of crops determined by the demands of agrarian markets, rather than by agronomic rationality. The consequent reduction in biodiversity encouraged the appearance of plagues and plant illnesses and the use of phytosanitary products, which had been quite limited until then. The application of this type of chemical remedy caused a vicious circle in which breaking the trophic chains (the disappearance of beneficial insects that controlled insect plagues is the result of insecticide use), along with advances in crops and homogeneous varieties in large expanses of land, brought about the growing use of these substances to control plagues and illnesses. This explains the fact that the use of phytosanitary products was multiplied by 20.3 between 1960 and 2008 (Table 6.1), as there has been no proportional increase in yields per unit of area or land cultivated.

But the most spectacular advance, the most defining for this industrialization phase, was the mechanization of most agrarian tasks. The power of tractors was multiplied by 30.5, harvesters by 63, and the small machinery by 176. This began with large farms and cereals and extended to vegetable gardens and greenhouses, affecting practically all the smaller ones. Table 6.1 also includes irrigation installations or collective and individual infrastructures built to store superficial waters and elevate underground water, one of the keys of industrialization. Nevertheless, both mechanization and the provision of irrigation infrastructures caused a significant increase in the demand for final energy in the agrarian sector. As seen in Table 6.1, fuel consumption was multiplied by 2 and electricity consumption by 19.

Finally, it is worth noting how forced cultivation under plastic in its different forms has developed in Spain, mostly greenhouses for growing vegetables and tunnels for fruit production, a reflection of the production specialization of Spanish agriculture in general. The surface devoted to these crops went from over 1,960 in 1975 to almost 112,000 hectares, mainly greenhouses. These surfaces, with various cultivation cycles in the same year, reach high levels of production that have conditioned and contributed to the lowering of the prices received by farmers and further reducing the profitability of open air crops, and especially, production in the rainfed crops of the inner peninsula.

Thus a vast amount of energy had to be injected into the agroecosystems to maintain continuous growth in agrarian production. It can therefore be observed that since 1960 the embodied invested has been multiplied by 5.05, an increase much higher than that experienced by crop production. This model, based on the growing use of input from outside the sector has had and continues to have very important economic consequences: the continued increase in costs from outside the sector has made farmers increasingly dependent on markets, has caused a reduction in agrarian income, and has destroyed employment in a mad race to save on labor costs in order to compensate for the fall in income. The agrarian income in real terms has decreased by 30% since 1961, while other branches of activity have continuously increased their income. The fact is that agriculture stopped being profitable some time ago, even with the grants from the Common Agricultural Policy (González de Molina et al., 2016).

From the start of mechanization, the best way to counteract the fall in agrarian income has been to increase labor productivity by replacing manpower with machines. For example, since 1990 the number of people employed in the agrarian sector has gone from 1,286,000 to 881,000 in 2012, so that the agrarian income per worker employed has gone from 10,699€ to 12,981€ in 2012, increasing by 12.45% (MAGRAMA, 2014). This destruction of employment is not new and has been a constant in the sector since the early twentieth century. According to the National Institute of Statistics (www.ine.es), the agrarian active population has fallen dramatically to a tenth of that of 1950. At that time it provided almost half of the country's employment, while in 2010 it provided approximately 4%. Numbers have dropped from almost five million farmers in the early twentieth century to under half a million at present, many of whom can only practice agrarian activities on a part-time basis.

This loss of profitability also has its biophysical counterpart. If we transfer this sustained increase in costs from outside the sector to energy magnitudes, the balances between input and final product show that this process has also been highly inefficient, with Spanish society investing large amounts of money to obtain food, fibers, and the raw materials it consumes. Moreover, agricultural products travel very long distances before reaching consumers' tables and require a vast logistic infrastructure.

Processed food has overtaken fresh food and increasingly more foods are consumed outside the home. New and more sophisticated "artifacts" powered by gas and electricity now intervene in human nutrition, and have increased the energy cost of nutrition. For example, in Spain the primary energy invested in the wide and complex process of nutrition was 1408 PJ in 2000, almost 27% of the 5240 PJ calculated for total primary energy consumption. This amount includes not only the energy costs in the agrarian sector strictly speaking, but also those from transport, processing, assembly and packaging, sale in food establishments, and the costs of conservation and preparation in homes. The agrarian sector is responsible for a little over a third of the total consumption of primary energy; transport of food (17.43%), industrial processing (9.83%), packaging (10.63%), sale (9.61%), and conservation and consumption (18.35%) of the other two thirds. At present, Spain requires more than 1408 PJ to satisfy its endosomatic metabolism, while the energy in the food consumed barely provides 235 PJ. For every energy unit consumed in the form of food, six have been spent on production, distribution, transport, and preparation (these and other complementary data can be seen in Infante-Amate and González de Molina, 2013).

6.6 ENERGY EFFICIENCY OF SPANISH AGRICULTURE FROM AN ECONOMIC POINT OF VIEW

Ecological economic indicators have been proposed to assess the energy efficiency and profitability of agrarian systems and of economic activity in general from a societal perspective. The energy return on investments (EROI) measures the efficiency of the productive use of energy from a social perspective, evaluating the returns received by society for each energy unit invested in agrarian production (Tello et al., 2015; Guzmán and González de Molina, 2015). The first EROI we have considered is known as *final EROI* and measures the total energy invested to get the socialized biomass; that is, the vegetable and animal biomass produced for human consumption. The formula is as follows:

Final EROI (FEROI) = socialized biomass (SB)/(reused biomass (RuB) + external inputs (EI))

Nevertheless, this index could be broken down in two more EROIs, depending on where the attention is focused, whether on external input used to obtain the socialized biomass or on the amount of biomass reused to this end. Therefore, the *internal final EROI* (IFEROI) measures socialized biomass in relation with reused biomass as follows:

$$IFEROI = SB/RuB$$

and *external final EROI* (EFEROI) measures the socialized biomass in relation with external inputs (mainly, but not always, fossil fuels) as follows:

$$EFEROI = SB/EI$$

Table 6.6 summarizes the main energy indicators for Spanish agriculture during the last century. The NPP_{act} for all Spanish agroecosystems grew moderately in energy terms (29%). However, the growth of *socialized biomass* increased to a

Table 6.6 Main Energetic Magnitudes of Spanish Agriculture (TJ) (1900 = 100)

	1900		1960		2008	
	TJ	%	TJ	%	TJ	%
NPP_{act} (a+c+d+e)	4,366,701	100	4,800,059	110	5,625,189	129
Socialized vegetable biomass (SVB) (a)	400,170	100	425,184	106	505,661	126
Socialized vegetable biomass (Cropland)	164,879	100	234,616	142	341,303	207
Socialized vegetable biomass (Forestland)	235,290	100	190,569	81	164,358	70
Socialized animal biomass (SAB) (b)	9,333	100	20,157	216	105,869	1134
Socialized biomass (SB) (a + b)	409,503	100	445,341	109	611,530	149
Reused biomass (RuB) (c)	501,739	100	746,671	149	854,664	170
Unharvested biomass (UhB) (d)	3,235,392	100	3,300,618	102	3,798,384	117
Aboveground unharvested biomass	1,540,562	100	1,482,719	96	1,825,377	118
Belowground unharvested biomass	1,694,829	100	1,817,899	107	1,973,007	116
Recycling biomass (RcB) (c + d)	3,737,130	100	4,047,289	108	4,653,048	125
Accumulated biomass (AB) (e)	229,401	100	327,586	143	466,480	203
External inputs (EI) (g)	23,723	100	92,690	391	510,260	2151
Total inputs consumed (TIC) (c + d + g)	3,760,854	100	4,139,979	110	5,163,308	137

Source: Author data.

greater extent (49%), as did the investment of biomass required to obtain it (*reused biomass*) (70%). This benefited livestock, since *reused biomass* was largely used to feed constantly growing numbers of livestock (see Figure 6.4b).

External inputs (*EI*) went through the roof, increasing twentyfold. The use of synthetic chemical fertilizers increased substantially, although their participation in the total *EI* was relatively low (15%) in 2008 (see Tables 6.1 and 6.6). This modest percentage must be linked with a phenomenon inherent to semiarid agroecosystems typical of the Mediterranean: the lack of rainfall means that the application of more fertilizer is of limited utility in terms of increasing NPP_{act} in the absence of optimum hydric conditions. In energy terms, the introduction of mechanical technologies has played a greater role, now accounting for 25% of *EI* (see Tables 6.1 and 6.6). However, the importing of animal feed saw the biggest growth, now representing 37% of *EI* (see

Tables 6.2 and 6.6). This fact, added to the growing importance of *reused biomass*, is the main cause for the significant loss of energy efficiency in Spanish agroecosystems.

Ultimately, Spanish society has invested a considerable amount of energy to obtain a supply of biomass with an increasingly large animal component. *Socialized vegetable biomass* from cropland has doubled in energy terms, but *socialized animal biomass* has by far and away increased the most, increasing by over elevenfold. In other words, the rise in productivity achieved by Spanish agroecosystems between 1960 and 2008 was largely invested in producing food for livestock. The well-known inefficiency of converting plant biomass into animal biomass has been transferred to the whole of Spain's agrarian sector, as shown by the proposed economic EROIs.

FEROI has fallen significantly (Table 6.7). The industrialization of Spain's agrarian sector has brought about a considerable loss of efficiency (over 40%). In other words, the growth of agrarian production has been made possible by the multiplication of energy uses. In this respect, traditional organic agriculture in Spain was more efficient than the industrial version in relation to the energy inputs invested by society. As expected, EFEROI has dropped even more dramatically, from a production of 17.3 J of socialized biomass for every joule invested from outside the sector, to 4.8 J in 1960 and 1.2 in 2008. The key to increasing yield per land area unit and increasing agrarian production has been the huge increase in energy incorporated into production (Smil, 2013), coming from fossil fuels and biomass. The important depressor effect on EFEROI brought about by importing external biomass inputs into agroecosystems has been demonstrated in other case studies (González de Molina and Guzmán, 2006; Guzmán and Alonso, 2008).

Ostensibly, traditional organic agriculture had to invest a huge amount of biomass to make its own reproduction possible, but in the case of Spain, IFEROI is also more inefficient for industrial agriculture than for traditional organic agriculture, having fallen by 12% since 1900. This latter phenomenon is to some extent unexpected. Traditional organic agriculture, given the high land cost required to replenish fertility and produce the energy required for the production process, is assumed to be more inefficient than industrial agriculture in terms of the investment of internal energy (Guzmán and González de Molina, 2009). In theory, the availability of external inputs should save on the amount of land required for production, or should decrease the investment of *reused biomass* (Guzmán et al., 2011). As a consequence, the IFEROI and even FEROI should increase. This tendency has been shown in another case study, applying this methodology, in which the opposite process was observed, in other words, a process of *agricolization* (see Chapter 5). However, in the case of Spain, this has not occurred. This is due to the fact that the increase in productivity achieved between 1960 and 2008 has largely been invested in feeding livestock.

Table 6.7   Evolution of EROIs from an Economic Point of View

	1900	1960	2008
Final EROI (FEROI)	0.78	0.53	0.45
External final EROI (EFEROI)	17.3	4.8	1.2
Internal final EROI (IFEROI)	0.82	0.60	0.72

Source: Author data.

6.7 EROIs OF SPANISH AGRICULTURE FROM AN AGROECOLOGICAL PERSPECTIVE

As we have seen in Chapter 2, the EROIs from an agroecological point of view inform us of the real productivity of the agroecosystem, not just the part that is socialized. Furthermore, they inform us on the reinvestment in the fund elements, that is, in the structure of the agroecosystem to sustain basic ecosystem services. The four EROIs proposed are (see Chapter 2) as follows:

1. The $NPP_{act} EROI$ explains the real productive capacity of the agroecosystem, whatever the origin of the energy it receives (solar for the biomass or fossil for an important portion of the EI).

NPP_{act} EROI = NPP_{act}/total inputs consumed, being total inputs consumed (TIC) = RcB + EI = RuB + UhB + EI

2. The *agroecological final EROI (AE-FEROI)* = *SB/TIC* gives a more exact idea of the total energy investment required to obtain *socialized biomass.*
3. From an agroecological point of view, the relationship between this indicator and the final EROI is of great interest.

$$\text{Biodiversity EROI} = 1 - \frac{\text{AE-FEROI}}{\text{FEROI}} = \text{UhB/TIC}$$

It can reach a minimum of 0, when all of the recycled biomass is reused, indicating agroecosystems with very significant human intervention, and a maximum value of 1 when there are no external inputs and no biomass is reused by society. This would be the case in natural ecosystems without human intervention.

Furthermore, this indicator allows us to explore the hypothesis of land sparing versus land sharing from the perspective of energy, since it links the productivity of the system with the biomass available for wild heterotrophic species.

4. We have included an agroecological EROI that tells whether the energy added to the system is contributing to store energy as accumulated biomasss (AB). Accumulated biomass can be considered as a fund element related to the ecosystem services provided by forests, but not only by them. The growth in AB can also be due to accumulation in croplands or grasslands (hedgerows, shade trees), providing ecosystem services for the agrarian activity. Biomass is also accumulated in living tissues of woody crops, growing when there is an expansion of these crops. In all cases, biomass accumulation contributes to carbon sequestration:

Woodening EROI = AB/TIC

According to the data in Table 6.8, the loss of efficiency in industrial agriculture with regard to traditional Spanish organic agriculture might have additional causes. As we said, agroecological EROIs allow us to detect whether the degradation of fund elements is undermining the productivity of agroecosystems. NPP_{act} EROI remained steady up until the 1960s. However, after that point it fell by 6%, coinciding with the industrialization of

Table 6.8   Evolution of EROIs from an Agroecological Point of View

	1900	1960	2008
NPP_{act} EROI = NPP_{act}/TIC	1.16	1.16	1.09
AE-FEROI = *SB*/*TIC*	0.11	0.11	0.12
Biodiversity EROI	0.86	0.80	0.74
Woodening EROI	0.06	0.08	0.09

Authors' calculations based on agricultural statistics cited.

Spanish agriculture. This decline occurred in spite of the injection of energy and water received. In a semiarid country such as Spain, the 82% increase in irrigated land area between 1960 and 2008, combined with the growth of invested energy (*TIC*) (25% more between 1960 and 2008) should have had the opposite effect. However, high rates of erosion (Gómez and Giráldez, 2008, Vanwalleghem et al., 2011), the decrease in organic soil matter, salinization and the overexploitation of water resources (European Commission, 2013), and the loss of agrarian biodiversity (Garrido, 2012; MAPA, 1995) are responsible for this decline. Ultimately, the deterioration of fund elements (soil, water, biodiversity), caused by industrial agriculture itself, is taking its toll.

AE-FEROI has grown by 9%. This increase might explain the fairly widespread notion that Spanish agriculture significantly increased productivity over the course of the twentieth century. However, as shown here, in reality total productivity did not grow; rather, growth was achieved in the part of productivity appropriated by society in relation to total inputs consumed.

The biodiversity EROI decreases by 15% indicating de-growth in unharvested biomass in relation to total inputs consumed, which entails a lower level of relative energy availability for wild heterotroph organisms, particularly in cropland (Figure 6.8). In this space, a major decline was observed for unharvested biomass, both below and aboveground. The fall in belowground unharvested biomass would help to explain why half agrarian plots of land in Spain have an organic carbon content of less than 1% currently (Rodríguez Martín et al., 2009).

In other words, the allocation of a growing amount of farming production to feeding cattle, a fundamental component of reused biomass, has a negative impact on biodiversity. This effect would not be compensated by the abandonment of pastureland and woodland (Figure 6.8), questioning the strategy of land sparing. The disassociation of the agroecosystem in areas of intensive production and abandoned and/or protected areas has not brought about a significant increase in the trophic energy available for transfer from plants to other levels in the trophic networks of ecosystems.

This argument adds to those put forward in other research, showing that the intensification of traditional agriculture has led to losses in biodiversity owing to the loss of ecological heterogeneity at multiple spatial and temporal levels (Benton et al., 2003; Firbank et al., 2008; Guzmán and González de Molina, 2009; Lindborg and Eriksson, 2004; Schuch et al., 2012; Vos and Meekes, 1999). Furthermore, it supports the strategy of land sharing, at least in countries where traditional agriculture has played a major role in shaping the landscape (Barral et al., 2015; Ramankutty and Rhemtulla, 2012; Wehrden et al., 2014).

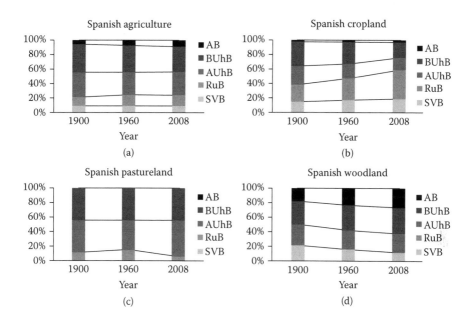

Figure 6.8 Evolution of NPP$_{act}$ (TJ) by its use in relative terms in (a) Spanish agriculture, (b) Spanish cropland, (c) Spanish pastureland, and (d) Spanish woodland. (SVB: socialized vegetable biomass; RuB: reused biomass; AUhB: aboveground unharvested biomass; BUhB: belowground unharvested biomass; AB: accumulated biomass.)

Last, woodening EROI grows 48% from 1900 to 2008. At first sight, this reforestation process would support a *land sparing* strategy, as agricultural modernization would have allowed the growth in forest area, much of it legally protected (e.g., 40% of Spain's total forest areas are protected, according to MAGRAMA, 2014). However, we have to put these values in context. AB increased from 229,401 TJ in 1900 to 466,480 TJ in 2008. Of this growth (237,079 TJ), approximately 10% corresponded to the expansion of woody crops, mainly olives (23,079 TJ). An additional 30% was due to the decrease in biomass extraction from forest (SB). Therefore, this growth in AB is not a side effect of industrialized agriculture, but rather it is mainly due to the substitution of firewood with fossil fuels in the Spanish economy. The remaining 60% of AB growth would be due to forest surface growth in areas freed from agricultural activities. However, this growth is due to the abandonment of spaces (pastureland and dryland) largely devoted to animal feeding, which have been largely substituted by imported feed, mainly from Latin America (Soto et al., 2016). It is likely that this partial externalization of the land cost of Spanish livestock has caused important deforestation in these areas with growing agricultural frontiers, a process that has not been studied here. In fact, estimated greenhouse gas (GHG) emissions from deforestation (land use and land-use change [LULUC] emissions) caused by Spanish feed imports in 2004 ranged between 20 and 64 Tg CO$_2$-eq in three different scenarios, which can be compared to an estimated emission from the Spanish livestock sector (excluding LULUC emissions) of 48 Tg CO_2-eq (Leip and Weiss, 2010). On the other hand, scientists using NASA satellite data have found that the size of the clearings used for crops has averaged twice the size of clearings used for pasture in the Amazon (Bettwy, 2006). Thus, discounting the *disaccumulated* biomass in other agroecosystems would probably lead us to negative results, but this issue should be addressed by further research.

6.8 CONCLUSIONS

The industrialization of Spanish agriculture allowed agrarian production to grow during the twentieth century, especially livestock production. This growth was based on the injection of large quantities of external energy in the form of fossil fuels and biomass from 1960 onward. The importing of biomass has been essential to sustain a model of intensive livestock farming that is decoupled from the territory, leading to the abandonment of pastureland and dryland.

The application of the proposed economic and agroecological EROIs to the case study has informed us, first, of the return on energy investments for society and, second, of processes that affect the fund elements of agroecosystems and their capacity to generate flows of ecosystem services. From a social point of view, the return was highest in the early twentieth century, when considering the total energy invested by society (final EROI) or the external or internal inputs separately (external versus internal final EROI). In short, Spanish society has obtained decreasing returns on the energy invested throughout the process of agrarian industrialization.

Considering economic EROIs by themselves, the loss of energy efficiency could be said to have been caused by the increase in livestock production. However, the agroecological EROIs show that it is also the result of the degradation of fund elements. The 6% fall in NPP_{act} EROI points in this direction. In fact, the agroecological EROIs tell us of other key processes that undermine the sustainability of the agroecosystem. The biodiversity EROI ratio alerts us to the low return to nature in the form of *UhB* available to aboveground and underground wildlife, especially in cropland. This low return in cropland is not compensated by the abandonment of pastureland and forestland, questioning the hypothesis of *land sparing* for the purpose of sustaining biodiversity, rather than *land sharing*. Furthermore, cropland soil suffered a drastic reduction in the flows of biomass required to maintain its quality.

Finally, the growth in forestland area and the loss of forest functionality has allowed a certain increase in accumulated biomass in this space. However, analyzed globally, the woodening EROI shows that this structural improvement in the agroecosystem through reforestation was collateral and marginal, with the bulk of energy investment aimed at maintaining livestock numbers.

CHAPTER 7

Energy Return on Investment in Traditional and Modern Agricultures
Coffee Agroecosystems in Costa Rica from an Agroecological Perspective (1935–2010)

Juan Infante-Amate, Wilson Picado, and Gloria I. Guzmán

CONTENTS

7.1 INTRODUCTION

This chapter compares energy flows in traditional and modern agrarian systems by examining the case of coffee growing in Costa Rica, a tropical crop managed typically using an agroforestry approach. This comparison evaluates the energy efficiency of each style of management by means of economic and agroecological energy return on investments (EROIs). This also allows us to understand the functional change this crop underwent during the process of industrialization. The socio-ecological transition (SET) in agriculture, as well as altering the energy efficiency of agroecosystems by incorporating technologies based on abiotic external inputs (EIs) (see also, Leach, 1976; Naredo and Campos, 1980), has had (and continues to have) another fundamental effect on agroecosystems: the change in the social and ecological functionality of the crop. Whereas in preindustrial economies, agriculture provided the majority of goods and services required by society, in industrial econo-mies its functions have been increasingly determined and limited by the appearance of replacement goods, and its objective has been practically reduced to the produc-tion of saleable fruits (Fischer-Kowalski and Haberl, 2007; González de Molina, 2010). Furthermore, the ecological functioning of the crop has undergone profound alteration: the internal flows that sustained traditional agroecosystems have been partially replaced by external energy flows, largely fossil in origin. Consequently, the environmental services provided by agroecosystems have also been altered (Guzmán and González de Molina, 2015; Tello et al., 2016). The future scenario of fossil fuel scarcity (Murray and King, 2012) requires increasingly efficient agrarian systems to be designed that are less dependent on these sources of energy. This, in turn, requires agriculture to regain its multifunctional role, reevaluating its productive potential that, in the case of coffee, goes beyond the bean (see also Haberl et al., 2011).

This chapter offers information about energy flows between 1935 and 2010. For the years between c. 1935 and c. 2010, we provide information that includes the shade layer, including productive flows and recycling within the plantation. In this respect, the results of our analysis offer new information to complement studies that have already highlighted the loss of efficiency in modern systems relative to traditional ones (Schroll, 1994; Cleveland, 1995; Carpintero and Naredo, 2006; Infante-Amate, 2014). It also provides unprecedented evidence when comparing the energy balances of tropical crops, since previous studies have not incorpo-rated the shade factor into their study of energy balances. The majority of these studies, carried out in coffee-growing countries such as Nicaragua (Cuadra and Rydberg, 2006), Brazil (Giannetti et al., 2011a, 2011b; Turco et al., 2012; Flauzino et al., 2014; Muner et al., 2015), and also Costa Rica (Marozzi et al., 2004, Mora-Delgado et al., 2006), have focused solely on comparing organic and conventional coffee production, without looking at the different structures and functioning of coffee agroecosystems (Photograph 7.1). Our chapter pays particularly close attention to this analysis and, in this regard, conducts a more rigorous comparison between traditional management (analogous to organic farming) and more modern management (industrialized), thereby making a relevant contribution to debates about the energy efficiency of organic agriculture (see also Smith et al., 2015). Finally, analyzing the multifunctionality of coffee crops helps us to understand not only the ecological rationality of farming approaches (Toledo, 1995), but also it provides valuable knowledge for the design of more sustainable coffee production. Agroecology considers this type of knowledge (*traditional ecological knowledge*, see also Berkes et al., 2000) as a basic instrument for the achievement of agrarian sustainability.

7.2 METHODOLOGY AND SOURCES

The methodology followed in this chapter follows the proposals of Tello et al. (2015, 2016) for estimating economic EROIs, and Guzmán and González de Molina (2015) for agroecological EROIs as summarized in Figure 7.1, and as we have seen

Photograph 7.1 Organic shade coffee plantation in Costa Rica.

in Chapter 2. We have followed the proposal developed by Infante-Amate (2014) to differentiate between the end uses of agrarian production from an energy perspective and to evaluate the change in social functionality. In general, we have used the conversion factors for energy values included in Appendix I, in the case of outputs, and in Appendix II in the case of inputs. In the case of outputs, we have focused particularly on the differences between traditional and modern management approaches, and in the case of inputs, we have taken a historical perspective to estimate the different direct and indirect input contents.

Table 7.1 provides available information and estimated indicators for different periods. Despite the difficulty of compiling so much information over such long time, in some cases we have been able to make annual estimations between 1935 and 2010. This is the case with coffee bean production, coffee plant waste (pruning), and external inputs. This way, we can estimate a specific EROI that relates bean production and external inputs (EI), which is the most commonly used measure in the literature; and another additional EROI that considers beans and prunings, in relation with EI. We have only been able to reconstruct reliably the role of the shade layer for two historical moments: 1935 (example of traditional management) and 2005 (example of modern management). Hence, it has only been possible to estimate total production for those years: actual net primary productivity (NPP_{act}), and its components (socialized vegetable biomass [SVB], recycling biomass [RcB], unharvested

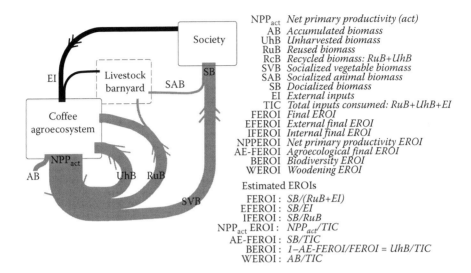

NPP_{act} *Net primary productivity (act)*
AB *Accumulated biomass*
UhB *Unharvested biomass*
RuB *Reused biomass*
RcB *Recycled biomass: RuB+UhB*
SVB *Socialized vegetable biomass*
SAB *Socialized animal biomass*
SB *Docialized biomass*
EI *External inputs*
TIC *Total inputs consumed: RuB+UhB+EI*
FEROI *Final EROI*
EFEROI *External final EROI*
IFEROI *Internal final EROI*
NPPEROI *Net primary productivity EROI*
AE-FEROI *Agroecological final EROI*
BEROI *Biodiversity EROI*
WEROI *Woodening EROI*

Estimated EROIs

FEROI : *SB/(RuB+EI)*
EFEROI : *SB/EI*
IFEROI : *SB/RuB*
NPP_{act} EROI : *NPP_{act}/TIC*
AE-FEROI : *SB/TIC*
BEROI : *1–AE-FEROI/FEROI = UhB/TIC*
WEROI : *AB/TIC*

Figure 7.1 Summary of energy flows studied, study limit and estimated energy return on investments (EROIs). (Authors' data, based on Guzmán, G. I. and González de, M. M., *Agroecology and Sustainable Food Systems*, 2015, 39(8), 924–952; Tello, E. et al., Social Ecology Working Paper 156, IFF—Social Ecology, available at https://www.uni-klu.ac.at/socec/inhalt/1818.htm, accessed on January 15, 2015; Tello, E. et al., *Ecological Economics*, 2016, 121, 160–174, doi:http://dx.doi.org/10.1016/j.ecolecon.2015.11.012.)

biomass [UhB], reused biomass [RuB], and accumulated biomass [AB]), and socialized animal biomass (SAB) as well as the input flows into the agroecosystem (total inputs consumed [TIC]).

In relation to the sources used, bean production and land area figures are taken from official sources in Costa Rica and FAOSTAT (FAO, 2016). On the basis of these, we have estimated the production of by-products (Rodríguez and Zambrano, 2010), prunings (Romijn and Wilderink, 1981), and the herbaceous layer (Romero, 2006). The shade layer is more complex: we have estimated the shaded land area in each period, as well as the dominant species. We have verified that Inga (*Inga*), Poró (*Erythrina poeppigiana*), Musaceae, and timber-yielding species (chiefly the Eucalyptus) have always provided canopy covering to around 90% of the total shaded area (Dirección General de Estadística y Censos [DGEC], 1953; ICAFÉ, 2007); hence, on the basis of our review of the literature—chiefly derived from CATIE, the Tropical Agricultural Research and Higher Education Center—we have estimated the net primary productivity of each species, as well as their recycling on the farm and their final uses. In the case of the Poró tree, of particular note are the works of Montenegro (2005), Romero (2006), and Merlo (2007). In the case of the Inga tree, we turned chiefly to Jiménez and Martínez (1979) and Salazar and Palm (1987). With regard to Musaceae, Farfán-Valencia (2005) provided a complete summary. For timber-yielding species, there is a complete study available on this subject (Detlefsen and Somarriba, 2012) along with the frequently used estimations of Beer et al. (1998, p. 151).

Table 7.1   Summary of Available Information and Estimated Indicators by Period

	Available in Annual Series (1935–2010)	Available for 1935 (Traditional Management) and 2005 (Modern Management)[a]
Information by production layer	Coffee layer (*Coffee bean, leaves, and firewood*) Herbaceous layer	Shade layer (*Fruits, firewood, timber, and leaves from four major species*)
Energy flows	External inputs Coffee bean output Coffee plant output	Total inputs consumed, recycled biomass, reused biomass, unharvested biomass, accumulated biomass, socialized biomass (vegetable, animal, and total) Shade layer output
Energy return on investment (EROIs) economic point of view	Coffee bean *external final EROI* (EFEROI) = *Coffee bean output/external inputs* (see Figure 7.3e) Coffee plant EFEROI = *Coffee plant output/external inputs* (see Figure 7.3f)	Final EROI (FEROI) External final EROI (EFEROI) Internal final EROI (IFEROI) (see Figure 7.5 and Table 7.2)
EROIs agroecological point of view		Net primary productivity EROI Agroecological final EROI Biodiversity EROI Woodening EROI (see Figure 7.5 and Table 7.2)

Source: Author data.
[a] Same information as in column two, plus the data offered in this column.

In the case of inputs, the most important estimation, since it has the greatest influence on the results, is that of fertilizers applied to the coffee plants (Cámara Nacional de Cafetaleros [CAFETICO], 1992; ICAFÉ, 2007, 2010) together with other estimations regarding the application of fertilizers in agriculture as a whole in the region (ICAITI, 1967; OAS, 1970; Rojas, 1979; FAO, 2016). A summary can be found in López and Picado (2012). In terms of pesticides, we have used a similar system (Ramírez, 2011; FAO, 2016) completed with studies that provide the percentage amount of pesticides added per crop in the country (Maltby, 1980; Hilje et al., 1987, 1989). Information about labor and machinery comes from Picado (2000) and Renjifo (1992), validated by other similar studies (Duque-Orrego and Dussán, 2004).

7.3 COSTA RICAN COFFEE IN THE GLOBAL COFFEE SYSTEM

When we talk about coffee, we are talking about one of the most widely consumed drinks in the world, and perhaps the second most important product in global commerce after oil, estimated on the basis of the final purchase price value (Grigg, 2002; Pendergrast, 1999; Ponte, 2002). Up to the eighteenth century, its production and consumption was confined to very specific parts of Africa and Asia (Pendergrast, 1999). Since then, fired by the first global groundswell, its production spread to other parts of the world, and its consumption became increasingly popular, chiefly in the Americas and Europe, where it is now more popular than tea (Grigg,

2002). Its consumption (availability) is 1.2 kg of coffee beans (green) per inhabitant per year (FAO, 2016), which is roughly equivalent to a total consumption of 2500 million cups a day (Dicum and Luttinger, 1999).

Figure 7.2a shows the geographical expansion of coffee worldwide, which also informs us of the agroclimatic requirements of coffee: it is confined to tropical

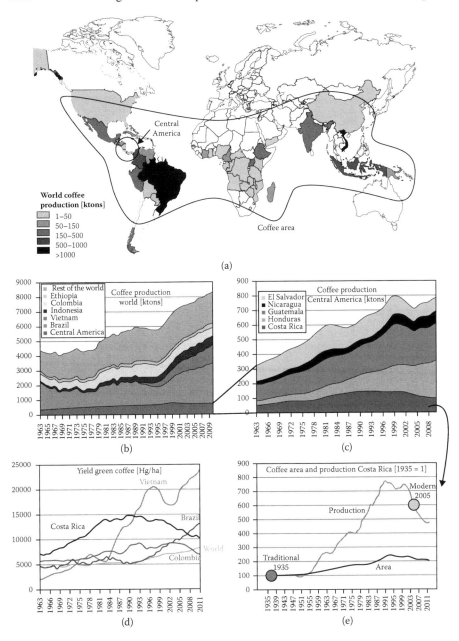

Figure 7.2 Contextual indicators for coffee in Costa Rica and the world (see text).

climates. Currently, Brazil (2.78 Tg) and Vietnam (1.16 Tg) account for over half of global production. Figure 7.2b also shows the main coffee-producing areas in the world: as well as Brazil and Vietnam, Colombia, Ethiopia, and Central America. Among them, they accounted for between 55% and 75% of global production in the last half century. In this period, production has almost doubled, from 4.40 Tg (millions of tons) in 1961–1965 to 8.38 Tg in 2008–2012, according to the FAO (2016). During this time, coffee-producing areas have continued to grow, particularly Vietnam, which has boomed since the early 1990s. No more than 10% of global food production enters the global market; in the case of coffee, this figure rockets to 80%. It is a paradigmatic example of a cash crop, the consumption of which has continued to grow right into the present day (FAO, 2016).

Central America encompasses five coffee-producing countries: Costa Rica, Honduras, Guatemala, El Salvador, and Nicaragua, which have accounted for a tenth of global coffee production more or less consistently over the past half century. The trend has been one of continued growth, with production figures rising from 341 to 792 Gg (Figure 7.2c). Although its participation in the global total is not as striking as countries such as Brazil or Vietnam, it is a paradigmatic example of a coffee-producing region since, of the main producing areas in the world, it is the only one where coffee generally represents the country's primary crop, and its percentage of cultivated land area is the highest in the world, reaching an average of 30% in the region (FAO, 2016).

Within Central America, Costa Rica has undergone a process of stagnation and a certain decline since the end of the 1990s, whereas Guatemala and Honduras have continued to grow at a constant rate, consequently gaining greater relative weight. The interest in the case of Costa Rica lies in the fact that it was the region's biggest coffee-producing country between the late nineteenth century and well into the twentieth century, but also because it led a process of industrial transition toward coffee-growing intensification from the 1950s onward, at unprecedented levels on a global scale. In just three decades, it trod an accelerated path to transition that altered the management approach and functionality of coffee agroecosystems: in the 1980s, it became the most productive country (based on land area units) in the world (Renjifo, 1992, p. 34). Section 7.4 aims to shed some light on this process, which led to stagnation since the 1990s, explained by several factors: a drive to produce quality coffee, production regulations, abandonment of more productive areas in the Central Valley owing to urbanization in the area, aging plantations, and, above all, falling coffee prices (Deugd, 2003; Castro et al., 2004). This has meant that Costa Rican Coffee has not moved toward an additional stage of hyperintensification in its management of this crop, as has occurred elsewhere, such as Vietnam, and in certain areas of Brazil and Colombia (Deugd, 2003), characterized by total mechanization of management and harvesting, with unprecedented levels of production never seen in history before (Fortunel, 2000; Agergaard et al., 2009).

In short, the evolution of coffee production in Costa Rica between 1935 and the present day reveals a history of agrarian intensification, in the transition from traditional to modern systems (Figure 7.2e), characterized by an increase in land productivity and labor in terms of commercialized beans. But this evolution seems much more complex if we incorporate a biophysical perspective, bearing in mind that these are tropical systems—the remaining biomass plays a crucial role—with a very marked multifunctional vocation.

7.4 INTENSIFICATION OF COFFEE AGROECOSYSTEMS

The history of agroecosystem intensification on a global scale, regardless of the approach used to analyze it, has several common but inevitable features: change in crop variety, chemical fertilization, pesticides, mechanization and so on (Evenson and Gollin, 2003). All of this led to an increase in agricultural productivity (see also Bindraban and Rabbinge, 2012; Federico, 2008), although it is not clear whether this change was accompanied by a general increase in the productivity of total biomass (González de Molina, 2010; Krausmann et al., 2008a, 2008b).

Analyzing this process in the specific case of coffee production, we see a similar tale, extensively detailed from a historical perspective by other authors (see also Hall, 1976; Pérez, 1977; Esguerra, 1991; Renjifo, 1992; Samper et al., 2001; López and Picado, 2012; Viales and Montero, 2010). As we noted in Section 7.2, between 1935 and 2010, it has been possible to reconstruct complete series of external inputs including fertilization, pesticides, machinery, and human labor. There is no documentation—or too little to be significant— regarding the use of fuels, electricity, or other installations that require the use of indirect energy.

Among the inputs described, as shown in Figure 7.3a,b, fertilization has played a particularly relevant role, accounting for between 63% and 90% of total energy requirements (in terms of embodied energy). Although up until the 1950s, organic fertilization represented between 30% and 50% of the energy consumption derived from fertilization, today it represents only around 10%. Energy dependency derived from chemical fertilization, which is now the predominant factor, was much lower in the 1930s (barely 30%) and nonexistent a few decades prior to that. The other inputs pertain to human labor, which was initially the second most important item: between 1935 and 1955 it accounted for between 20% and 35% of energy inputs, with a downward trend that has seen recent figures plummet to between 5% and 10% of consumption. From the 1960s onward, the second most important item became the pesticides, with average consumption of around 12%.

It is certainly true that, beyond their relative participation, they have all grown constantly in the period studied with the exception of between 1988 and 1993 and, intermittently, over the course of the twenty-first century. In both cases, this can be explained by the crisis of prices and, in recent years, also having to complete with urban land, and the aging of plantations. Chemical fertilization was already present in 1935, although in a very limited way: it is estimated that only 5.1% of coffee plantations were fertilized at that time using external inputs, of which only 2.2% used chemical fertilizers. Today, these percentages have risen to 90% and 89%, respectively. The use of pesticides was first documented in 1916 with the use of inorganic products (Hilje et al., 1987, 1989). However, they were not used in a significant way until the 1950s, and coffee has never been a crop that is treated using high doses of

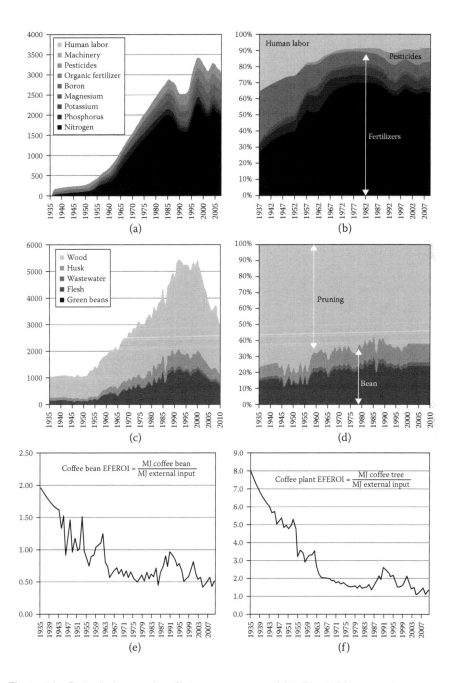

Figure 7.3 Embodied energy in coffee agroecosystems (a) in TJ and (b) in percentage terms. Coffee plant production (parts of the bean, prunings) (c) in TJ and (d) in percentage terms. (e) Coffee bean external final EROI (EFEROI) and (f) coffee plant EFEROI (f). (Author's own.)

chemical products relative to other crops. Between 1970 and 2000, different estimations indicate that consumption was probably between 5% and 10% of the country's
total pesticide consumption, even though it was the most prominent crop. Banana
crops accounted for 57% of pesticide usage in 1993, whereas coffee, which covered a
greater land area, consumed just 7% (Santos et al., 1997, p. 18). In the present day, in
52% of coffee growing land, weeds are managed manually (ICAFÉ, 2007).

Consumption relating to human labor and machinery has been lower in absolute
terms and is becoming increasingly less significant in relative terms. Although the
total number of man-hours has increased, from demand for 30.2 million hours in
1935 to 82.6 million in 2010 (peaking at 120.8 in the year 2000), the growth of this
factor has been lower than the other inputs. Greater growth has been due to harvesting, where technological improvements have not been implemented: labor productivity has remained constant, but with the increase in coffee production, the demand
for labor has also increased. Mechanization of coffee production is very limited in
Costa Rica: small machinery for the manual application of treatments, fertilizers,
and weed management. This country has not made the transition to hyperintensive
models, the emblematic example of which is Vietnam in recent years (Fortunel,
2000; Agergaard et al., 2009).

The changes described in the management of coffee agroecosystems led to a
formidable change in production (Figure 7.3c,d). In the period studied, between 1935
and 2010, the production of coffee beans increased from 117 Gg (263.8 TJ) to 500 Gg
(1131.5 TJ), peaking in 1989 at 948 Gg (2144.6 TJ). Up to 1955, production remained
relatively stable (c. 110 Gg, 250 TJ). It also remained stable between 1985 and 2000
(c. 900 Gg, 2000 TJ), after which it fell to c. 500 Gg (1131 TJ). To put it another way,
up until 1955 production maintained stable figures under traditional management
approaches, and in barely 30 years, between 1955 and 1985, it underwent substantial
expansion that was sustained up until the year 2000, which marked a turning point
as the sector entered its crisis period.

The trend observed for coffee bean production is similar to that of inputs from an
energy perspective as we can see in Figure 7.3a,c: a pattern of rapid growth between
1955 and 1985, subsequent stagnation, and final downturn into the sector's current
crisis. However, comparing the production of coffee beans and external inputs, identified in Table 7.1 as coffee bean EFEROI, we can see a progressive decline. External
inputs have grown at a much higher rate than coffee production, although the general
tendency is similar. In 1935 it was 1.97, meaning that for each unit invested, 1.97
of coffee beans would be obtained, whereas in 2010 this had fallen to 0.50. Energy
efficiency measured in such terms is four times lower.

Figure 7.3c shows not only the evolution of coffee bean production but also the
production of prunings. These results show how the productivity of coffee grew by
4.29 between 1935 and 2010, whereas pruning products only grew by 1.97. In other
words, this tendency informs us of the evolution of the harvest index; how the bean
represents a growing part of the biomass appropriated in industrialized agricultures.
In the case of coffee, this index rose from 0.44 to 0.64 (measured in terms of fresh
matter) between 1935 and 2010. This change is related with the change in plant
variety observed in the region, moving away from the Arabiga (Typica) variety that

accounted for almost all plantations at the start of the twentieth century (Viales and Moreno, 2010) toward the Caturra variety, which currently represents 75% of the coffee-growing land area (ICAFÉ, 2007). Having said this, if we estimate coffee plant EFEROI, including prunings in the input (numerator), we see that in this case it has fallen from 8.0 to 1.3, an even greater decline than coffee bean EFEROI since external inputs have grown at a faster rate than coffee beans but also substantially quicker than the prunings of the coffee plant.

These indicators of energy efficiency, which compare coffee bean production and the whole of the socialized biomass (SB) of the coffee plant with external inputs, are the most commonly used in the literature (particularly the first one). A little later, we will compare these with the results of other studies that have examined the energy balances of coffee systems in this way. However, such estimations neglect the complexity of a crop that is predominately an agroforestry system. The inclusion of the shade layer profoundly alters the flows produced and recirculated within the system, directly affecting its ecology, but also the products obtained and used by coffee-growing communities. Section 7.5 aims to show this productive complexity before analyzing its impact on EROIs.

7.5 COFFEE AS AN AGROFORESTRY SYSTEM

Coffee is, above all, an agroforestry system in which the coffee plant is inserted among dozens of different species of trees that play a fundamental role in the energy circuits of this agroecosystem (see also Muschler, 1999). Their relevance is such that it is futile to analyze total consumed inputs (TCI) in the system without taking account of the production and recycling of shade-providing trees (see also Romero, 2006; Merlo, 2007). Their effects on coffee agroecosystems are multiple and have been widely studied: recycling of nutrients, water and temperature regulation, allelopathy and improvements in biodiversity, and even increases in production (Perfecto et al., 1996; Beer et al., 1998).

In general terms, coffee agroforestry systems have been losing space in the SET in a process that has been analyzed with great concern owing to its environmental implications (Perfecto et al., 1996; Tscharntke et al., 2011). In Costa Rica, this phenomenon has been observed, although to a lesser extent than in the burgeoning regions of hyperintensive coffee production. Even so, in the last few decades of the twentieth century, shaded areas underwent a gradual decline in terms of surface area (Perfecto et al., 1996). From almost total occupation prior to the industrialization of the sector, current statistics estimate that between 75% and 88.9% of coffee plantations are currently shade grown (ICAFÉ, 2007). There has also been a change in the structure of the species used to provide shade, which explains part of the functional change noted in coffee agroecosystems, as we will see later. In 1950, of all the shade-providing plants and trees planted in Costa Rica's coffee plantations, Inga occupied 61.7%, Musaceae accounted for 17.5%, and the occupation of timber-yielding species was insignificant (DGEC, 1953). By 2005, it is estimated that Inga had fallen to 18.4%, having been replaced by the Poró Tree, with 25.2%, and with a growing presence of timber-yielding species (particularly Eucalyptus and Laurel), which now account for 11.1% of the shade-providing species (ICAFÉ, 2007). The reasons behind this change are related in some way to the nature of the SET.

The productive changes derived from the SET focused on increasing land and labor productivity, especially for the saleable fruit or bean in the majority of crops. In other words, the first way in which the SET can be observed in agriculture is through the industrialization of the management approaches described previously, or to put it more directly, through the green revolution. However, the SET has forced another major change in the world's agrarian systems: social functional change. The transition is from a system that supplied the majority of goods required to sustain a family and, consequently, which was capable of providing multiple uses, to another much more simplified system, which focuses on the production of grains or fruits destined for human or animal consumption (Fischer-Kowalski and Haberl, 2007; González de Molina, 2010; Singh et al., 2012; Infante-Amate, 2014). Traditional uses have been lost (or at least their secular importance has) as providers of multiple goods: construction materials, energy goods, medicinal goods, lighting, packaging, and so on. In the case of coffee production, this process is perfectly recognizable, and a great deal of it can be explained by the central role of the shade layer.

Figure 7.4 provides a flowchart showing the final destinations of the production of coffee agroecosystems in traditional and modern systems. There are four main changes that in turn tie in with the process of the SET.

First, production destined to supply human consumption, in other words, the production of coffee beans, has doubled, from 531 to 1197 kg d.m. ha^{-1} year^{-1} (Figure 7.4). This fact is linked with the process of intensification in the management of agroecosystems and the growing commercialization of the crop and its export dynamics since the mid-twentieth century. The bean is, ultimately, the most lucrative part of the agroecosystem's products and, consequently, much of the struggle in managing agrarian systems focuses on increasing production of this part.

Second, the fuel for which the destination usage represented 75.5% of the total in 1935. This figure points to the multifunctional nature of a paradigmatic cash crop in the same direction as olive groves (Infante-Amate, 2014) and vines (Infante-Amate and Parcerisas, 2013). The total production flows in agriculture show that many of them, other than the fruit itself, played a key role in domestic sustenance. The question is, to what extent does the production of firewood in a coffee agroecosystem have palpable importance in sustaining the rural farming communities of Costa Rica? The percentage of homes that used firewood in the country dropped from 67% to 39% between 1963 and 1984 (DGEC, 1965, 1974, 1985). This information suggests an evident decline, explained by the arrival of electricity (the percentage of homes with electricity increased from 25% to 49%). However, it also suggests that until well into the twentieth century, the majority of the country continued to depend on firewood. Lemckert and Campos (1981), in their extensive analysis, studied consumption by small landowners in the country: 50% of the country's farms. Although 60% stated that they had electricity, the majority continued to use firewood (86% regularly and 91% occasionally). They noted that 83% of farmers used firewood from their own farm or given to them by another member of the community, and only

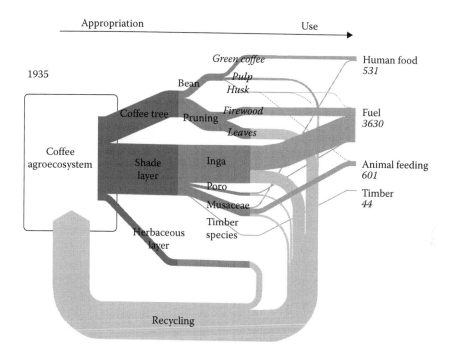

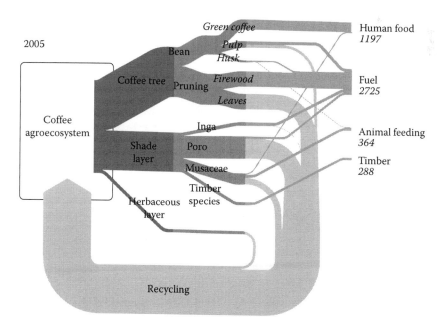

Figure 7.4 Appropriated flows distinguishing between the end uses of coffee agroecosystems in Costa Rica. Figures expressed in kg d.m. ha^{-1} year^{-1}. (Authors' data, based on the methodology described in Chapter 2 of this book and in Section 7.2 of this chapter.)

22% were unable to supply all the firewood they needed. In other words, three out of every four small landowners fulfilled their energy demands with firewood from their own farm or community. Where did this firewood come from? In all regions, the response was largely from Inga trees or from coffee plant prunings (Lemckert and Campos, 1981, p. 37). Ugalde (1982) noted the important consumption of pruned wood among the benefits of coffee growing. Sisson (1979), Domínguez (1979), and Gewald and Ugalde (1981), in a study that encompassed countries in Central America, highlighted the same criterion, along with the country's energy balances (see also CNE, 1976). By 2005, the percentage represented by prunings with regard to total usage had fallen to 59.6% (Figure 7.4).

Third, the weighting of products aimed at supplying animal feed requirements (principally animal feed derived from Musaceae) has also declined, although their total weighting is not very significant. This fact is linked with the increased availability of external feed, as well as the industrialization of the country's livestock sector. However, the figures are too low to draw any strong conclusions in this regard.

Finally, although it is similarly insignificant in terms of total biomass, but it is qualitatively important, consumption of wood has risen from 44 to 288 kg d.m. ha^{-1} year^{-1}, marking a growth trend that suggests continued growth in the future. This phenomenon is being studied in detail and is suggested as a plausible option for the future, while impacting on the maintenance of the shade layer, but through the generation of a second saleable product, wood or timber, which could be as profitable as coffee in its association with species such as Eucalyptus (Detlefsen and Somarriba, 2012). However, as we will see, replacing leguminous shade-providing trees with nonleguminous species, with a high growth rate, could further compromise the sustainability of the agroecosystem. In other words, the growing demand for wood and timber for industrial purposes is also conditioning coffee production and could foreseeably condition it even further.

The available data inform us, among other things, of the major presence of the shade layer in traditional and modern coffee agroecosystems in Costa Rica. Average production ranges between 2.5 and 10 Mg of dry matter per hectare including wood, prunings, and fruit. The majority of this production is recycled on the plantation (leaves and branches that fall to the ground or are pruned and are reused). Section 7.6 analyzes the energy functioning of coffee systems in Costa Rica, including the shade factor.

7.6 ENERGY EFFICIENCY: AN AGROECOLOGICAL PERSPECTIVE

Section 7.4 (see Figure 7.3e,f) provided information about energy efficiency related with coffee bean production and the socialized production of coffee plants with external inputs. Using the proposal developed by Guzmán and González de Molina (2015) and detailed in Chapter 2, which integrate not only bean production and prunings, but also the other biomass (including root biomass) and the internal circulation of energy flows, we obtain more complex efficiency indicators.

Figure 7.5 summarizes these flows for the years 1935 (traditional management) and 2005 (modern management). Table 7.2 summarizes the economic and

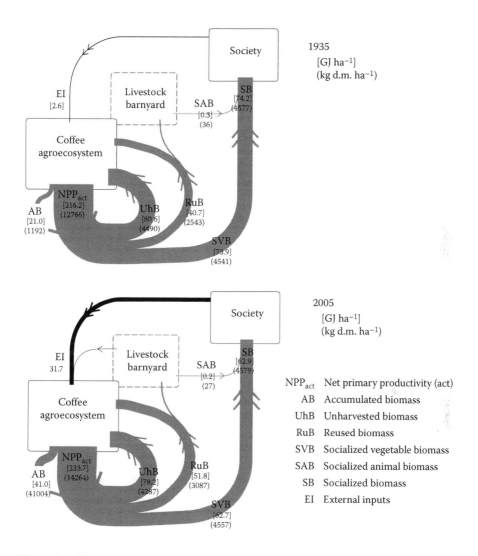

Figure 7.5 Energy flows of coffee agroecosystems in Costa Rica in 1935 (traditional management) and 2005 (modern management) [GJ ha⁻¹], kg d.m. ha⁻¹. (Authors' data, based on the methodology described in Chapter 2 of this book and in Section 7.2.)

agroecological EROIs derived from the relationships between flows described in Figure 7.5. The first conclusion to highlight is that, in all cases, energy efficiency has fallen. Running counter to the trend observed with regard to labor and land productivity, which always increases in the industrial transition and also in the case of coffee (Infante-Amate and Picado, 2016), the data on energy efficiency suggest the degradation of these indicators in all cases. FEROI is the most recurrent indicator and relates SB with RuB and EI. It fell from 1.71 in 1935 to 0.75 in 2005.

Table 7.2   Summary of Estimated EROIs for Coffee in Costa Rica

	1935		2005	
Proposed EROIs from an Economic Point of View				
Coffee bean external final EROI	1.97	100	0.50	25
Coffee plant external final EROI	8.0	100	1.3	16
Final EROI = SB/(RuB + EI)	1.71	100	0.75	44
External final EROI = SB/EI	28.38	100	1.98	7
Internal final EROI = SB/RuB	1.82	100	1.21	67
Proposed EROIs from an Agroecological Point of View				
NPP_{act} EROI = NPP/TIC	1.74	100	1.44	83
Agroecological FEROI = SB/TIC	0.60	100	0.39	65
Biodiversity EROI = UhB/TIC	0.65	100	0.48	74
Woodening EROI = AB/TIC	0.17	100	0.25	150

Source: Author data, based on the methodology described in Chapter 2 of this book and in
          Section 7.2 of this chapter.
Note: See Figure 7.1 and Table 7.1 for further details on EROI composition.

In other words, for every unit of energy invested in the system, the return to society
fell by almost half. The reasons are varied. One of them is the change in productive orientation. In 1935, SB was 18% higher in energy terms relative to the figure
obtained in 2005 (Figure 7.5) and the composition was different. Whereas in 1935,
coffee beans represented 8% of SB, in 2005 they represented 24%. This change in
orientation toward a less energetic product such as the coffee bean partly justifies
the fall in this indicator. However, such a major reduction is not explained by this
argument alone. The fundamental reason was the drastic decline in the efficiency
of external inputs, from 28.4 to 2.0 (EFEROI). The exponential growth of external
inputs (1218%) yields a negative return. However, in this case, it does not manage to
decrease the investment society makes in the form of reused biomass, which is also
growing (up 27%), pushing IFEROI down by 33%. In other words, the change in productive orientation and the specialization of the coffee plant induced by the market
has yielded a very negative behavior in terms of the energy return for society.

As noted in the Introduction, the literature about energy balances in coffee production is limited. In general, it is difficult to make comparisons bearing in mind the
different criteria used by different studies when examining EROIs, for example, in
the selection of flows (using gold coffee, or green coffee, or the rest of the socialized
biomass, etc.) and the conversion of these flows into energy (e.g., the literature takes
factors of between 0.2 MJ h^{-1} and almost 200 MJ h^{-1} for the case of labor) (Murphy
et al., 2011; Aguilera et al., 2015). According to the literature review provided by
Infante-Amate and Picado (2016), documenting 15 examples of studies with information about energy balances in coffee systems, we find extremes between fairly
nonintensive family-based organic management systems (1.5 GJ ha^{-1} of IE) and
very intensive management approaches in Brazil (943 GJ ha^{-1}). The EROIs reached
by these studies indicate lower efficiency in the case of conventional management,

generally lower than unity; greater efficiency for mixed management, with few external inputs, with an average EROI of 1.94; and, finally, the greatest efficiency in organic systems with an average of 3.41. These data are not representative and only refer to the relationship between coffee produced and external inputs. However, they provide evidence in the case of coffee of the trend reiterated in the literature of greater efficiency in organic or traditional systems (see also Smith et al., 2014) and help to contextualize our results.

Regarding agroecological EROIs, NPP–EROI has fallen from 1.74 to 1.44. As we set out in Chapter 2, a drop in the total productivity of the agroecosystem with regard to total inputs consumed (TIC) would indicate processes of deterioration in the agroecosystem, which would diminish its capacity to transform energy received into biomass. This deterioration could be induced by various processes of degradation, which might have synergistic effects. Castro-Tanzi et al. (2012) attributed changes in soil chemistry and nutrient retention capacity, along with the reduction of NPP_{act} in coffee plants today in Costa Rica, to the N saturation hypothesis. According to these authors, based on their own data and on other studies, the excess in nitrogen applied to industrialized coffee plantations (they estimate 212 kg N ha^{-1}) may induce a state of N saturation, since N recovery efficiency in these plantations is less than 40%. The excess N would deplete the soil exchangeable Ca and increase Al3+ toxicity. Moreover, lower soil exchangeable Ca concentrations reduce Ca assimilation by the plant, which in turn becomes more susceptible to fungal diseases. Additionally, the greater number of stems per plant in industrialized coffee plantations might also play a role in the infectious process of fungus. More stems per plant create denser coffee plantations that may in turn affect the movement of air and increase the relative humidity inside coffee plantations. This in turn could produce ideal conditions for higher infection rates and severity of fungus outbreaks. Continual and intensive application of fungicides may eventually lead to the development of resistance of fungal agents to fungicides and contamination of soils with copper.

Agroecological FEROI has experienced an even sharper drop, with a lower return for society than all the energy that circulates around the agroecosystem. This relative decline in human appropriation does not benefit biodiversity that sees a drastic fall in the availability of biomass in absolute (Figure 7.5) and relative terms (Table 7.2). This lower availability of biomass for wildlife, together with the increase in energy investment in pesticides in 2005, contributes to the loss of biodiversity documented in coffee-growing regions (Philpott et al., 2008).

Last, woodening EROI has increased. The increase in the conversion of total inputs consumed (TIC) into annually accumulated biomass is not due to the greater presence of woody biomass per area unit at any given time, which would be similar. Rather it is due to a lower useful lifespan of plantations and trees in the shade layer. The growth rate of trees follows a sigmoidal curve, which implies that once the mature stage is reached, growth slows down. Therefore, removing the plantations before this turning point increases the average biomass accumulation rate. This fact, together with the introduction of a rapid growth species such as the Eucalyptus in recent times, accelerated the annual accumulation of biomass in the agroecosystem in 2005. The renewal of coffee plants in 1935 occurred approximately 40 years after they were initially planted (Samper et al., 2000), in contrast to the current useful lifespan of 20 years (ICAFÉ, 2010).

7.7 AGROECOLOGICAL DESIGN OF COFFEE PLANTATIONS IN THE TWENTY-FIRST CENTURY

The twenty-first century has brought with it major socioenvironmental challenges that call into question the process of agrarian modernization and challenge us to come up with proposals to transition toward more sustainable agrofood systems (Gliessman, 1998; Guzmán et al., 1999; Méndez et al., 2015). The deterioration of agroecosystems and the services they provide following decades of modernization (Gliessman, 1998); population growth (Alexandratos and Bruinsma, 2012); the depletion of nonrenewable natural resources that sustain modern agrarian production, such as oil or phosphorus minerals (Cordell et al., 2009; Koppelaar and Weikard, 2013; Murray and King, 2012); and climate change (Lal et al., 2011): all these factors place humanity at a crossroads that requires us to produce more socialized biomass, with fewer nonrenewable resources, while investing in the recovery of the fund elements of agroecosystems. As we expounded previously, studying the functioning of preindustrial agroecosystems could offer guidelines as to how we could tackle these challenges, since they had to adapt to the growing demands of the population with very limited external inputs.

The case of coffee production presented in this chapter shows us that in 1935, coffee agroecosystems were capable of producing greater quantities of socialized biomass in absolute terms, with a very low dependence on external inputs, and generating internal flows of biomass that enabled them to maintain the fund elements, soil, and biodiversity of the agroecosystem in good condition. That year, the energy composition of socialized biomass was firewood (88%), coffee beans (8%), fruits produced by Musaceae (bananas and plantains) (3%), wood (1%), and meat (0.003%). By 2005, the composition had varied to the detriment of firewood (67%) and bananas and plantains produced by Musaceae (2%), in favor of coffee beans (24%) and wood (7%). Meat production has stayed the same in terms of proportions (0.003%). From this perspective, the coffee agroecosystem has maintained apparently high levels of diversification, which would not justify the drastic decline in economic and agroecological EROIS. So what is the key? How can we explain these major declines in efficiency?

Possibly, the concept of functional biodiversity, emphasized strongly in agroecology (Gliessman, 1998; Altieri and Nicholls, 2007), might help to explain it. In 1935, two leguminous species (*Inga* sp. and *Erythrina* sp.) accounted for 111 GJ of biomass produced (51% of NPP_{act}), whereas in 2005 they contributed just 83 GJ (35% of NPP_{act}). The inclusion of leguminous plants in polyculture is a classic strategy of traditional agriculture to generate low-entropy internal loops, which increase recycling and socialized biomass, since they enrich the agroecosystem with nitrogen. The replacement of these species with other nonleguminous ones, either coffee plants and/or other timber-yielding species, such as eucalyptus, does not guarantee the performance of these functions. Consequently, the increasing precariousness of this internal loop necessarily entails the incorporation of external inputs to sustain the productivity of the agroecosystem. Paradoxically, this inputting of nitrogen of fossil origin sparks a decline in the rate of N_2 fixation by leguminous species that remain in the agroecosystem (Cannavo et al., 2013), thereby reinforcing the need to incorporate industrially manufactured nitrogen.

Beer (1988) argued that leguminous shade-providing trees in coffee plantations fix relatively low levels of nitrogen (60 kg N/ha) and stated that other functions, such as the production of aboveground recycling biomass, could be more important in terms of maintaining soil fertility and the supply of nutrients to the coffee plants. However, according to our estimations, there is no difference between the two systems in terms of aboveground recycling biomass. In 2005, 3433 kg ha^{-1} above-ground biomass was recycled, provided by shade-giving trees, in comparison to 3040 kg ha^{-1} in 1935. Furthermore, Beer also referred to modern-day coffee plantations that incorporate nitrogenous chemical fertilizers, which have eliminated the need for symbiotic fixing. But this situation, from an energy perspective, is very costly, as reflected in Figure 7.3a,b, and in the decline in the EROIs (Table 7.2).

The replacement of internal biomass flows with chemical fertilization has also generated major problems in terms of soil degradation, as shown previously, which are hampering its capacity to produce biomass, as reflected in the NPP–EROI. Furthermore, the relative decline in flows of unharvested biomass affects biodiversity, which is also damaged by the use of pesticides, a process reflected in the biodiversity EROI. Another deteriorated fund element is water owing to nitrate contamination. Cannavo et al. (2013) measured nitrate run-off losses of 157.2 kg N ha^{-1} year^{-1}, in shade-grown coffee plantations in Costa Rica fertilized chemically with 250 kg N ha^{-1}. Other authors (Babbar and Zak, 1995; Harmand et al., 2007) have also pointed to major nitrogen run-off losses, although the amounts are lower.

In short, it is essential to maintain territorial equilibrium, even within polyculture, between species dedicated to providing flows of biomass and nutrients that sustain production, and species that offer greater market value. The rupture of this balance forces us to incorporate external energy flows (Guzmán and González de Molina, 2009; Guzmán and Foraster, 2011) and leads to the gradual deterioration of fund elements.

Given that the driving force for change is the market, measures need to be developed that guarantee adequate earnings for coffee growers, without sacrificing the maintenance of high levels of ecosystem functioning. Other authors have argued the need for considering trade-offs among income, biodiversity, and ecosystem functioning during agroforestry intensification in tropical areas. To this end, they have determined an appropriate crop:shade layer ratio, and proposed the establishment of market mechanisms that help growers maintain this relationship (Steffan-Dewenter et al., 2007). The results of the diachronic research carried out here underline the importance of the composition of the shade layer, beyond the simple crop:shade ratio. They also showed that, in the case of coffee production in Costa Rica, this compromise between profitability and environmental services has not been achieved. The abandonment of this crop in recent years suggests low profitability levels for growers and the deterioration of fund elements.

7.8 CONCLUSIONS

The intensification of coffee agroecosystems from a biophysical perspective can be explained by a series of concatenated factors: new more productive varieties that require the addition of external inputs and the increase of which leads to an increase in coffee bean productivity, to the detriment of other types of biomass. In terms of energy, we see that most energy requirements stem from the addition of chemical fertilizers. This is due to the limited use—in comparison with other crops—of pesticides and the low mechanization levels of coffee growing in Costa Rica. This process took place between 1955 and 1985, followed by a period of stabilization (in terms of intensity) and then crisis in the sector, as the country did not move toward hyperintensive models in the way that other countries have.

Our research corroborates the pattern reiterated in the literature regarding the loss of energy efficiency in the transition from traditional to modern systems, expressed in terms of classic economic EROIs. Coffee is not a product with special energy value from a nutritional perspective, but it is a crop that has increasingly demanded greater inputs of energy. Traditionally, coffee, beyond a system used to provide beans used to make a steaming hot beverage, has been a formidable provider of goods and services for coffee-growing communities. In traditional systems, prior to the energy transition, it had to meet many other needs through the pruning of coffee plants and shade-giving trees for fuel, as well as to supply wood and timber used domestically, and for animal feed, human food, and so on. The SET has caused this system to lose its multifunctional nature as it has increasingly specialized in coffee bean production, and more modestly, in wood.

This transition has weakened internal loops of biomass and led to the loss of functional biodiversity. Specifically, leguminous species in the shade layer have become less important in NPP_{act} as a whole, which has only been made possible by inputting very high levels of industrially manufactured nitrogen, which in turn weaken the symbiotic fixation capacity of the leguminous species that remain in the agroecosystem. These changes are keys to the loss in energy efficiency observed in coffee plantations. Economic EROIs have fallen dramatically, affected by the decline in socialized biomass brought about by increased specialization, but above all as a result of the huge increase in EI. Agroecological EROIs (NPP–EROI and biodiversity EROI) have also suffered a major decline, resulting from the deterioration of fund elements (soil, water, and biodiversity). Only woodening EROI yields positive values, which are not due to the increased productive capacity of the agroecosystem, but rather to the shortening of the useful lifespan of plantations, which fell by half between 1935 and 2005.

From an agroecological perspective, it is fundamental to recuperate leguminous species to provide functional biodiversity in the design of sustainable agroecosystems for the twenty-first century. These species are capable of generating low-entropy internal loops that effectively increase recycling and socialized biomass, by enriching the agroecosystem with nitrogen.

CHAPTER 8

Organic Farming
Between the Relocation of Energy Flows and Input Replacement

Gloria I. Guzmán and Marta Astier

CONTENTS

8.1 INTRODUCTION

Organic farming (OF) has experienced major growth around the world in recent decades. In 2014, 43.7 million hectares were farmed organically worldwide. That same year, the number of organic farms stood at 2.26 million. In both cases, year-on-year growth in the twenty-first century (1999–2014) reached double-digit figures

(10% and 18%, respectively). This growth in terms of land area covered and number of farming businesses has also been accompanied by growth in the organic food market, with 12% year-on-year growth in the same period, reaching US$80 billion in 2014 (Willer and Lernoud, 2016).

Organic farming (OF) constitutes a more sustainable alternative to industrialized agriculture in environmental terms. The environmental benefits of OF have been broadly studied in recent decades in very diverse agroecosystems, paying particularly close attention to the impact on biodiversity, soil and water quality, energy efficiency, and greenhouse gas (GHG) emissions. The majority of the studies reviewed clearly show that the richness and abundance of species in a wide range of taxa (insects, birds, small mammals, reptiles, etc.) tend to be greater on organic farms than conventional farms present in the same area (Bengtsson et al., 2005; Hole et al., 2005; Norton et al., 2009).

As for the quality of water and soil, OF significantly reduces pesticide and nitrate contamination and pollution in both natural resources, and chemically, biologically, and physically improves the soil, leading to more effective water usage. In the long-term, tests carried out by the Rodale Institute in the United States, water volumes percolating through soil were 15%–20% higher in organic-diversified systems relative to conventional ones, with more groundwater recharge and less runoff (Rodale Institute, 2011). Long-term tests conducted under diverse agroclimatic and cultural conditions demonstrate the potential of OF in this regard (Mäder et al., 2002; Raupp et al., 2006; Meco et al., 2010; Rodale Institute, 2011).

A great deal of agricultural land presents very low levels of organic matter. Hence, such areas can be considered as a potential major global carbon sink (Smith, 2004). There are relatively few comparative studies that examine conventional versus organic farming with regard to GHG balance, and there is a much lower consensus in this regard. In very simplified terms, studies show that OF reduces emissions when quantified by area unit, but the results yield greater variability when calculated by product unit (Haas et al., 2001; Flessa et al., 2002; Mondelaers et al., 2009; Skinner et al., 2014). From one point of view, the disparity of results in terms of product unit is linked with the complexity of the processes involved in the net balance of emissions. These processes are strongly affected by soil and climate conditions, as well as by management and type of products (Weiske et al., 2006; Chirinda et al., 2010; Aguilera et al., 2015a; Martin and Willaume, 2016). The diversity of results is also due to methodological aspects: the emission factors applied (de Boer, 2003; Aguilera et al., 2013a) and the system limits defined (Wood et al., 2006; Thomassen et al., 2008;) can profoundly alter the results. Moreover, soil carbon balance is usually not taken into account in agricultural GHG balances, despite OF has been shown to promote the soil carbon sink function by enhancing carbon storage (Gattinger et al. 2012; Aguilera et al. 2013b).

With regard to energy efficiency, studies show that OF consumes less external energy, particularly nonrenewable energy, to obtain the same product, with some exceptions (Smith et al., 2015). In relation to crops, greater efficiency in the use of nonrenewable energy is due, above all, to the replacement of chemical fertilizers (the synthesis of nitrogenous compost is highly costly in energetic terms) with organic fertilization (Alonso and Guzmán, 2010; Rodale Institute, 2011). For livestock, ruminant production systems tend to be more energy efficient under organic management due to the production of forage in grass–clover leys (Smith et al., 2015). Productive orientation influences this differential, reducing it to zero or even making it negative when fertilization accounts for a small portion of the total nonrenewable energy consumed (e.g., sheltered horticulture) and/or there is a strong downturn in yield (kilogram per hectare) with regard to conventional production and/or an energy-intensive technique is used in OF (e.g., flame weeding) (Pimentel et al., 1983; Alonso and Guzmán, 2010; Smith et al., 2015).

However, the majority of these studies approach this issue with a typical "black box approach." In other words, they only consider external energy flows imported into the agroecosystem and the energy contained in biomass outputs that have commercial value. Obviously, the finding that OF offers greater efficiency in terms of nonrenewable energy is very important. Among other things, it grounds the institutional development of measures to support this model of farming with a view to reducing problems linked to the consumption of nonrenewable energy.

However, this approach also has some significant limitations. First, the "black box approach" ignores the importance of internal energy flows in the form of biomass when it comes to sustaining the fund elements of the agroecosystem (biodiversity, soil, accumulated biomass [AB], and water). Improving these fund elements is, at one and the same time, a goal and an achievement of OF, as we have seen before. We believe it is important to understand the role played by internal energy flows in these achievements. Second, input–output analysis cannot explore the quantity and quality (high-low entropy) of the internal loops operating in the system. As they are not taken into account, information that could be used to improve the internal functioning of the system is not generated. Last, by only considering biomass with commercial value as the output, it overlooks the multifunctional nature of agrarian activity for many farmers and rural communities, both in the past and in the present day. These socialized biomass (SB) flows, which are noncommercial in nature (firewood, timber, edible and medicinal wild plants, etc.), have been and continue to be fundamental in the social reproduction of rural communities.

Therefore, to resolve these limitations, methodological tools are required that are capable of opening up the black box and exploring the internal energy functioning of agroecosystems (Tello et al., 2016). The use of such tools could be particularly relevant in the case of OF. Often, farms certified as organic embarked on the conversion process based on intensive management of external inputs (EIs) within contexts of industrialized farming. This means that the internal mechanisms for replenishing fertility and supplying energy that enabled traditional agriculture to function are broken at the level of individual farms and also on a local scale (Guzmán and González de Molina, 2009; Tello et al., 2016). When faced with this situation, farmers develop a host of strategies ranging from the partial internationalization of energy and material flows (sowing leguminous crops, composting of waste from the farm, etc.) to the complete replacement of inputs. In this latter case, they acquire organic inputs (biomass) in the market that they need to continue producing. Regrettably, the input replacement strategy is all too common and is part of what has come to be known as the conventionalization of OF (Rosset and Altieri, 1997; Zoiopoulos and

Hadjigeorgiou, 2013). The inability to relocate energy and material flows might well be limiting the provision of ecosystemic services in OF.

In this chapter, we aim to apply economic and agroecological energy return on investments (EROIs) to the comparative study of organic and conventional farms, with a view to improving understanding of the energy functioning of the agroecosystems studied and making contributions capable of redirecting management approaches, if necessary, toward greater levels of sustainability. To this end, we have chosen a sample of olive and avocado groves, where input–output energy analysis has already previously been applied (Guzmán and Alonso, 2008; Astier et al., 2014). From this perspective, the study presented now aims to complement the studies published previously, and we will endeavor to explain the advances offered by this new approach.

Olive crops are of immense socioeconomic importance in Mediterranean countries and are currently undergoing global expansion. According to Food and Agriculture Organization (FAO), in 2014, they occupied 10.3 Mha worldwide, 40% more than in 1990 (FAO, 2016). For olive crops, we selected groves in the area of Sierra Mágina, which lies within Jaén, a province with a long tradition in olive growing. This province represents 23.3% of all Spanish olive groves, which in turn account for 24.4% of olive crops worldwide (MAGRAMA, 2016). Avocado is also one of the most important economic crops in Mexico and this country is the largest producer, consumer, and exporter of this fruit in the world. Mexico produced 1,231,000 Mg of avocado in 2009 that accounted for 32% of the global production (SE, 2012). In the State of Michoacán alone, there are more than 153,000 ha under avocado, and the production area increases steadily—often at the expense of native forests—more than doubling in the last 6 years (Morales Manilla et al., 2012). The growing importance of these crops justifies the interest in applying our methodological proposal to them.

8.2 DESCRIPTION OF THE AREAS AND CROPS STUDIED

The areas studied are representative of the crops chosen (Figure 8.1). The olive groves of Sierra Mágina can be grouped into two categories: traditional rainfed cultivation on moderate to steep slopes and traditional irrigated cultivation on moderate slopes. For avocado, we selected six avocado orchards in four localities within the Cupatitzio Watershed that is located within the main avocado producing region in the surroundings of Uruapan (19°25 N, 102°03 W) in the state of Michoacan in central Mexico. Table 8.1 shows the general characteristics of the areas and crops studied.

8.2.1 Primary Data Collection of Olive Groves

Information about management practices was obtained via personal interviews conducted within the interviewees' own environment (at home, on the land, and/ or at the olive oil mill), to obtain detailed information on management techniques, types of machinery, and inputs used. Generally, certain aspects of management were also discussed and verified with workers from the olive oil mill, especially if tasks were carried out jointly (irrigation, aerial crop spraying, etc.), and with

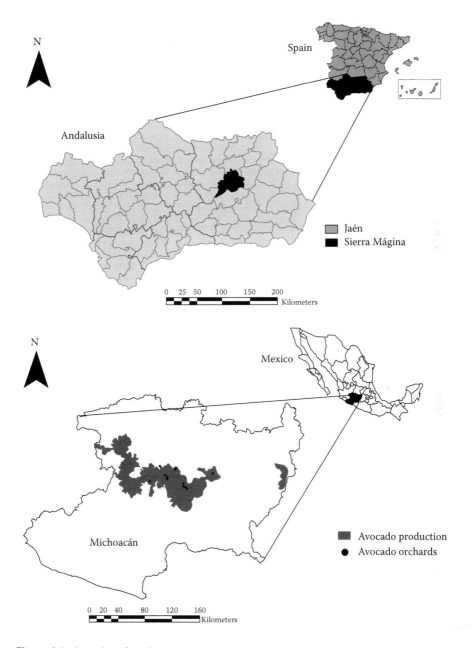

Figure 8.1   Location of study areas.

irrigation communities. The data for olive production for the last 4 years (3 years in some cases) were mainly obtained from olive oil mill records.

The interviews were conducted between 2002 and 2003. The organic olive growers were selected according to how long they had been producing organically,

Table 8.1   Characteristics of Areas and Crops Studied

	Olive Grove	Avocado
Localization	Sierra Mágina comarca (Spain)	Cupatitzio Watershed (Mexico)
Soils	Basic, with moderate to steep slopes	Andosoils, acidic, with moderate to steep slopes
Rainfall (mm)	400–600	1000–1500
Climate	Continental Mediterranean	Temperate to warm-subhumid
Trees per hectare	70–90	132
Rainfed Irrigated	100–130	132
Main varieties	Picual	Hass
Integration of cattle	Low	Zero
Watering regime		
Rainfed	38%	50%
Irrigated	62%	50% (only few weeks)

as it took time to establish management practices and to overcome a possible downward turn in production following the switch to organic farming. These growers had all been operating for between 4 and 10 years. The total number of cases in Sierra Mágina consisted of 31 organic farms (rainfed farming: 13 and irrigated farming: 18) and 30 conventional farms (rainfed farming: 10 and irrigated farming: 20). Conventional growers were chosen according to their proximity to organic ones, usually those with neighboring plots, to ensure similar agroclimatic conditions, and with the same farming regime (rainfed or irrigated). The technical characteristics are represented in Table 8.2, showing the percentage of farms that use each agricultural practice in each study area according to management type (conventional or organic).

8.2.2 Primary Data Collection of Avocado Groves

Three organic and three conventional groves that produce avocados intended for export were selected randomly in the Cupatitzio Watershed. These groves were selected using four criteria: (1) they had to be representative of their management system (organic or conventional), (2) they had to have been operating for 6–12 years since establishment, (3) they had to be approximately 20 ha in size, and (4) the producers had to be collaborative. All the groves were planted with *Persea americana* "Hass" trees, the most important avocado crop in the region and in the province. Data were collected for two production cycles (2010 and 2011) through personal interviews and the farmers' working and input schedules. The main agricultural practices conducted in the organic and conventional avocado groves during the period studied are presented in Table 8.2.

Table 8.2   Technical Characteristics of the Farms Analyzed in the Two Areas (%)

	Olive Grove		Avocado	
	Con	Org	Con	Org
Agricultural Practices				
Soil cultivation	60	90	Oc.	Oc.
Groundcovers	0	68	Oc.	Oc.
Herbicides	87	0	100	0
Green manure	0	0	0	0
Manure/compost	Oc.	100	100	100
Chipping of pruning cuttings	0	0	100	100
Synthetic soil fertilization	100	0	100	0
Foliar fertilization/ fertigation	100	100	100	100
Disease control	100	100	100	100
Pest control	100	100	100	100
Pruning frequency	100	100	100	100
Cattle grazing	0	13	0	0
Harvesting Manual	23.9	65	100	100
Branch shaker	76.1	35	–	–
Trunk shaker	0	0	–	–

Notes: Con, conventional; Oc., occasional; Org, organic.

8.3 CALCULATION OF NET PRIMARY PRODUCTIVITY AND SOCIALIZED ANIMAL BIOMASS

8.3.1 Olive Groves

In the case of olive groves, actual net primary productivity (NPP_{act}) was calculated on the basis of olive production. Using this data, we can calculate the quantity and type of pruning using the algorithm developed by Civantos and Olid (1982). This algorithm is highly suited to this case study, since it is based on the Picual olive variety and the type of pruning typically carried out in the Sierra Mágina area. It offers information about the amount of large woodcuttings taken, twigs, and leaves removed from the tree when pruning. Large pieces of wood are taken off the farm to be used as firewood and timber. Such cuts, therefore, represent socialized vegetable biomass. Twigs and leaves remain on the farm, where they are burned or incorporated mechanically into the soil, constituting reused biomass (RB). In our case study, they are burned on all the farms analyzed (Table 8.2). We considered that the quantity of leaves left on the trees is five times greater than the leaves removed by pruning and which fall to the ground completely every 2.5 years (Della Porta, 2015).

These naturally recycled leaves constitute aboveground unharvested biomass (UhB). To transform the different types of biomass into dry weights, we used the data provided by Ferreira et al. (1986).

When calculating belowground unharvested biomass, we multiplied the annual root:shoot partition coefficients, 0.30 for rainfed olive groves and by 0.21 for irrigated olive groves (Table AI.2), by the sum total of the aboveground unharvested biomass plus pruned material. In this species of tree, characteristic of semiarid climates, root growth is proportionally greater under rainfed conditions. This enables a larger volume of soil to be explored in periods of drought (Connor and Fereres, 2005).

Finally, the amount of aboveground biomass accumulated annually in trees was calculated using the total biomass of a hectare of olive grove in adult state (trunk and main branches) (45,000 and 50,000 kg fresh matter [fm], for rainfed and irrigated olive groves, respectively) and divided by the useful lifespan of the plantation (100 years for these traditional olive groves). The humidity level of this wood is 30% (Table AI.3). Annual accumulated belowground biomass was calculated considering a root:shoot ratio (kg dm:kg dm) at the end of planting of 0.21 for irrigated olive groves and 0.30 for rainfed olive groves (Table AI.2).

The aboveground biomass of the groundcovers in the different types of olive groves (rainfed or irrigated) was measured directly in the field over several years, taking an average in each case for use in our calculations (Foraster et al., 2006a, 2006b). Calculation of the belowground biomass of groundcovers was based on a root:shoot ratio (kg dm:kg dm) of 0.8 (Table AI.2).

The level of livestock integration in olive groves is low, but in some organic farms (Table 8.2), groundcover is used to feed ovine livestock. We calculated livestock load on these farms on the basis of the metabolizable energy (ME) of fodder (7.8 MJ ME·kg^{-1} dm based on Patón et al., 2005 and Díaz Gaona et al., 2014), considering that only 50% of the biomass produced in the herbaceous stratum of the olive grove can be consumed for grazing to be sustainable, and using the annual needs of a standard sheep (45 kg live weight with 1.5 lambings a year) as the benchmark animal (3464 MJ ME·yr^{-1} and sheep^{-1}). The annual production of this standard sheep would be 24 kg of live weight meat at farm gate.

The gross energy of olives is 11.55 MJ·kg^{-1} fm, and the gross energy of meat is 7.0 MJ·kg^{-1} fm. For other vegetable biomass, the conversion is 17.57 MJ·kg^{-1} dm (Table AI.5).

8.3.2 Avocado Groves

Information regarding avocado production (socialized vegetable biomass) and pruning was provided through interviews and direct measurement, respectively. In the case of avocados, pruned branches are milled and added to the soil as mulch (4±0.73 Mg dm·ha^{-1}) (Astier et al., 2014). Therefore, it counts as reused biomass. Accumulated aboveground biomass (106.4 Mg·dm·ha^{-1}) was taken from Ordóñez et al. (2008). To calculate the amount of biomass accumulated annually, it was divided by the useful lifespan of the plantation (40 years). Naturally recycled leaves constitute aboveground unharvested biomass, calculated on the basis of a leaf:shoot ratio of 0.072 (Rosecrance and Lovatt, 2003). Annual accumulated belowground biomass was calculated considering a root:shoot ratio (kg dm:kg dm) at the end of planting of 0.25 (Table AI.2). The mass of roots returned annually to the soil (belowground unharvested biomass) was based on a root/litter fraction of 0.2 of the total aboveground biomass (Shepherd et al., 1996).

The biomass of the groundcovers is 0 in conventional farms and on two organic farms that carry out strict controls of groundcover. For the third organic farm, we estimated aboveground biomass production of 2 Mg·dm·ha^{-1}. Calculations of the belowground biomass of groundcovers were based on a root:shoot ratio (kg dm:kg dm) of 0.71 (Astier, 2002).

There is no integration of livestock in the avocado groves studied; hence socialized animal biomass (SAB) is zero.

The gross energy of avocados is 8.16 MJ·kg^{-1} fm. For other biomass, the conversion is 17.57 MJ·kg^{-1} dm (Table AI.5).

8.4 CALCULATION OF EXTERNAL INPUTS

External inputs (EIs) include human labor, as well as all the inputs (fertilizer, pesticides, machinery, compost, etc.) that originate outside the agroecosystem. They can be divided into industrial inputs (chemical fertilizers, machinery, etc.) and non-industrial inputs (manure, human labor, etc.). The allocation of energy to each type of input is summarized in the following sections.

8.4.1 Nonindustrial Inputs

We estimate energy in human labor as dietary energy consumption (2.2 MJ·h^{-1}) (see a discussion in Chapter 4).

The method for calculating the gross energy of manure is based on the energy balance partitioning of livestock animals. Starting from gross and metabolizable energies in feed, we can estimate the amount of energy that is rejected as feces, methane, and urine (the nonmetabolizable fraction of the gross energy), and the energy that is metabolized into retained energy and heat production (see a discussion in Chapter 4). To standardize calculations, we considered a standard adult sheep (45 kg live weight, grazed, with 1.5 lambings a year) as the benchmark animal. We also considered that the metabolizable energy of the feed consumed by this sheep is 70% of the gross energy available in said feed. In short, the gross energy of the feed consumed by said sheep would be 4948 MJ and the metabolizable energy would be 3464 MJ. Therefore, the difference (1484 MJ) would be the energy attributed to the annual excretions of the animal.

To transform the different manures or composts imported by farmers into "standard sheep" excretions, we considered the nitrogen provided by different products on the basis of the bibliography consulted or the product labels. We calculated that a "standard sheep" annually produces 500 kg of feces and 250 kg of urine, with a nitrogen content of 0.7% and 1.7%, respectively (Urbano Terrón, 1992, p. 386).

Losses considered as a result of handling this waste are 44.6% (see Aguilera et al., 2015c). Therefore, the nitrogen available as fertilizer is 4.29 kg·N per standard sheep. In short, for every 4.29 kg of nitrogen imported onto the farm in the form of manure or compost, we have calculated a cost of 1484 MJ.

In the case of compost, we added in the cost of producing the compost (0.48 MJ·kg^{-1}) (Astier et al., 2014). Last, we added the amount of gross energy consumed in transporting manure or compost to the farm. In the case of olive groves, the distance traveled was estimated to be 50 km, whereas in the case of avocado groves, it was estimated to be 250 km. The gross energy consumed in transportation is 4.21 MJ·Mg^{-1}·km^{-1} (Table AII.11.4).

The gross energy of other nonindustrial inputs is as follows: wheat bran 19.4 MJ·kg^{-1} and soy oil 38.96 MJ·kg^{-1}.

8.4.2 Industrial Inputs

Table 8.3 shows the gross energy values for the industrial inputs used on the farms.

8.5 CALCULATION OF EROIs

Having transformed biomass and external inputs into energy, we calculated the following EROIs, as specified in Chapter 2:

Proposed EROIs from an economic perspective:

$$\text{Final EROI (FEROI)} = SB/(RuB + EI)$$
$$\text{External final EROI (EFEROI)} = SB/EI$$
$$\text{Internal final EROI (IFEROI)} = SB/RuB$$

Proposed EROIs from an agroecological point of view:

$$\text{NPP}_{act}\ \text{EROI} = NPP_{act}/TIC$$
$$\text{Agroecological final EROI (AE-FEROI)} = SB/TIC$$

Biodiversity EROI:

$$\text{Biodiversity EROI} = 1 - \frac{\text{AE-FEROI}}{\text{FEROI}} = UhB/TIC$$

Woodening EROI:

$$\text{Woodening EROI} = AB/TIC$$

where *socialized biomass* (*SB*) = socialized vegetable biomass (*SVB*) + socialized animal biomass (*SAB*); *RuB* = reused biomass; *EI* = external inputs; *RcB* = recycling

Table 8.3   Gross Energy of the Industrial Inputs Used

	Unit	Gross Energy	Reference
Synthetic Fertilizers			
N fertilizers average	MJ·kg^{-1} N	73.3	Table AII.5.13
P fertilizers average	MJ·kg^{-1} P$_2$O$_5$	18.5	Table A.II.6.6
K fertilizers average	MJ·kg^{-1} K$_2$O	14.4	Table AII.7.6
Organic and Mineral Fertilizers			
N biofertilizers	MJ·kg^{-1} N	12.8	Table AII.5.8 through Table AII.5.9
Ground rock	MJ·kg^{-1} P$_2$O$_5$	13.1	Table AII.6.5
KCl	MJ·kg^{-1} K$_2$O	12.4	Table AII.7.5
Synthetic pesticides	MJ·kg^{-1} a.i.	447.0	Table A.II.8.1
Organic Pesticides			
Extracts and teas	MJ·L^{-1}	0.27	Astier et al. (2014)
Microorganisms	MJ·kg^{-1}	77.2	Astier et al. (2014)
Irrigation			
Trickle average 100 m lift	GJ·ha^{-1}·yr^{-1}	36.0	Table AII.9.26
Fuels			
Gasoline	MJ·L^{-1}	44.6	Tables AII.1.1 and AII.1.9
Diesel	MJ·L^{-1}	48	Tables AII.1.1 and AII.1.9
Machinery	MJ·ha^{-1}	Several	Table AII.4; Astier et al. (2014); Guzmán and Alonso (2008)

Note: a.i., active ingredient.

biomass = reused biomass + unharvested biomass; *UhB* = unharvested biomass; *TIC* = total inputs consumed = RcB + EI; *AB* = accumulated biomass.

8.6 RESULTS AND DISCUSSION

8.6.1 Net Primary Productivity in Olive and Avocado Groves

Figure 8.2 shows the net primary productivity (NPP$_{act}$) of avocado and olive groves. There are very few differences between organic and conventional olives groves in the case of rainfed olive production. The lack of rainfall limits the production of biomass in semiarid climates and reduces, or even cancels out, response to synthetic fertilization (Meco et al., 2010). In addition, in these systems, olive growers are fearful that groundcover will compete with olive trees for water. This fear leads organic olive growers to exercise greater control over groundcover than their peers working

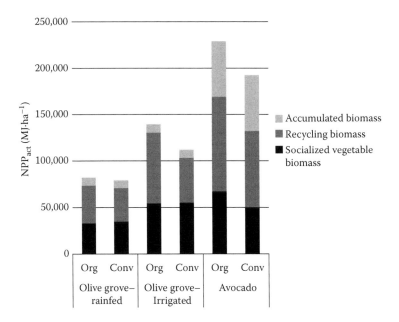

Figure 8.2 Composition of actual net primary productivity (NPP$_{act}$) in the olive and avocado groves analyzed.

on irrigated groves. However, in irrigated olive groves and avocado groves, NPP$_{act}$ is clearly higher in organic production than in conventional farming. The causes for this increase are different in these two cases. In irrigated olive groves, it occurs due to the groundcover that organic olive growers allow to grow between the olive trees in rainy seasons. The growth of these plants means that recycling biomass (reused plus unharvested biomass) is 1.59 higher than in conventional irrigated olive groves. However, on avocado groves, organic growers allow very little groundcover to grow in between the trees, and the increase in NPP$_{act}$ is due to the increase in fruit production (1.32 times higher) more than to the presence of grass or weeds, which means that recycling biomass is 1.25 higher. Among farmers and technicians, there is the belief that weeds keep pests. Since the early twentieth century, the U.S. government imposed phytosanitary restrictions on fresh avocado imports to combat seed and stem weevils, an avocado seed moth. This pest, however, has been eradicated since the late 1980s.

Figure 8.3 details the components of NPP$_{act}$ and their transformation into socialized animal biomass. In addition to the herbaceous component mentioned previously, only the meager presence of livestock in 13% of the organic olive growing area (Table 8.2) differentiates between the flows of biomass and energy in organic and conventional production. On average, organic rainfed olive groves could maintain 0.04 sheep a year, which would generate 1.1 kg of meat at farm gate, allowing for 0.19 kg·N·ha^{-1} to be recycled, after losses. Organic irrigated olive groves, on the other hand, could maintain 0.20 sheep a year, which would generate 4.7 kg of meat at farm gate, and would enable 0.85 kg·N·ha^{-1} to be recycled, after losses. In organic avocado groves,

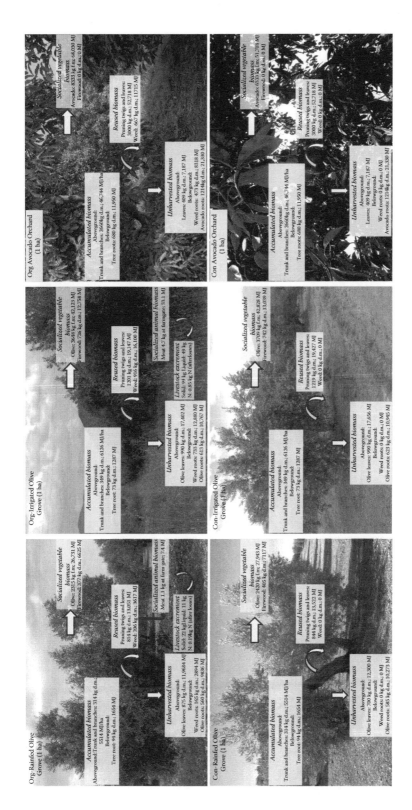

Figure 8.3 Net primary productivity and socialized animal biomass in the different crops studied.

regulations governing certification prohibit the integration of livestock on these plantations (personal communication of producers). In both cases, official regulations have also been flagged as a cause for the conventionalization of organic agriculture (Zoiopoulos and Hadjigeorgiou, 2013).

From the perspective of the social multifunctionality of agrarian production, this signifies a very moderate advance. In organic and conventional avocado groves, only the fruit itself is extracted as socialized biomass. In olive groves, olive output accounts for 77%–80% of the socialized biomass, the rest being wood (20%–23%). Livestock production on irrigated and rainfed organic olive groves represents just 0.1% and 0.02%, respectively, of socialized biomass.

As for the fruit, the main commercial portion of NPP_{act}, the results show a slight lower figure for organic farming in comparison with conventional olive production (5% in rainfed olive groves and 2% in irrigated groves) and a substantial increase of 32% for organic avocados.

Organic production has embarked on a timid process of diversification, more accentuated in irrigated organic olive groves. However, it is still a far cry from the quantity and quality of internal loops of biomass and energy achieved by traditional production, as you will see in Chapter 7 dedicated to coffee, or in González de Molina et al. (2014), in relation to olive groves. Consequently, they are still systems that require copious energy flows to be imported to sustain their functioning, as discussed in the following section.

8.6.2 External Inputs

The importation of inputs is slightly lower in organic production (Figure 8.4). It ranges between 82% for rainfed organic olive groves and 96% for organic avocado production, in comparison with their conventional equivalents. However, the composition is very different. Organic agriculture proportionally consumes a greater quantity of nonindustrial inputs. Hence, human labor increases by 3%–31%; and manure–compost by 84%–276%, with regard to conventional production of a similar category. Conventional farms, on the other hand, consume more industrial inputs. Industrial fertilizers in conventional production increase by 153%–8602% and pesticides by 113%–2619%. With regard to machinery, the difference is not as marked and there is no clear tendency. It is 24%–31% higher on organic olive groves and 7% lower on organic avocado groves.

As a consequence of the different composition, the amount of fossil energy invested varies with the different management approaches. This aspect has been examined by Guzmán and Alonso (2008) and Astier et al. (2014).

The different strategies used by organic olive and avocado growers when converting their farms are also interesting to note. Organic olive growers have intensified their use of machinery, particularly in the management of soil and grasses/weeds, sacrificing total energy investment in fertilization. Organic avocado growers have intensified their energy investment in fertilization, slightly decreasing their energy investment in machinery. In the case of Spain, other studies have highlighted the low replenishment of nutrients observed in extensive organic crop production, such as olive groves, owing to the high cost of organic fertilization (Alonso et al., 2008). This high cost is related with the chronic shortage of biomass in the Mediterranean and the lack of infrastructures to recover and compost existing biomass, which increases the land cost of replenishing nutrients locally (Guzmán et al., 2011). In addition, there is a shortage of machinery for mechanical spreading of manure/compost on farms.

8.6.3 Energy Return on Investments

Of the three systems analyzed, the greatest changes in the flows of internal biomass during the switch over to organic production have been seen in irrigated olive groves. In avocado crops, the amount of biomass reused by producers increased by just 22%, whereas the total recycling biomass increased by just 25%. In rainfed olive groves, the changes were even less patent. Organic producers increased reused biomass by 23% and recycling biomass by 14%. However, irrigated olive groves increased reused biomass by 81%, and in total, recycling biomass increased by 59% (Table 8.4). Such a marked increase reduces the internal efficiency of the system by almost a half, halving the energy return on reused biomass in the form of socialized biomass (IFEROI). This decline in internal efficiency is not compensated by a significant increase in external energy (EFEROI), since high energy investments caused by irrigation and the increased use of machinery in organic crops minimizes the impact on savings achieved by replacing synthetic with organic fertilizers. As a result, the energy return of both flows in the form of socialized biomass drops to 85% (FEROI). For the same reason, the return on total inputs consumed in socialized biomass (AE-FEROI) falls to 80% (Table 8.4).

The conversion of rainfed olive groves is based on a decline in external energy flows, specifically those dedicated to fertilization and pest control (Figure 8.4). Given that water shortages are the main limiting factors of these agroecosystems, the drastic decline in these inputs has a reduced effect (5%) on socialized biomass. Their impact can even be compensated on organic farms by the greater availability of water for olive trees, owing to the control of runoff water losses provided by plant cover and the increased accumulation of water in the soil resulting from an increase in organic matter (Durán Zuazo et al., 2009; Gómez et al., 2009). As a result, the return on external energy investment (EFEROI) is positive, and the return on investment made by society (FEROI) decreases slightly (4%), at a similar level to socialized biomass. Although internal flows are only moderately increased in organic rainfed olive groves, internal efficiency (IFEROI) and total efficiency (AE-FEROI) decrease, albeit to a lesser extent than on irrigated olive groves (Table 8.4).

The transformation from conventional to organic avocado production, on the other hand, substantially increases the energy return of energy invested for society (FEROI and AE-FEROI), internally (IFEROI) and externally (EFEROI). In this case, the system can successfully respond to the change in strategy implemented by organic growers during the conversion. The fall in *EI* is similar to the levels noted in irrigated organic olive groves (4%–5%), but the composition is clearly different (Table 8.4 and Figure 8.4). In this case, the external investment of energy in fertilization increases by 33%, whereas in the conversion of irrigated olive groves it

Table 8.4 Energy Indicators of the Olive and Avocado Grove Analyses

	Olive Grove–Rainfed			Olive Grove–Irrigated			Avocado		
	con	org	org/con %	con	org	org/con %	con	org	org/con %
NPP_{act}[a]	78,324	81,786	104	110,787	138,076	125	191,631	227,991	119
Socialized biomass[a]	35,060	33,363	95	55,867	54,914	98	51,703	68,030	132
Reused biomass[a]	13,522	16,668	123	19,427	35,247	181	52,718	64,434	122
Unharvested biomass[a]	22,573	24,594	109	28,601	41,049	144	28,516	36,834	129
Recycling biomass[a]	36,096	41,263	114	48,028	76,296	159	81,235	101,268	125
Accumulated biomass[a]	7,168	7,168	100	7,413	7,413	100	58,693	58,693	100
External inputs[a]	19,407	15,936	82	62,070	58,895	95	95,786	91,863	96
Total inputs consumed[a]	55,502	57,198	103	110,198	135,191	123	177,020	193,131	109
Proposed EROIs from Economic Point of View									
FEROI	1.06	1.02	96	0.69	0.58	85	0.35	0.44	125
EFEROI	1.81	2.09	116	0.90	0.93	104	0.54	0.74	137
IFEROI	2.59	2.00	77	2.88	1.56	54	0.98	1.06	108
Proposed EROIs from an Agroecological Point of View									
NPP_{act} EROI	1.41	1.43	101	1.01	1.03	101	1.08	1.18	109
AE-FEROI	0.63	0.58	92	0.51	0.41	80	0.29	0.35	121
Biodiversity EROI	0.41	0.43	106	0.26	0.30	117	0.16	0.19	118
Woodening EROI	0.13	0.13	97	0.07	0.05	81	0.33	0.30	92

Notes: AE-FEROI, agroecological final energy return on investment; EFEROI, external final energy return on investment; IFEROI, internal final energy return on investment; EROIs, energy return on investments; NPP_{act}, actual net primary productivity.

[a]Megajoule per hectare.

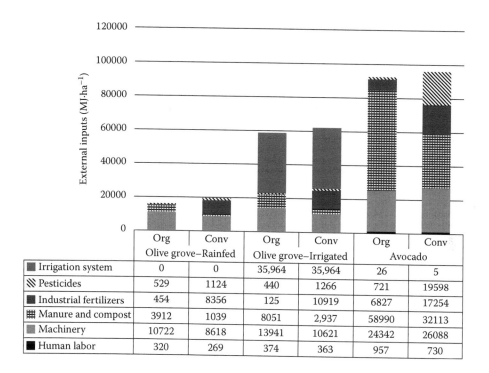

	Org	Conv	Org	Conv	Org	Conv
	Olive grove–Rainfed		Olive grove–Irrigated		Avocado	
Irrigation system	0	0	35,964	35,964	26	5
Pesticides	529	1124	440	1266	721	19598
Industrial fertilizers	454	8356	125	10919	6827	17254
Manure and compost	3912	1039	8051	2,937	58990	32113
Machinery	10722	8618	13941	10621	24342	26088
Human labor	320	269	374	363	957	730

Figure 8.4 Composition of external inputs used in the case studies.

decreases by 41%. Another notable difference can be found in the baseline situation of these agroecosystems prior to their conversion. On the one hand, conventional avocado production also integrates high quantities of organic fertilizer. On the other hand, the amount of recycled biomass is significantly higher in absolute terms, since rainfall is not such a limiting factor in this case. Consequently, these agroecosystems can potentially be in better condition to respond to the change in fertilization strategy owing to a greater quality and more biologically active soil than is present in conventional olive groves. Put another way, the greater the level of soil degradation, the greater investment of energy is required to achieve a sufficient improvement to increase returns in the form of socialized biomass. The distance between the pathways of degradation and restoration is known as the hysteresis of land rehabilitation (Tittonell et al., 2012).

In this respect, the growing value of NPP_{act} EROI indicates that the rehabilitation of fund elements has begun in all the cases studied; mildly in the case of olive groves and more definitely in the case of the avocado groves, where better starting conditions allow for a greater response to the intensification of organic fertilization in terms of biomass production. In the case of organic olive groves, the recovery of fund elements will be slower. However, numerous studies have shown improvements in the physical, chemical, and biological quality of the soil, including a higher organic matter content and a net balance of carbon sequestration in organic olive groves (Benítez et al., 2006;

Castro et al., 2008; García-Ruiz et al., 2009; Parras Alcántara et al., 2015; Aguilera et al., 2015b). In the cases examined here, this improvement is based more on the increase in recycling biomass (5,167 and 28,268 MJ·ha^{-1} extra, respectively in rainfed and irrigated organic olive groves) than on organic fertilization (2,872 and 5,114 MJ·ha^{-1} extra, respectively). However, in avocado crops, the improvement would be fostered chiefly by external flows of organic fertilizer (26,877 MJ·ha^{-1} extra) in comparison with the energy contribution of plant cover (20,033 MJ·ha^{-1} extra).

Taken overall, organic conversion has in all cases increased internal flows of energy. The consumption of *EI* has fallen consistently, particularly in terms of industrial *EI*. This partial internalization of energy flows through the recirculation of biomass has led to an increase in biodiversity EROI (Table 8.4). In other words, the recirculation of biomass in these systems has generated an increase in the availability of food for wildlife. This, combined with the decreased pressure exerted by pesticides, explains the improvement in biodiversity associated with organic approaches to management in these crops (Jerez-Valle et al., 2015; Villamil et al., 2016).

Finally, woodening EROI declines in all cases. In these agroecosystems, accumulated biomass is not modified, since the woodening component is not increased and, consequently, the environmental services provided by this component similarly do not increase. Hence, the increase in total inputs consumed leads to a proportional decline in this indicator in all cases.

In short, opening up the black box of the agroecosystems studied by looking at EROIs shows the benefits offered by switching to organic farming. Efficiency in terms of external inputs and nonrenewable energy (see Guzmán and Alonso, 2008 and Astier et al., 2014) has increased, but, above all, the fund elements of agroecosystems managed organically have begun to be rehabilitated. However, the analysis shows that there is still a great deal of room for improvement in these agroecosystems. In both crops, groundcover has not been adopted by all organic producers. This limits the average biomass generated by these systems. The spread of this practice would afford substantial improvements. In the case of olive groves, the decline in energy returns toward society (FEROI) could be reduced, stimulating greater internalization of biomass flows by increasing other low-entropy internal loops, for example, by sowing leguminous plants, palliating the fertilization deficit they suffer. In avocado production, both measures would be adequate to reduce imports of manure and compost. Imports that entail a high consumption of fossil fuels, since a third of energy investment represented by manure and compost in avocado production correspond to transportation. This cost could be reduced through the integration of livestock and the composting of agroindustrial waste on a local scale.

8.7 CONCLUSIONS

Switching over to organic farming has modified energy flows in two ways: first, by increasing internal flows of biomass through the adoption of groundcover by some organic producers, and, more exceptionally, through the introduction of grazing; and second, by modifying flows of external inputs in terms of quantity and quality. The proposed EROIs are capable of showing these variations and also of reflecting a significant change in tendency in the state of the fund elements of the agroecosystem (land, biodiversity, and accumulated biomass).

Opening up the black box has allowed us to go further in seeing how switching to organic farming improves the return toward society of external energy flows (total and nonrenewable) (EFEROI and nonrenewable external final EROI [NR-EFEROI]). This has allowed us to verify the hypothesis that conversion generates changes in the internal energy flows of the agroecosystem. These changes promote improvements in the fund elements and help to explain the ecosystemic services detected by numerous studies that have compared organic and conventional farming. In the case of olive groves, groundcover becomes a key element that, by generating a low-entropy internal loop, stimulates the recovery of these fund elements. However, in avocado groves, the development of internal loops is insignificant. In this case, the importing of organic fertilizer is the catalyst for change. From an energy perspective, both forms of organic management could be improved.

Consequently, the application of the proposed EROIs shows that there is still a great deal of room for improvement in organic management from an agroecological perspective, moving further away from the analyzed agroecosystems of conventional farming and making them even more sustainable. This margin for improvement is directly related with the increase in low-entropy internal loops and an even more substantial improvement in the fund elements of the olive and avocado agroecosystems. In short, EROIs, especially agroecological ones, provide a very useful tool not only to discover how far removed a certain agroecosystem is from conventional farming when it seeks to switch over to organic farming, but also it indicates the way to improve its management to make it more sustainable.

ACKNOWLEDGMENT

The authors express their gratitude to the project PAPIIT N-210015 funded by Programa de Apoyo a Proyectos de Investigación e Innovación Tecnológica (PAPIIT), Universidad Nacional Autónoma de México.

CHAPTER 9

Energy in Agroecosystems
A Tool for the Sustainable Design of Extensive Livestock Farms

Gloria I. Guzmán, Eduardo Aguilera, Leticia Paludo Vargas, and Romina Iodice

CONTENTS

9.1 INTRODUCTION

The global livestock sector has undergone huge growth in recent decades, motivated by the growing human population and by the increase in per capita animal product consumption across large swathes of the planet (Alexandratos and Bruinsma, 2012). This dynamic has been called into question on account of its contribution to serious socioenvironmental problems, such as climate change (Herrero et al., 2016), deforestation and the loss of biodiversity (Steinfeld et al., 2006; Thornton and Herrero, 2010; Erb et al., 2016), and competition between food for human consumption and the production of animal feed (Schader et al., 2015). One of the keys to explaining the impact of livestock farming on these problems is its territorial expanse. Managed grazing occupies more than 33 million km^2 or 25% of the global land surface, making it the single most extensive form of land use on the planet (Asner et al., 2004), and the production of livestock feed uses 33% of agricultural cropland (Steinfeld et al., 2006). The high land costs of livestock farming are a result of its low efficiency as an energy converter, meaning that it requires huge quantities of phytomass to produce relatively small amounts of animal biomass. Consequently, at a global level, livestock contributes to only 15% of total food energy and 25% of dietary protein (FAO, 2009).

This low energy efficiency is due to the fact that a double energy transformation must occur. First, solar (and fossil) energy is converted into phytomass by photosynthesis. Second, when the phytomass is fed to the animal, a major share of energy intake is spent on keeping up body metabolism, and only a small portion is used to produce meat, milk, or eggs. However, there is major variability in energy and land consumption among livestock species and types of production system. FAO (2009) categorizes livestock production systems into grazing, mixed farming, and industrial systems. Grazing and mixed farming systems have their own territorial basis, and ruminants usually play an essential role on them. In grazing systems, ruminants graze mainly grasses and fodder. In mixed farming systems, cropping and livestock rearing are linked activities. In these systems, the livestock consumes, among other things, crop by-products and stubble. Industrial systems are defined as those systems that purchase at least 90% of their feed from other enterprises. In these systems, monogastric animals and the consumption of feed play a major part. In territorial terms, grazing systems occupy greater land areas, but the biomass produced in said territory is not apt for direct human consumption. At the other extreme, we have industrial systems. Proportionally they occupy smaller territories, but this territory has an agricultural vocation. This leads to considerable competition with producing food for direct human consumption. In terms of fossil energy consumption, this increases substantially from grazing systems to industrial systems, with mixed farming systems falling somewhere in between the two (Pimentel, 2004; Veermäe et al., 2012).

The inefficiency of livestock as an energy converter and the increase in animal products in the human diet mean that the performance of these farms must be improved to increase energy returns on investment in agroecological and economic terms. Several strategies have been suggested. They largely fall into three categories:

1. Increase feed conversion efficiency; in other words, the productivity of socialized animal biomass. This may include improved feed quality (e.g., digestibility, protein, and mineral contents), optimally matching animal feed requirements, and breeding and herd management. This path has been particularly successful when applied to the rearing of monogastric animals in industrial systems, but progress has also been made with ruminants (Pimentel, 2004; Herrero, 2013).
2. Improve the energy efficiency of the production of pasture, fodder, and feed. Techniques such as the phosphorus fertilization of pastureland, the inclusion of leguminous plants in pastureland and crop rotations, rotational grazing, the elimination or substitution of inputs and labor with a high energy cost, and so on, allow the energy efficiency of livestock feed production to be increased (Pimentel, 2006; Rodale Institute, 2011; Latawiec et al., 2014). Consequently, this decreases the energy intensity of the animal product (milk, meat, eggs, etc.).

3. Redesign livestock agroecosystems through the balanced integration within the territory of pastureland, crops, and woodlands/forests. Integrated crop/livestock systems, where animals are fed only from grassland and by-products from food production, effectively decrease the consumption of fossil energy (Schader et al., 2015). The same occurs when animal numbers are in balance with the quantity of feed crops grown on the farm, reducing dependence on imported feed and the associated fossil energy use by growing a mixture of grain and fodder crops (Pimentel, 2006; Malcolm et al., 2015). In another study, Giambalvo et al. (2009) showed that higher proportions of permanent pastures to total farm area contributed to improving the energy efficiency of livestock farms in Sicily. Furthermore, designing integrated crop–livestock–forestry systems has also been proposed as a means of improving energy efficiency, as one way of contributing to agrarian sustainability (Latawiec et al., 2014; Murgueitio et al., 2015).

These three strategies are not mutually exclusive. For example, switching from conventional to organic farming can increase the efficiency of fossil energy owing to changes in the design of the agroecosystem (e.g., partial substitution of grain crop land areas with pastureland and fodder); along with changes in the way they are managed, applying techniques that incorporate lower levels of fossil energy (e.g., replacing chemical with organic fertilization) (Pimentel, 2006; Veermäe, 2012).

The aim of this chapter is to illustrate, by means of an example, the utility of the energy perspective when it comes to the sustainable design of extensive livestock farms, in line with strategies 2 and 3. To this end, we selected a series of extensive farms dedicated to breeding ruminant livestock, either for meat or dairy production, presenting major differences in terms of the structure of their respective agroecosystems (from highly specialized to highly diversified in their structural components) as well as their levels of input intensification (from highly self-sufficient to highly dependent on external inputs). For this purpose, we chose 36 farms, of which 23 are pastureland farms known as *dehesas* located in Andalusia, in the south of the Iberian Peninsula (Photographs 9.1 and 9.2). *Dehesas* are structurally very complex farms. They are agroforestry systems, usually combining pastureland, woodland, and cropland, whose main production consists of extensive livestock farming. Of the 23 *dehesas* selected, 15 were certified organic. The remainder was conventionally managed. A priori, organic certification presupposes a lower level of intensification, with lower stocking rates and a higher level of self-sufficiency in terms of livestock feed. We also selected eight farms in Rio Grande do Sul (Brazil), which were structurally very simple, based on natural pastureland with no trees (Photograph 9.3). Of these, four were meat farms, three were dairy (milk), and one was mixed. Presumably, different farming orientations should also infer differences in management intensity. Finally, five farms were selected to the north of Buenos Aires province (Argentina) (Photograph 9.4), which were structurally simple but more intensive, because they focused on the production of milk, and also because four of them cultivated grasslands as the main source of livestock feed (Table 9.1). As we shall see, for the 36 farms, we calculated the economic and agroecological energy return on investments (EROIs), and we also explored which of the factors in terms of structure, farm management, and livestock type, impacted on the configuration of these indicators. The locations of the farms are shown in Figure 9.1.

Photograph 9.1   Panoramic view of *dehesa* pastureland in Andalusia (Spain).

Photograph 9.2   Landscape integrating *dehesa*, cropfields, and livestock in Andalusia
(Spain).

Photograph 9.3   Meat cattle grazing on natural pastures in Rio Grande do Sul (Brazil).

Photograph 9.4   Dairy cattle grazing on cultivated grassland in the province of Buenos Aires (Argentina).

Table 9.1 Characteristics of the Farms Studied

Case	Study Area	Total Surface Area (ha)	Natural Wooded Pasture (ha)	Natural Treeless Pasture (ha)	Cropland (Grain and Forage Crops) (ha)	Crop Species	Sown Pasture (ha)	Livestock Species	Number of Mothers	Number of Other Components of the Herd	Certification	% Raised in Barns
1	North Buenos Aires province (Argentina)	4.92	0	0	0	—	4.92	Dairy cattle	6	5 calves, 1 heifer[a]	No	43
2		9.10	0	0	0	—	9.10	Dairy cattle	16	1 stud, 2 heifers, 15 calves	No	43
3		11.11	0	0	0	—	11.11	Dairy cattle-carne	7	8 heifers, 9 calves	No	43
4		14.87	0	0	0	—	14.87	Dairy cattle	9	1 stud, 1 heifer, 7 calves	Organic–biodynamic	43
5		46.42	0	45.42	0	—	1.0	Dairy cattle	21	1 stud, 6 heifers, 16 calves	No	0
6	Alegrete-Rio Grande do Sul (Brazil)	20.0	0	20.0	0	—	0	Meat sheep	200	6 studs, 50 ewes, 112 lambs	No	0
7		40.0	0	40.0	0	—	0	Meat sheep	55	2 studs, 45 ewes, 28 lambs	No	0
8		20.0	0	20.0	0	—	0	Meat sheep	18	1 stud, 34 lambs, 5 ewes[a]	No	0
9		30.0	0	30.0	0	—	0	Meat sheep	16	1 stud, 4 ewes, 10 lambs	No	0
10		20.0	0	20.0	0	—	0	Dairy cattle	6	1 heifer[a]	No	0
11		30.0	0	30.0	0	—	0	Meat sheep/ Dairy cattle	15/20	2 studs/3 ewes/ 5 lambs; 15 heifers	No	0
12		17.0	0	17.0	0	—	0	Dairy cattle	2	1 heifer	No	0

(Continued)

Table 9.1 (*Continued*) Characteristics of the Farms Studied

Case	Study Area	Total Surface Area (ha)	Natural Wooded Pasture (ha)	Natural Treeless Pasture (ha)	Cropland (Grain and Forage Crops) (ha)	Crop Species	Sown Pasture (ha)	Livestock Species	Number of Mothers	Number of Other Components of the Herd	Certification	% Raised in Barns
13		11.5	0	11.5	0	–	0	Dairy cattle	9	3 heifers/9 female calves	No	0
14	Andalusia (Spain)	300	150	0	150	Wheat/ sorghum/ sunflower/ barley	0	Meat cattle	138	22 heifers/ 110 calves	No	2
15		50	25	0	25	Wheat/ sorghum/ sunflower/ fodder barley	0	Meat cattle	29	1 stud/17 heifers/ 25 calves	No	17
16		40	16	0	24	Wheat/ sorghum/ sunflower/ broad beans	0	Meat cattle	20	1 stud/11 heifers/ 15 calves	No	0
17		300	120	0	165	Wheat/ sunflower/ fodder barley	15	Meat cattle	116	19 heifers/ 95 calves	No	25
18		130	45	0	85	Wheat/ sorghum/ sunflower/ broad beans	0	Meat cattle	56	9 heifers/ 36 calves	No	4
19		150	75	0	75	Wheat/barley/ broad beans/ fodder barley	0	Meat cattle	120	1 stud/73 heifers/ 100 calves	No	17

(Continued)

Table 9.1 (*Continued*) Characteristics of the Farms Studied

Case	Study Area	Total Surface Area (ha)	Natural Wooded Pasture (ha)	Natural Treeless Pasture (ha)	Cropland (Grain and Forage Crops) (ha)	Crop Species	Sown Pasture (ha)	Livestock Species	Number of Mothers	Number of Other Components of the Herd	Certification	% Raised in Barns
20		125	50	0	75	Wheat/ sorghum/ sunflower/ fodder barley	0	Meat cattle	55	1 stud/ 33 heifers/ 38 calves	No	0
21		53	13	0	35	Wheat/ sunflower/ fodder barley	5	Meat cattle	20	1 stud/ 11 heifers/ 15 calves	No	29
22		140	90	0	33	Wheat/fodder barley	17	Meat cattle	86	14 heifers/ 55 calves	Organic	23
23		104	80	0	12	Broad beans/ fodder barley	12	Meat cattle	52	1 stud/ 31 heifers/ 38 calves	Organic	38
24		200	100	0	100	Wheat/ sorghum/ Sunflower/ broad beans	0	Meat cattle	100	1 stud/ 60 heifers/ 90 calves	Organic	0
25		240	95	0	145	Wheat/barley/ broad beans/ fodder barley	0	Meat cattle	69	11 heifers/ 50 calves	Organic	33
26		70	30	0	40	Wheat/ sorghum/ Sunflower/ broad beans/ fodder barley	0	Meat cattle	41	1 stud/ 24 heifers/ 30 calves	Organic	21
27		100	80	0	20	Wheat/ sorghum/ Barley/fodder barley	0	Meat cattle	67	11 heifers/ 50 calves	Organic	17

(Continued)

Table 9.1 (Continued) Characteristics of the Farms Studied

Case	Study Area	Total Surface Area (ha)	Natural Wooded Pasture (ha)	Natural Treeless Pasture (ha)	Cropland (Grain and Forage Crops) (ha)	Crop Species	Sown Pasture (ha)	Livestock Species	Number of Mothers	Number of Other Components of the Herd	Certification	% Raised in Barns
28		400	200	0	200	Wheat/ sorghum/ sunflower/ broad beans/ fodder barley	0	Meat cattle	90	15 heifers/ 60 calves	Organic	25
29		50	33	0	15	Wheat/ sorghum/ broad beans	2	Meat cattle	33	5 heifers/ 25 calves	Organic	38
30		212	110	0	62	Wheat/fodder barley	40	Meat cattle	69	11 heifers/ 55 calves	Organic	33
31		100	50	0	44	Wheat/ sorghum/ barley/fodder barley	6	Meat cattle	70	1 stud/ 42 heifers/ 50 calves	Organic	35
32		420.0	350	70	0	—	0	Sheep	800	30 studs/ 100 ewes/ 1000 lambs	Organic	0
33		342.5	180	120	42.5	Oats	0	Sheep	520	16 studs/ 58 ewes/ 800 lambs	Organic	0
34		198.0	39.5	158.5	0	—	0	Meat cattle	55	2 studs/ 33 heifers/ 47 calves	Organic	0
35		100.0	25.0	75.0	9	Alfalfa and vetches	0	Dairy goats	320	16 studs/40 cull females/50 kids	Organic	43
36		97.0	38.8	58.2	0	—	0	Meat sheep/ dairy goats	208/6	6 studs/ 17 ewes/ 115 lambs; 1 stud	Organic	0

a Cull females aggregated.

The research described in this chapter is, therefore, an exploratory exercise of the potential of EROIs as a tool to improve the energy design of livestock farms, in terms of structure as well as management. Under no circumstances, we assume that the sample is representative of each region, or of each type of management. To do that, the research design would have required a random stratified sampling weighted according to various criteria, and would have encompassed a very large number of farms, which exceeds the scope and aims of this chapter.

Therefore, the research goals outlined were as follows: (1) to understand how territorial design and management intensity can affect energy flows and the provision of ecosystemic services and (2) to explore how farming businesses can be improved in terms of energy sustainability, with a view to producing an adequate return for socialized biomass and socialized animal biomass, and maintaining the fund elements of the agroecosystem.

From a methodological perspective, to achieve the goals set, we divided our task into four stages. First, we selected the farms and obtained primary information through surveys and direct sampling. In the second stage, we calculated the EROIs. In the third stage, we identified the variables that differentiate the farms, establishing homogeneous groups of farms in terms of those variables. The techniques used were principal component analysis (PCA) and cluster analysis to classify the farms into homogeneous segments. In the fourth stage, we established the relationship among the homogeneous groups of farms based on their principal components (cluster), having calculated the EROIs. The results obtained are set out and discussed in Section 9.4, where we put forward some strategies in terms of territory design and management that could improve the energy sustainability of farms.

Figure 9.1 Global distribution of study areas.

9.2 METHODOLOGY

9.2.1 Gathering Primary Information

Primary information about the characteristics of each farm and livestock management was obtained by surveying farmers and local experts or zootechnicians. The surveys were conducted between 2013 and 2015. The questionnaire comprised two principal blocks: a technical section to gather descriptive data on the area, structure, and livestock management regime of the farm; and an economic section to collect data on external inputs (labor, fertilizers, seeds, etc.) and output generated by the system (milk, animals, timber, firewood, grain, etc., sold or consumed by the family). Table 9.1 shows the basic characteristics of the farms studied. In terms of herd composition, in cases where there were no cull females, a useful lifespan of 6 years was considered for dairy cows (16% restocking rate) and 25% for adult ewes and nanny goats. This way, the territorial cost of restocking was internalized. However, we did not internalize stud animals in cases in which the farmers decided to use artificial insemination techniques. This technique means that males do not have to be kept with the herd, and considerably reduces the territorial cost of reproduction, taking it down to insignificant levels.

Productivity and the fodder value of cultivated grasslands and trees (if any) for ruminant animals were directly measured on the farms in Argentina. This information was obtained for Brazil through surveys with experts and zootechnicians, and finally, in Spain, it was based on the previous studies. Table 9.2 summarizes the data for different cases.

9.2.2 Calculating Net Primary Productivity and EROIs

The actual net primary productivity (NPP_{act}) of pasturelands with and without trees, and of cultivated grasslands was calculated on the basis of aboveground dry biomass (Table 9.2) considering a root:shoot ratio of 0.8 (Table 9.3). In the case of holm oak trees, it was calculated using annual biomass distribution coefficients based on the production of acorns (Almoguera Millán, 2007, p. 93, 98). For crops, converters from Appendix I were used (Table 9.3), multiplied by the harvested portion (data taken from surveys conducted with livestock farmers).

At this point, we were able to calculate NPP_{act} (kg dm) per standard hectare for the different farms. The standard hectare differed from one farm to another, since it represented the percentage of each usage (crop, wooded pastureland, treeless pastureland, and cultivated grassland) obtained by means of surveys (Table 9.1) We then calculated reused biomass (kg dm). In the case of Argentine farms (No. 1–5), it was measured directly on-site (Iodice, 2013). For the other farms, the reused biomass of crops was obtained by surveying the farmers. In the case of oats, barley, and broad beans, these crops are usually used as feed for livestock, whereas in the case of wheat, sunflowers, and sorghum, the grain produced is mostly sold, and any waste or by-products are used as livestock feed on the farm itself. To calculate the

Table 9.2 Productivity and Metabolizable Energy of Pasture and Forestry Resources That Can Be Used by Ruminants

	Treeless		Wooded					
	Natural Pasture or Cultivated Grassland[a]		Pasture[a]		Acorn[b]		Twigs and Leaves[b]	
Farm	kg dm/ha	MJ ME/kg dm	kg dm/ha	MJ ME/kg dm	kg fm/ ha	MJ ME/kg fm	kg dm/ha	MJ ME/kg dm
1	7,209	9.46						
2	10,602	9.39						
3	9,899	10.01						
4	10,971	9.44						
5	5,050	8.65						
6–13	4,500	9.5						
14–36	3,986	5.9	1,821	7.8	399	7.3	840	5.82

Notes: 1–5: Direct measurement (Iodice, R., *Estudio del metabolismo social y la salud del suelo en cinco producciones familiares tamberas en transición agroecológica de la cuenca del río Luján*, Buenos Aires, Argentina, Master's thesis, Universidad Internacional de Andalucía, Baeza, Spain, 2013).

6–13: Surveys with experts who advise the farms (not published).

14–36: In the *Dehesa* system (known as Montado in Portugal) of the Iberian Peninsula, *Quercus* trees (chiefly holm oaks) are scattered around the territory. The average number of trees in Andalusia is 42 trees/ha (Costa Pérez, J.C. et al., *Dehesas de Andalucía. Caracterización ambiental*, Sevilla, Spain, Consejería de Medio Ambiente, Junta de Andalucía, 2006.). Between the trees, herbaceous pasture and bushes grow. Depending on the animal species (cows, sheep, goats, or pigs), different uses are made of the different resources.

Productivity of pasture in the *dehesa* (López Díaz, M.L. et al., Matorralización de la *dehesa*: Implicaciones en la productividad total del sistema, Actas VI Congreso Forestal Español, Ed, Sociedad Española de Ciencias Forestales, Vitoria-Gasteiz, Spain, 2013). Metabolizable Energy of *dehesa* pasture (Patón, D.J. et al., Calidad nutritiva del pastizal mediterráneo de ecosistemas de la reserva de la biosfera de Monfragüe, Actas XLV Reunión Científica de la SEEP, Sesión, Ecología y Botánica de Pastos, pp. 869–874, 2005). Productivity of holm oak, acorns (Pulido, G F. et al., *Evolución, ecología y conservación*, Junta de Extremadura, Mérida, Spain, 2007; López Díaz, M.L. et al., Matorralización de la *dehesa*: Implicaciones en la productividad total del sistema, Actas VI Congreso Forestal Español, Ed, Sociedad Española de Ciencias Forestales, Vitoria-Gasteiz, Spain, 2013. Of leaves: Calculated by means of biomass partitioning coefficient based on acorn production (Almoguera Millán, J., *Modelo Dehesa sobre las relaciones pastizal-encinar-ganado*, Trabajo Fin de Carrera, Universidad Politécnica de Madrid, Madrid, Spain, p. 93, 98, 2007). Metabolizable energy, of acorns (Fundación Española para el Desarrollo de la Nutrición Animal (FEDNA), *Tablas FEDNA de composición y valor nutritivo de alimentos*, 2010). Of leaves (Robles, A.B. et al., *Pastos, clave en la gestión de los territorios: Integrando disciplinas*, Junta de Andalucía, Sevilla, Spain, pp. 31–51, 2008).

Productivity of treeless pasture: Calculated using the algorithm proposed by Le Houerou, H.N., and Hoste, C.H., *J. Ran. Man.* 30, 181–9, 1977; developed for the Mediterranean basin, with rainfall levels of 500 mm—Robles, A.B. et al., *Pastos, clave en la gestión de los territorios: Integrando disciplinas*, Junta de Andalucía, Sevilla, Spain, pp. 31–51, 2008. Metabolizable energy (Boza, J. et al., Impacto ambiental en las explotaciones ganaderas del extensivo mediterráneo, In: F. Férnandez-Buendía et al. [eds.], *Globalización medioambiental. Perspectivas agrosanitarias y urbanas*, MAPA, Madrid, Spain, pp. 257–268, 2000; Robles, A.B. et al., *Pastos, clave en la gestión de los territorios: Integrando disciplinas*, Junta de Andalucía, Sevilla, Spain, pp. 31–51, 2008).

[a] Aboveground actual net primary productivity (NPP$_{act}$) of pasture.

[b] Production of forestry biomass (acorn, leaves, and thin twigs) that could potentially be used by ruminants.

Table 9.3 Converters Applied to Calculate NPP_{act}

	Residue:Product Ratio (kg fm:kg fm)	Root:Shoot Ratio (kg dm:kg dm)	Fruit Biomass Partitioning Coefficient	Leaf Biomass Partitioning Coefficient	Aboveground Biomass Partitioning Coefficient (Excluding Leaves and Fruit)	Root Biomass Partitioning Coefficient
Grazing pasture and cultivated grasslands		0.80				
Wheat	1.36	0.20				
Barley	1.20	0.21				
Oats	1.43	0.40				
Sunflower	2.30	0.18				
Sorghum	1.69	0.09				
Broad beans	1.56	0.60				
Alfalfa (fodder)		1.20				
Vicia sp. (fodder)		0.56				
Holm Oak		0.84[a]	7.4	17.12	24.62	50.87

Source: Appendix I (this volume).
[a] Root:shoot ratio for an adult tree.

biomass reused of pasturelands, acorns, and holm oak trees, we used the following method: (1) We calculated the metabolizable energy requirements of the herd based on the coefficients in Table 9.4. (2) We deducted from these requirements the metabolizable energy of feed bought by farmers and from their own crops (grain and/or waste/by-products) consumed by their livestock. (3) The difference in metabolizable energy was then attributed to pasturelands (and, if applicable, to acorns and holm oak leaves) according to the metabolizable energy of these food stuffs (Table 9.2). Socialized vegetable biomass includes harvested firewood and the biomass of commercialized crops (normally grains). The aboveground accumulated biomass of trees was calculated based on the total aboveground accumulated biomass of a hectare of *dehesa* pastureland (Almoguera Millán, 2007) and divided by the average number of years until replenishment (100 years in the case of holm oak trees). Finally, unharvested biomass was calculated. To obtain the aboveground unharvested biomass, we subtracted reused biomass, socialized vegetable biomass, and aboveground accumulated biomass from aboveground NPP_{act}. To obtain the belowground unharvested biomass, we subtracted the belowground accumulated biomass from the root NPP_{act}.

Finally, Table 9.5 provides the gross energy values for biomass and industrial inputs for conversion into energy. With regard to human labor, a value of 2.2 MJ/h was used. These values are based on the energy content of consumed food (Chapter 4).

Having transformed the biomass and industrial inputs into energy, we then calculated the following EROIs as specified in Chapter 2.

Proposed EROIs from an economic point of view:

1. Final EROI (FEROI) = $SB/(RuB + EI)$
2. Crop FEROI (Crop-FEROI) = SVB *from cropland*$/(RuB + EI)$
3. Forestry FEROI (For-FEROI) = SVB *from forestland*$/(RuB + EI)$
4. Livestock FEROI (Liv-FEROI) = $SAB/(RuB + EI)$
5. External final EROI (EFEROI) = SB/EI
6. Internal final EROI (IFEROI) = SB/RuB

Proposed EROIs from an agroecological point of view:

7. NPP_{act} EROI = NPP_{act}/TIC
8. Agroecological final EROI (AE-FEROI) = SB/TIC
9. Biodiversity EROI =

$$1 - \frac{AE - FEROI}{FEROI} = \frac{UhB}{TIC}$$

10. Woodening EROI

$$\text{Woodening EROI} = \frac{AB}{TIC}$$

where SB = socialized biomass; SVB = socialized vegetable biomass; SAB = socialized animal biomass; RuB = reused biomass; EI = external inputs; RcB = recycling biomass = reused biomass + unharvested biomass; UhB = unharvested biomass; TIC = total inputs consumed = $RcB + EI$; AB = accumulated biomass.

Table 9.4   Livestock Metabolizable Energy Requirement (MJ/year)

	Grazed	Kept in Barns
	MJ/year	MJ/year
Dairy Cattle		
Cow (4 L/day)	38,758	32,299
Cow (8 L/day)	48,326	40,273
Cow (10 L/day)	52,312	43,595
Cow (15 L/day)	62,276	51,898
Adult cow, not gestating or lactating	21,652	18,044
Cull heifer	19,668	16,390
Stud	32,748	27,291
Calf (<1 year)	11,909	9924
Dairy Goats		
Mothers	5,759	4,738
Adult goat, not gestating or lactating	4,291	3,301
Cull nanny goat (1–2 years)	3,545	2,727
Stud	4,828	3,714
Kid (<1 year)	2,204	1,696
Meat Sheep		
Mothers	3,463	
Adult sheep, not gestating or lactating	2,644	
Ewe (1–2 years)	2,908	
Stud	3,966	
Lambs (<1 year)	1,586	
Meat Cattle		
Mothers	27,498	22,916
Adult cow, not gestating or lactating	21,652	18,044
Cull heifer (<3 years)	19,668	16,390
Cull heifer (2–3 years)	23,276	19,623
Stud	32,478	27,066
Calf (<1 year)	11,909	9,924

Sources: Adult sheep, not gestating or lactating (Martín Bellido, M. et al., *Metodología para la determinación de la carga ganadera de pastos extensivos,* INIA, Madrid, Spain, 1986; Pulido García, F., and Escribano Sánchez, M., *Arch. Zoot.*, 43, 163, 239–49, 1994). Adult goat not gestating or lactating (González Rebollar, M. et al., *Congresos y Jornadas*, 30, 31–45, 1993; Flores Mengual, M.P., and Rodríguez Ventura, M., *Nutrición Animal*, Universidad de las Palmas de Gran Canaria, Las Palmas, Gran Canaria, 2013). Adult meat cow not gestating or lactating (García-González, R., and Marinas, A., *Pastos del Pirineo*, CSIC, Madrid, Spain, pp. 229–53, 2008). Adult dairy cow not gestating or lactating (Flores Mengual, M.P., and Rodríguez Ventura, M., *Nutrición Animal*, Universidad de las Palmas de Gran Canaria, Las Palmas, Gran Canaria, 2013). Correction factors for the rest of the herd (based on Martín Bellido, M. et al., *Metodología para la determinación de la carga ganadera de pastos extensivos*, INIA, Madrid, Spain, 1986).

Table 9.5 Gross Energy of Biomass and Industrial Inputs (MJ·unit^{-1})

Biomass	Unit	Gross Energy	Industrial Inputs	Unit	Gross Energy
Wheat (grain)	kg fm	13.84	*Fertilizers*		
Wheat (straw)	kg fm	15.23	Ammonium nitrate	kg N	62.0
Barley (grain)	kg fm	15.63	DAP	kg N	66.9
Barley (straw)	kg fm	15.18	NPK	kg N	113.9
Sorghum (grain)	kg fm	15.98	Urea	kg N	67.8
Sorghum (straw)	kg fm	15.29	Single superphosphate	kg P_2O_5	25.2
Oat (grain)	kg fm	15.18	NPK	kg P_2O_5	20.3
Oat (straw)	kg fm	15.94	Pesticides	kg a.i.	447.0
Barley, Green	kg fm	4.36	*Machinery*		
Maize (grain)	kg fm	14.44	Cultivating	ha	614.0
Triticale (grain)	kg fm	15.77	Plowing	ha	1129.0
Soybeans (grain)	kg fm	18.20	Seeding	ha	250.0
Pea, dry, fodder varieties	kg fm	18.44	Fertilizing	ha	171.0
Broad bean (grain)	kg fm	18.52	Spraying	ha	68.0
Broad bean (straw)	kg fm	15.57	Harvest	ha	935.0
Alfalfa	kg fm	4.92	*Irrigation*		
Vetch, green	kg fm	3.39	Sprinkler average 50 m lift	ha·year	34,800
Sunflower seed, with shell	kg fm	15.65			
Acorn	kg fm	18.33			
Meat[a]	kg fm	7.00			
Goats' milk	kg fm	3.07			
Cows' milk	kg fm	3.01			
Other biomass	kg dm	17.57			

Source: Appendices I and II of this book.
Notes: a.i., active ingredient; DAP, diammonium phosphate; NPK, complex NPK fertilizers.
[a] Live weight at the farm gate.

9.2.3 Statistical Analyses

Multivariate statistical methods are often used in the characterization and classification of farms, usually on the basis either of their structural characteristics (Maseda et al., 2004), of the productive characteristics (Bernués et al., 2004), or technical–economic variables. Principal component analysis (PCA) is a multivariate statistical method that transforms a set of variables Z1, Z2, …, Z*j* into a new, smaller set of uncorrelated variables Y1 (CP1), Y2 (CP2), …, Y*j* (CP*j*) that are arranged in descending order of variance (Liberato et al., 1999). Each principal component is a linear combination of all the original variables, independent of each other, and estimated, with the aim of retaining the maximum of information and total variation contained in the original data. The goal of this procedure is that a few of the first principal components contain most of the variability of the original data, thus reducing the amount of variables to be studied while retaining most of the original information. PCA can also unveil relationships not previously identified, contributing to a better interpretation of the collected data (Baker et al., 1988).

9.2.3.1 Selecting the Original Variables

We selected the original variables based on our group reflections regarding which aspects pertaining to the structure of the farms (land uses, composition of the herd, etc.), livestock feed (composition of the feed: grazing, feed produced internally, or imported, etc.), and the management of livestock and pastureland (confinement in barns, use of pastureland, etc.) might be relevant when it comes to explaining the values of the economic and agroecological EROIs. We selected the following nine quantitative variables:

Complexity of land uses: We found five types of land use on the farms studied: natural treeless pastureland, natural wooded pastureland, cultivated grassland, cropland, and woodlands. We assigned a value of 1 for each use, and for each farm the number of uses was added together and divided by 5. This indicator would range, therefore, between 0.2 (one use only) and 1 (five types of land use).

Livestock orientation index: This is the ratio of socialized biomass on the farm that comes from livestock farming (milk, calves, etc.), in terms of gross energy. Theoretically, this indicator reaches values of between 0 and 1. The value would be 0 for farms that did not produce any livestock products, and 1 for farms that did not have any agricultural or forestry production. None of the farms studied presents a value of 0, although some of them were close, given that they are all dedicated partially or exclusively to livestock production.

Forestry orientation index: This is the ratio of socialized biomass on the farm that comes from forests or woodlands, in terms of gross energy. This indicator reaches values of between 0 and 1. The value would be 0 for farms that do not harvest any timber or firewood. The value increases as forestry products gain more weighting in terms of socialized biomass.

Agricultural orientation index: This is the ratio of socialized biomass from the farm that comes from crops, in terms of gross energy. This indicator reaches values of between 0 and 1. The value would be 0 for farms that do not sow any crops or which only sow crops that are consumed exclusively by their own livestock. The value increases as agricultural products gain more weighting in terms of socialized biomass.

Intermediate values in agricultural, forestry, and livestock orientation indexes indicate diversified agroecological systems.

Stocking rate: This was measured in terms of livestock units (LUs) per hectare, bearing in mind that an LU is equivalent to 33,915 MJ of metabolizable energy from food per year (250 kg barley/month, where the metabolizable energy from a kilogram of barley is 11,305 MJ/kg) (Almoguera Millán, 2007; San Miguel Ayanz, 2009).

The values used to calculate stocking rates for each type of animal are shown in Table 9.4. The values ranged from 0.1 to 2.5 LU ha^{-1}. To standardize this indicator on a scale from 0 to 1, the stocking rate of each farm was divided by the maximum stocking rate encountered (2.5). Standardization is necessary to ensure that all variables can be compared and no errors are committed on account of differences in scale.

Grazing intensity: This is the ratio of consumption of aboveground NPP$_{act}$ of pasturelands by livestock. Theoretically, the value should range between 0 for farms with no grazing livestock and 1 for farms with high pressure from grazing, which would lead to the total consumption of pastureland aboveground biomass by livestock. Grazing intensity is of tremendous interest in terms of evaluating the sustainability of rangeland ecosystem management. Golluscio (2009) suggested that domestic herbivores may consume only a proportion of aboveground net primary productivity, known as the harvest index, to make a sustainable use of rangeland ecosystems. However, it is not easy to establish an adequate generic value for this index in natural pasturelands. Milchunas and Lauenroth (1993) offered maximum values of 60%. Le Houerou and Hoste (1977) considered 50% for pasturelands in semiarid climates. But the value also depends on the fragility of the system and the existence of other ecological restrictions (e.g., conservation of protected spaces) (García-González and Marinas, 2008). These authors proposed a range of 20%–60%. For managed grasslands, this index is easier to establish, given that human intervention facilitates the reimplementation of useful grazing species. Iodice (2013) considered a value of 2/3 for cultivated grasslands.

Dairy orientation index: This is the proportion of socialized animal biomass that corresponds to milk production, in terms of gross energy. This indicator reaches values of between 0 and 1. The value will be 0 for farms that specialize in the production of meat, and 1 for those that specialize in the production of milk. Intermediate values would indicate a lower degree of farming specialization.

Self-sufficiency for livestock feed: This expresses dependency on external feed to maintain livestock. The value would be 0 for farms that base their livestock feed completely on imported feed, and 1 for farms that produce all the food required by their livestock. Biomass produced internally and imported biomass for livestock feed are both expressed as gross energy.

Level of livestock enclosure: This expresses the ratio of time spent by the livestock in barns or enclosures. It would be 0 for farms where livestock are free to roam outside all the time, and 1 for farms where livestock is kept permanently in barns.

9.2.3.1.1 Principal Component Analysis (PCA)

Principal component analysis (PCA) was applied to reduce the number of variables needed to categorize the farms. We checked that some of the nine variables selected were correlated. Some of these variables were not normally distributed, but PCA has proven to produce a good low-dimensional projection of the data even when it is not normally distributed (McGarigal et al., 2000; Jaška et al., 2015). We performed the PCA retaining those principal components with eigenvalues greater than 1 for further analyses (McGarigal et al., 2000).

9.2.3.1.2 Cluster Analysis

Cluster analysis was used to classify groups of farms that were similar to one another but different from others. The objective was to maximize intragroup homogeneity and intergroup diversity. Partitional clustering was performed employing K-means clustering. The K-means method is a prototype-based clustering technique that attempts to identify a user-specified number of clusters (K), which are represented by their centroids (Tan et al., 2006), and tries to minimize the sum of squared distances between cluster centers and cluster members. The previous PCA results were used as the input data for the cluster analysis to avoid both a large number of variables and the existence of strong correlations among them. We performed K-means clustering with two to seven clusters and selected the four-cluster results for further analyses, as they showed the minimum within-group variance and maximum between-group variance, being significant for the three components analyzed.

9.2.3.1.3 Correlation between EROIs and Principal Components

Partial correlations between the principal components and the different EROIs were studied using Pearson's correlation coefficient.

The statistical package STATISTICA v.10 (StatSoft, 2011) was used to perform all the analyses.

9.3 RESULTS

9.3.1 Principal Component Analysis

Based on the PCA results, the respective eigenvalues, and the percentages of explained variance (Table 9.6), we selected the first three components since they all have variances (eigenvalues) greater than 1, and between them they account for

Table 9.6   Principal Components (PCs), Eigenvalues (λi), and Variance Percentage Explained by Components (Simple Variance and Accumulated Variance) of Measured Variables

	λi	Simple Variance	Accumulated Variance
PC1	3.82	0.42	0.42
PC2	2.04	0.23	0.65
PC3	1.28	0.14	0.79
PC4	0.76	0.08	0.88

79% of the variance of the original variables, offering a satisfactory percentage (Malhotra, 2008). We can interpret these selected three PCs as follows:

PC1: Structural diversification

> PC1 explains 42.2% of the variance. It presents very high positive correlation coefficients with the variables complexity of land uses and agricultural and forestry orientation indexes. This PC is negatively correlated with livestock orientation index and with dairy orientation (Table 9.7). In short, this component is greater the more complex the livestock farms are and the more they combine forestry and agricultural production, decreasing the relative weight of livestock and dairy products in the composition of socialized biomass.

PC2: Intensification of livestock production

> PC2 explains 22.6% of the variance. It is related positively to the stocking rate of livestock and the level of livestock enclosure, and negatively with self-sufficiency for livestock feed (Table 9.7). The highest scores for this PC correspond to the most intensive dairy cattle farms in Argentina, and the lowest scores to the least intensive cattle farms in Brazil.

PC3: Degree of biomass appropriation

> PC3 explains 14.2% of the total variance of the model. It is related positively with grazing intensity and the agricultural orientation index, and negatively with forestry orientation (Table 9.7). An increase in grazing intensity and a change in land uses toward agriculture implies a higher degree of biomass appropriation by society. The opposite occurs if forestry orientation is increased. The highest scores for this PC correspond to conventional Spanish farms, and the lowest to organic Spanish farms.

Table 9.7   Factor Loading of the Variables with the Four Principal Components to Explain Total Variation

	Component 1	Component 2	Component 3	Component 4
Complexity of land use	0.91	0.13	−0.24	0.11
Livestock orientation index	−0.98	−0.10	0.06	−0.01
Forestry orientation index	0.72	−0.04	−0.54	−0.40
Agricultural orientation index	0.66	0.20	0.52	0.46
Stocking rate	−0.49	0.76	0.13	−0.15
Grazing intensity	0.40	0.46	0.62	−0.40
Dairy orientation	−0.81	0.18	−0.19	0.04
Self-sufficiency for animal feed	−0.02	−0.79	0.15	0.18
Level of livestock confinement	0.01	0.73	−0.44	0.41

9.3.2 Cluster Analysis

The cluster analysis yielded the most significant results for a four-cluster solution. The groups thus obtained were compared by analysis of variance (Table 9.8). Figure 9.2 shows the distribution of four groups of farms in relation to the principal components 1, 2, and 3.

The characteristics differentiating the clusters are as follows:

> *Group 1*: Organic *dehesas* with no or very little agricultural orientation. This group of eight holdings represents 53% of the Spanish organic farms. They are *dehesa* pastureland farms, in which there is no or very little agricultural orientation (4% on average), dedicated to meat or mixed production. This group presents medium–high values in terms of complexity and structural diversification (PC1), medium–low levels of intensification (PC2), and a very low level of biomass appropriation (PC3).
>
> *Group 2*: *Dehesas* with an agricultural orientation. This group of 15 holdings represents 100% of the Spanish conventional farms and 47% of the Spanish organic farms. They are *dehesa* pastureland farms dedicated to meat production, with an important agricultural orientation (58% on an average). This group presents high values with regard to structural diversification, since they have the three components (agriculture, livestock, and forestry) (PC1), an average level of intensification (PC2), and a high degree of biomass appropriation (PC3).

Table 9.8   ANOVA Results for Cluster Analysis

	Df Between	Df Within	F	Significance
PC1-Diversification	3	32	238.83	***
PC2-Intensification	3	32	45.46	***
PC3-Appropriation	3	32	32.36	***

Notes: ANOVA, analysis of variance; Df, degrees of freedom.
*$p < .05$, **$p < .01$, ***$p < .001$.

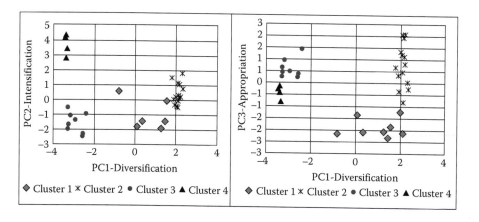

Figure 9.2   Distribution of the four clusters, in relation to the principal components 1, 2, and 3.

Group 3: Treeless natural pasturelands, with no livestock enclosures, and very high levels of self-sufficiency for livestock feed (nine cases). This group encompasses eight cases in Brazil and one case in Argentina. It includes holdings where the production orientation is meat, dairy, and mixed. This group presents low values with regard to the level of diversification (PC1) and the degree of intensity (PC2), and high levels of biomass appropriation (PC3).

Group 4: Cultivated grasslands with dairy cattle, partially kept in barns or enclosures, with an average level of feed self-sufficiency. This group encompasses four cases in Argentina. This group presents a very low level of diversification (PC1), very high production intensity (PC2), and an average level of biomass appropriation (PC3).

9.3.3 Correlation between EROIs and the Principal Components

Table 9.9 presents the correlation data between the three principal components and the EROIs. PC1, which establishes the structural diversification of each farm, has a major impact on economic and agroecological EROIs. There is a positive correlation between structural diversification and return of energy toward society, with regard to the energy consciously invested (FEROI) and to the total energy invested (AE-FEROI). This improvement in energy efficiency is not due to an increase in efficiency with regard to the external inputs of these agroecosystems (EFEROI), which is not affected (Table 9.9). It is due to the increased return with regard to internal flows of reused biomass (IFEROI). This is linked with the greater energy efficiency of vegetable production over animal production. In exclusively livestock agroecosystems, reused biomass is practically entirely consumed by livestock, whereas in more complex systems, livestock consumes only part of this reused biomass. However, it brings returns in the form of livestock products down, since a lower proportion of reused vegetable biomass is used as animal feed.

As for agroecological EROIs, it correlates positively with all of them with the exception of biodiversity EROI. This fall is related with the greater appropriation of biomass that takes place within agricultural spaces, which are present in the more complex farms included in our analysis. Clearly, woodening EROI benefits from the presence of the arboreal layer that allows biomass to be accumulated in the agroecosystem.

As for PC2, greater intensity negatively affects two agroecological EROIs and does not improve economic EROIs. The importing of external feed does not guarantee the greater availability of biomass for heterotrophic wild species (biodiversity EROI), if accompanied by a substantial increase in LUs that foster overgrazing. Similarly, it does not promote the production of vegetable biomass (NPP_{act}) within the agroecosystem, since the extra investment of external energy is channeled directly into feeding the animals, and increased livestock numbers can trigger processes of degradation in the agroecosystem.

Finally, with regard to PC3, the increased appropriation of biomass, through grazing and/or changes in land uses toward agricultural crops, decreases the availability of biomass for heterotrophic wild species (biodiversity EROI). Although it improves the availability of agricultural products for society in relation to the

Table 9.9 Correlation between the Principal Components Selected and the EROIs

	Mean	SD	r(X,Y)	R^2	t	p
PC1-Diversification	0.000000	2.42				
FEROI	0.28	0.20	0.77	0.60	7.11	***
Liv-FEROI	0.03	0.03	−0.70	0.48	−5.65	***
For-FEROI	0.12	0.14	0.62	0.39	4.65	***
Crop-FEROI	0.11	0.16	0.61	0.37	4.51	***
EFEROI	2.72	3.74	0.23	0.05	1.39	n.s.
IFEROI	0.34	0.25	0.81	0.65	7.94	***
NPP_{act} EROI	1.04	0.124	0.53	0.29	3.69	***
AE-FEROI	0.10	0.09	0.81	0.66	7.96	***
Biodiversity EROI	0.64	0.179	−0.53	0.29	−3.65	***
Woodening EROI	0.03	0.03	0.90	0.81	12.02	***
PC2-Intensification	0.000000	1.73				
FEROI	0.26	0.20	−0.03	0.00	−0.17	n.s
Liv-FEROI	0.04	0.03	0.23	0.08	1.67	n.s
For-FEROI	0.12	0.14	−0.18	0.038	−1.08	n.s
Crop-FEROI	0.11	0.16	0.06	0.00	0.37	n.s
EFEROI	2.72	3.74	−0.31	0.10	−1.91	n.s
IFEROI	0.34	0.25	0.09	0.01	0.51	n.s
NPP_{act} EROI	1.04	0.12	−0.38	0.14	−2.39	*
AE-FEROI	0.10	0.09	0.13	0.02	0.77	n.s
Biodiversity EROI	0.64	0.17	−0.48	0.23	−3.16	**
Woodening EROI	0.03	0.03	0.01	0.00	0.05	n.s
PC3-Appropriation	0.000000	1.26				
FEROI	0.26	0.20	0.08	0.00	0.44	n.s
Liv-FEROI	0.036	0.03	−0.07	0.00	−0.36	n.s
For-FEROI	0.12	0.14	−0.54	0.28	−3.66	***
Crop-FEROI	0.11	0.16	0.58	0.33	4.05	***
EFEROI	2.72	3.74	−0.32	0.10	−1.95	n.s
IFEROI	0.34	0.25	0.15	0.02	0.91	n.s
NPP_{act} EROI	1.04	0.12	0.06	0.00	0.34	n.s
AE-FEROI	0.10	0.09	0.43	0.18	2.76	**
Biodiversity EROI	0.64	0.17	−0.47	0.22	−3.07	**
Woodening EROI	0.03	0.03	−0.20	0.04	−1.16	n.s

Notes: AE-FEROI, agroecological FEROI; Crop-FEROI, crop FEROI; EFEROI, external FEROI; EROIs, energy return on investments; FEROI, final EROI; For-FEROI, forestry FEROI; IFEROI, internal FEROI; Liv-FEROI, livestock FEROI; n.s., nonsignificant; SD, standard deviation.
*$p < .05$,
**$p < .01$,
***$p < .001$.

social investment of energy (Crop-FEROI), it decreases the return in terms of forestry products (For-FEROI) and does not affect returns with regard to livestock (Liv-FEROI), or the total return (FEROI).

External FEROI is not affected by structural changes or by the management of livestock or pasturelands. This is because external inputs are composed almost entirely of feed for the livestock (Figure 9.3) and their conversion is similar regardless of the type of livestock farm.

Figure 9.4 represents the mean values of the four clusters with regard to economic and agroecological EROIs. Group 1 (organic *dehesas* with no or very little agricultural orientation) obtains the most balanced values for EROIs globally. This group presents high values with regard to economic EROIs (Figure 9.4a), always bearing in mind that they are livestock agroecosystems. These strong values are based on a high response to external inputs, with little dependence on them. This low dependence is due to the fact that stocking rates (0.39 LU·ha⁻¹) are very close to the optimum for *dehesa* pasturelands (0.33 LU·ha⁻¹) (Gaspar García et al., 2009). This allows them to achieve a high level (85%) of self-sufficiency in terms of animal feed, without needing to degrade the agroecosystem through overgrazing. This group also displays an acceptable response with regard to internal flows of reused biomass, basically grasses and pasture, acorns, and twigs from the holm oak trees, which feed the livestock. The conversion of invested energy into socialized animal biomass is relatively low (Figure 9.4b) comparative to treeless agroecosystems, but higher than Group 2. Furthermore, it presents good values for agroecological EROIs (Figure 9.4c, d). The values achieved for NPP$_{act}$ EROI and biodiversity EROI are particularly strong. These agroecosystems maintain NPP$_{act}$ EROI values greater than 1, producing more energy through photosynthesis than they receive (total input consumed) and, of this latter input, 77% is converted into unharvested biomass available for the sustenance of wild heterotrophic organisms. The high availability of unharvested biomass would help, alongside other factors (absence of biocides, complex territorial structure, diversity of plant layers), to support the important wildlife that lives in these agroecosystems.

The EROIs of Group 2 (*Dehesas* with an agricultural orientation) differ from those of Group 1 displaying a greater return on energy invested by society

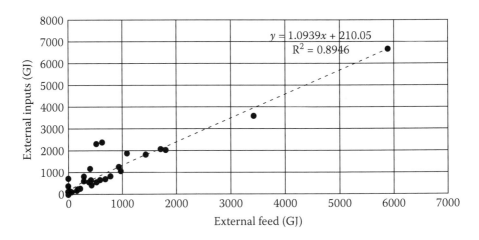

Figure 9.3 Correlation between energy invested as animal feed on the farms and the total energy of external inputs (GJ).

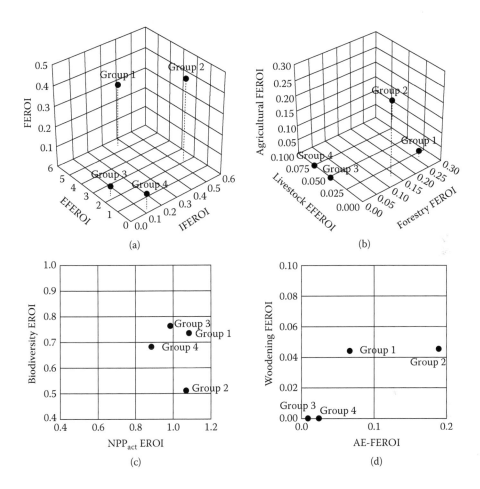

Figure 9.4 Mean values of the clusters for economic and agroecological energy return on investments (EROIs): (a) with regard to final EROI (FEROI), external final EROI (EFEROI), and internal final EROI (IFEROI); (b) with regard to crop FEROI (Crop-FEROI), forestry FEROI (For-FEROI), and livestock FEROI (Liv-FEROI); (c) with regard to actual net primary productivity (NPP$_{act}$) EROI and biodiversity EROI; and (d) with regard to AE-FEROI and woodening EROI.

(Figure 9.4a) and total energy consumed (TIC) (Figure 9.4d). This is due to the incorporation of agricultural activity that improves the return on internal flows of biomass in the form of agricultural products (Figure 9.4a,b), which is much more efficient than livestock conversion. However, the return in the form of socialized animal biomass declines (Figure 9.4b). This is due to the fact that a lower proportion of energy invested by society (EI and reused biomass) is aimed at livestock activity.

The stocking rates on these farms (0.55 LU·ha^{-1}) are 43% higher than in Group 1. This increase is based on the incorporation of feed produced internally and imported, but also on a higher grazing intensity. The percentage of aboveground NPP_{act} consumed rises from 40% in Group 1 (within the boundaries of sustainability), to 76% in Group 2. This grazing intensification not only endangers the renewal of pastureland itself (Le Houerou and Hoste, 1977; Milchunas and Laurenoth, 1993; García-González and Marinas, 2008); but it also ignores environmental restrictions (García-González and Marinas, 2008), damaging the sustenance of wild heterotrophic organisms. This is manifested in the major decline in biodiversity EROI. This increase in stocking rates and overgrazing in the *dehesa* pastureland has been fostered by subsidies awarded to livestock farmers as part of the Common Agricultural Policy (CAP) (Gaspar García et al., 2009).

Groups 3 and 4 present very low returns on energy to society (Figure 9.4a,b), but they are exclusively in the form of animal biomass (Figure 9.4b). Therefore, whereas the FEROI for these groups is divided by 4.0 and 6.2 with regard to Group 1, respectively, the Liv-FEROI is multiplied by 2.4 and 3.7. As we have seen in Chapter 7, the loss of functionality and consequent specialization of agroecosystems often entails the loss of energy efficiency. This process occurs when uses of plant biomass with higher gross energy are sacrificed, which is the case for firewood, for example. In the case of Group 4, efficiency is particularly low with regard to external inputs (EFEROI). In this group, for every GJ invested, society only obtains a quarter in return. This is due to the fact that these farms have very high stocking rates (1.73 LU ha^{-1}), without overgrazing, by outsourcing 31% of their livestock feed. Therefore, the global inefficiency of livestock conversion is passed onto the EI.

In terms of agroecological EROIs, these groups present values of less than 1 for NPP_{act} EROI, which implies that in spite of photosynthesis, these systems generate less energy than they receive. These agroecosystems have become, therefore, net consumers of energy, an unprecedented phenomenon until recently in agrarian history. Processes of degradation could be related with the high levels of biomass appropriation in Group 3, and the high stocking rates of Group 4. The absence of a hedgerow/arboreal layer also prevents energy from accumulating in the form of plant biomass (woodening EROI). However, both groups achieve high values for biodiversity EROI. Approximately 70% of the total inputs consumed remains available for wild heterotrophic organisms. In Group 3, this is due above all to the fact that TIC are very low. In the case of Group 4, it is made possible by the fact that around a third of livestock feed is brought in from off the farm, which prevents overgrazing (only 43% of the aboveground NPP_{act} of pastureland is consumed).

9.4 DISCUSSION

The contributions of this research could be discussed in terms of methodology and findings. As for the findings, it is important to note that this is an exploratory study, with a very limited sample of farms analyzed. Therefore, the results should be considered preliminary. However, we believe that they are useful and of interest since they shed light on some highly controversial discussion points. Reports about the growth of the human population and demand for livestock products in recent decades, combined with the significant contribution made by livestock to increasing greenhouse gases (GHGs), have placed this issue at the center of the debate. It is crucial to define the role livestock should play in the territory and in our diet, as well as the farming model that generates greater advantages for society (not only in terms of the greatest livestock product but also providing greater resilience to climate change) and less damage (deforestation, loss of biodiversity, greenhouse gas emissions, etc.).

9.4.1 Methodological Aspects

Studies of energy applied to livestock are usually based on the relationship between the energy output of the system in the form of livestock products and the energy inputs society invests to achieve this, either total energy (similar to our Liv-FEROI) or fossil energy (Pimentel, 2006; Veermäe et al., 2012). This black box approach has already been called into question (see Chapter 2) and we shall not spend any further time on it here. We only point out that making decisions with such precarious and partial information seriously damages complex livestock agroecosystems. In industrial systems, which operate like factories, the input and output flows of the system are basically the only ones that exist. But in complex agroecosystems, when such an approach is applied, it ignores the most powerful flows of energy that operate in the agroecosystem. As we saw in Chapter 2, the excluded flows are those responsible for sustaining the fund elements of the agroecosystems and the ecosystemic services. Making decisions that affect the design and/or management of these agroecosystems, without knowing how these would be affected, could quickly lead us to greater levels of unsustainability. It also prevents us from assessing other output flows that society also requires, such as agricultural and forestry.

In our case study, this becomes clearly manifest. Looking at Liv-FEROI, Group 4 followed by Group 3 would be the most efficient. But a more systemic vision allows us to see that they present a negative return for NPP_{act} (NPP_{act} EROI), indicative of processes of degradation. They also fail to develop structural diversity, since they do not integrate either agricultural or forestry spaces (crop and For-FEROIs, woodening EROI). We have seen in our case study the major impact of the *structural diversification (PC1)* of livestock agroecosystems on EROIs. Hence, the elimination of this diversity negatively affects the majority of them. Furthermore, greater structural diversity within livestock systems is often suggested as a solution to promote redundancy within such systems and therefore increase their adaptive capacity and reduce their vulnerability against climate change and variability (Gil et al., 2015; Martin and Magne, 2015; Murgueitio et al., 2015).

In short, a conventional approach to energy might possibly recommend moving toward simpler agroecosystems that are more specialized in animal production. But an agroecological approach takes us in precisely the opposite direction.

9.4.2 Results

The results show that some models of livestock farming could reconcile interests that in the current debate appear to be antagonistic. This is the case of the binomial livestock farming/deforestation, or livestock farming/loss of biodiversity (Steinfeld et al., 2006; Erb et al., 2016).

One interesting result of this preliminary study is that silvopastoral systems (Group 1) are energetically more adequate than agrosilvopastoral systems (Group 2). The introduction of crops into *dehesa* pastureland has reduced the majority of the EROIs, including Liv-FEROI. The reason is that feeding livestock with grazing pasture (herbaceous or woody) constitutes a lower entropy loop (see Chapter 2) than feeding them with crops (grains and by-products). This latter loop incorporates higher amounts of energy. This is in line with the findings of Giambalvo et al. (2009) for other Mediterranean livestock agroecosystems, where replacing pasture and pastureland with crops grown for cattle led to a decline in the energy indicators of these farms. Furthermore, it would also reinforce the strategy of feeding livestock with resources that do not compete with the human food supply (Schader et al., 2015). In our study, this silvopastoral model (Group 1) decreases Liv-FEROI comparative to other more intensive farming methods (e.g., Group 4) and would be along the lines of decreasing the animal component of the average human diet, as proposed by various authors (Pimentel and Pimentel, 2003; Erb et al., 2016), to minimize the environmental impact of livestock farming, but without completely relinquishing these food products, which offer high nutritional value.

These results would call into question certain models of livestock intensification as a means of minimizing environmental problems. Cardoso et al. (2016) investigated the impact of increasing animal productivity using fertilizers, forage legumes, supplements, and concentrates, on the emissions of greenhouse gases (GHGs) in five scenarios for beef production in Brazil. They concluded, along the lines of the land sparing hypothesis, that the great advantage of intensification was not directly associated with the emissions of enteric CH_4, N_2O emissions from excreta or fossil energy CO_2 from inputs production and transport, but in the reduction in area required to produce the same quantity of product. This would avoid the need to expand farming boundaries that would increase deforestation (Cardoso et al., 2016). It is not our intention here to discuss the land sparing hypothesis (see Latawiec et al. [2014], in reference to livestock systems). However, we wish to suggest that there could be various models of intensification. Some are based principally on the incorporation of more external inputs (fertilizers, supplements, concentrates, etc.), and others are based on an intensification of low-entropy internal loops (improving pastureland with leguminous plants, introducing woodlands and forests for fodder, producing firewood, and feeding with crop waste and by-products, etc.). This thesis is defended by other authors (Calle et al., 2013; Gama et al., 2014; Murgueitio et al., 2015) based on livestock systems in Latin America and elsewhere (Mosquera-Losada et al., 2004), allowing interests that until now have seemed irreconcilable: increasing livestock productivity per unit of land, sharing territory between trees and livestock, sustaining greater biodiversity values, and helping to mitigate greenhouse gas emissions. This second path appears to be more promising

according to our preliminary results. We believe that the application of the methodological proposal developed in this book to both models of livestock farming intensification could shed led on their sustainability in terms of energy. The need for greater
information to make adequate decisions had been highlighted by Erb et al. (2016) who
point to the "lack of critical knowledge on, for example, the role of management, different livestock species and biomass flows and their geographic location but also on the
interrelation between grazing and ecosystem processes."

Finally, we should point out that organic certification seems to encourage farms
to move toward greater levels of sustainability in energy terms. However, encourage-
ment is not the same as obligation. Almost half the organic farms studied remain in
Group 2, together with conventional farms. It would be advisable for organic certification to introduce energy criteria in its certification processes. It would also be
advisable for agricultural policies to introduce adequate criteria so that all livestock
farms can move toward greater levels of sustainability (Gaspar García et al., 2009;
Gil et al., 2015).

9.5 CONCLUSIONS

By applying PCA, we have been able to identify three principal components
(PC1: Structural diversification of the agroecosystem, PC2: Degree of livestock farming intensification, and PC3: Degree of biomass appropriation) that contain most of
the variability of the original data. *Structural diversification* has a major impact on
economic and agroecological EROIs. It offers a positive correlation with the energy
return to society, with regard to energy consciously invested (FEROI) and total energy
invested (AE-FEROI). It also correlates positively with all the agroecological EROIs,
with the exception of biodiversity EROI. This decline is related with the greater
appropriation of biomass that takes place in agricultural spaces. The *degree of livestock farming intensification* negatively affected biodiversity EROI and NPP_{act} EROI,
and did not improve the economic EROIs. Finally, the *degree of biomass appropriation* reduced the availability of biomass for wild heterotrophic species (biodiversity
EROI) and only improved one of the economic EROIs (Crop-FEROI).

Cluster analysis allowed us to classify the farms into four homogeneous segments. Of these, Group 1 (Organic *dehesa* pastureland farms with no or very little
agricultural orientation) obtained the most balanced values for the EROIs globally.

The results obtained support the proposal of intensifying livestock farming
through the enhancement of low-entropy internal loops in the agroecosystem, rather
than incorporating external flows of energy. This opinion, which reconciles different
uses within a single territory, seems capable of delivering greater energy efficiency,
sustaining fund elements (biodiversity and the arboreal layer), and supplying animal
proteins for a diet with a low proportion of animal products (but not lacking them).
However, these results need to be confirmed in subsequent research.

The results obtained confirm the usefulness of agroecological EROIs for a more
profound comprehension and evaluation of the energy functions in agroecosystems
from the point of view of sustainability.

CHAPTER 10

A Few Useful Conclusions for the Design of Sustainable Agroecosystems

CONTENTS

10.1 INTRODUCTION

Chapters 1 through 4 looked at the technical and methodological aspects of this innovative proposal aimed at evaluating energy efficiency in agrarian systems from an agroecological perspective. In Chapters 5 through 9, we applied this proposal to seven case studies. Three of them are diachronic studies conducted on different scales: national (Spain), municipal (Santa Fe), and crop (café), showing the changes that have occurred with the transformation from traditional to industrialized agroecosystems. The remaining four case studies are synchronic, and three of them apply to the crop scale: avocado (AVO), dry-farmed olives (OLIdry), and irrigated olives (OLIirri), while the last one was applied to a large number of livestock holdings (LIV). The aim was to illustrate the transition from industrialized agroecosystems to modern certified organic agroecosystems. In Chapters 5 through 9, we attempted to analyze in depth the use of energy and its efficiency in different situations and farming businesses, showing the robustness of the proposed methodology and the consistency of the results obtained. In this chapter, we draw more general conclusions that can provide a foundation for the sustainable design of agroecosystems based on energy analysis and the use of the proposed indicators. Indicators not only take into account the input and output flows of energy, but also the flows that circulate within. This chapter aims, first, to identify the major common trends highlighted by the different case studies; second, to explain the existence of common patterns or divergence; and third, to draw conclusions that can be used to design more sustainable agroecosystems from an agroecological perspective.

10.2 EVOLUTION OF EROIs FROM AN ECONOMIC POINT OF VIEW

Figures 10.1 through 10.4 show the results of the case studies that illustrate the transition from traditional to industrialized farming, and then from industrialized to certified organic farming. We have omitted intermediate points in the case of the diachronic studies, selecting only those from the start and end of the transition from traditional to industrialized farming to make the figures clearer. Similarly, we have excluded two cases of livestock farming (Groups 3 and 4; see Chapter 9), for which there is no "organic" equivalent.

Figure 10.1 shows the return on the energy investment made by society (final EROI [FEROI]). In all cases, the transition from traditional to industrialized farming has led to a decline in FEROI. In other words, society has been progressively receiving less energy with regard to the amounts invested during the process of agrarian modernization. This decline is due, principally, to the loss of efficiency with regard to *external inputs* (*EI*) (external final EROI [EFEROI]) (Figure 10.2), but not solely

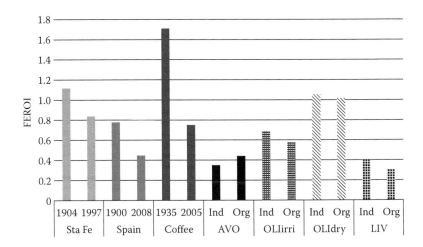

Figure 10.1 Evolution in energy returns to society in the transition from traditional to industrial farming, and from industrial to certified organic farming. *Notes*: Final energy return on investment (FEROI) = socialized biomass/(external inputs + reused biomass). The cases of Liv-org and Liv-ind correspond, respectively, to Groups 1 and 2 in Chapter 9. Group 2 encompassed all conventional farms and a few more intensively managed organic ones. Group 1 was made up of organic farms only. (Author data.)

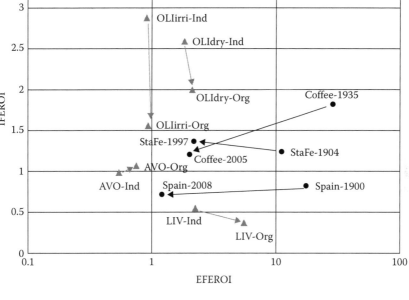

Figure 10.2 Evolution of the energy return toward society on external inputs (external final energy return on investment [EFEROI]) and reused biomass (internal final energy return on investment [IFEROI]) during the transition from traditional to industrial farming, and from industrial to certified organic farming. *Notes*: EFEROI = socialized biomass/external inputs; IFEROI = socialized biomass/reused biomass. (●) Transition from traditional to industrialized management. (▲) Transition from industrialized to certified organic. The cases of Liv-org and Liv-ind correspond, respectively, to Groups 1 and 2 from Chapter 9. Group 2 encompassed all conventional farms and a few more intensively managed organic ones. Group 1 was made up of organic farms only. The arrow point indicates the direction of the transition process. (Author data.)

so. In two of the three cases, there has also been a decline in efficiency with regard to *reused biomass (RuB)* (internal final EROI [IFEROI]) (Figure 10.2).

As for the transition from industrialized to organic farming, in three of the cases studied there has been a decline in FEROI, and this energy return increased in just one case. The three cases in which it has fallen have seen a major decline in returns related to *reused biomass (RuB)*, not compensated by the increase in efficiency in terms of *external inputs (EI)* (Figure 10.2). Only in the case of avocado growing, FEROI has been increased when the switch is made over to organic production, resulting from increased efficiency with regard to external and internal inputs.

In short, the transition to organic models increases efficiency with regard to *EI* (EFEROI) and, consequently, the transition toward industrialized models decreases this efficiency (EFEROI) (Figure 10.2).

However, the evolution of efficiency with regard to *RuB* (IFEROI) does not show such a clear tendency. To analyze these results, it is important to separate cases in which a substantial change has been seen in the composition of *socialized biomass* (*SB*), which could alter the results owing to a change in the numerator, from cases in which the nature of *SB* does not vary. This second group would encompass the cases of olive and avocado production, where the composition of the numerator is exactly the same. Therefore, the difference in response with regard to internal energy flows must be due to the condition and health of the agroecosystem itself. As we saw in Chapter 8, a notable difference can be found in the baseline situation of these agroecosystems prior to their conversion. On the one hand, conventional avocado production also integrates high quantities of organic fertilizer. On the other hand, the amount of recycled biomass is significantly higher in absolute terms, since rainfall is not such a limiting factor in this case. Consequently, these agroecosystems can potentially be in better condition to respond to the change in fertilization strategy (from chemical to organic) owing to a greater quality and more biologically active soil than that is present in conventional olive groves. Put another way, the greater the level of soil degradation, the greater investment of energy is required to achieve a sufficient improvement to increase returns in the form of *SB*. The distance between the pathways of degradation and restoration is known as the hysteresis of land rehabilitation (Tittonell et al., 2012).

These results do not support the hypothesis put forward by Tello et al. (2015) according to which potential improvements in FEROI are greater when the combination of EFEROI and IFEROI is unbalanced—that is, when the *EI:RuB* ratio is far from one, and the most abundant is partially replaced by the scarcest. This synergy between *EI* and *RuB* might have occurred at the start of modernization. During this period, the fund elements were in good health thanks to the accumulated and constant investment of energy flows (biomass) made previously to maintain them. Under these conditions, it was apparently possible to increase portion of *SB*, without observing negative consequences resulting from the decrease in biomass investment (*RuB*). As we saw in Chapter 5, the case of Santa Fe would exemplify this situation in the early twentieth century. In the present day, this synergy does not occur in industrialized countries, where a decline in *SB* has been documented when *EIs* are partially substituted by *RuB* in organic farming. Ponisio et al. (2015) reported an average fall in yield of 19.2% in organic farming, based on a meta-analysis of 115 studies. Organic farms (and therefore the farmers who adopt this approach) have to pay the cost of restoring the fund elements of agroecosystems, degraded by industrialized farming. This is a paradigmatic case of spatial and temporal circulation of energy through the internal loops of the agroecosystem, as discussed by Ho and Ulanowicz (2005), which turned traditional agrarian systems into low-entropy systems. Those circuits were destroyed by industrialization, and it is now necessary to invest energy in rebuilding them. It will take even longer before they begin to function again and enable intertemporal energy exchanges within the same agroecosystem. It is hard for organic farmers to take on this task individually without governmental support and the implementation of public policies to foster ecological restoration (González de Molina, 2013). Guzmán et al. (2011) proposed specific ways to translate the land cost of sustainability into monetary values that organic farming is currently bearing.

In the four remaining cases, changes in the composition of *SB* would justify the lower efficiency with regard to flows of *RuB*. In three of them, lower internal efficiency (IFEROI) is directly linked to the greater weighting of livestock farming in these agroecosystems. In Spain, livestock products went from representing 2.3% of *SB* in 1900 to 17.3% in 2008. In the case of the livestock systems studied, in the group Liv-ind, livestock farming has a weighting of 4.1%, which increases to 10.2% in the group Liv-org. In the case of Santa Fe, livestock farming represented 7.1% of *SB* in 1904 and just 4.0% in 1997. Given that livestock are inefficient energy converters, any increase in livestock weighting can be expected to be accompanied by a decline in IFEROI (which happens in the case of Spain and in the case of livestock farming) and any increase in agricultural weighting should be accompanied by an increase in IFEROI (which is true for the case of Santa Fe). The last case we need to analyze is the decline in IFEROI observed in the industrialization of coffee plantations. Here, livestock farming does not play a significant role. However, specialization in plant products with little calorific value (coffee) justifies the drop in IFEROI (see Chapter 8).

In conclusion, efficiency in terms of *EI* (EFEROI) is always higher in the organic agroecosystems studied, but efficiency in the conversion of *RuB* into *SB* (IFEROI) is influenced by the health and condition of the agroecosystem's fund elements, and by the composition of *SB*. Hence, increasing livestock production and relative specialization in plant products with little calorific value would also particularly lower this indicator. Finally, the evolution observed in terms of FEROI is the result of both components.

10.3 EVOLUTION OF EROIs FROM AN AGROECOLOGICAL POINT OF VIEW

Figure 10.3 shows the evolution of two agroecological EROIs (biodiversity EROI and actual net primary productivity [NPP$_{act}$] EROI) during the transition from traditional to industrial farming, and from industrial to certified organic farming.

10.3.1 Biodiversity EROI

In all cases, the energy return in the form of unharvested biomass (*UhB*) is greater in organic agroecosystems, both traditional and modern, freeing up greater proportions of phytomass for heterotrophic wildlife.

As we saw in Chapter 2, the availability of phytomass, together with the absence of biocides and the presence of a diversified land matrix, are pillars that support biodiversity in agroecosystems. These three conditions are found in traditional farming, and at least two of them (greater return in the form of *UhB* and less pressure from biocides) are also found in certified organic farming.

In the model of industrialized farming, society reduces *UhB* through various strategies. First, modern varieties have been selected to increase the harvest index (HI). In the cases of Santa Fe and Spain (Chapters 5 and 6), different converters (HI and

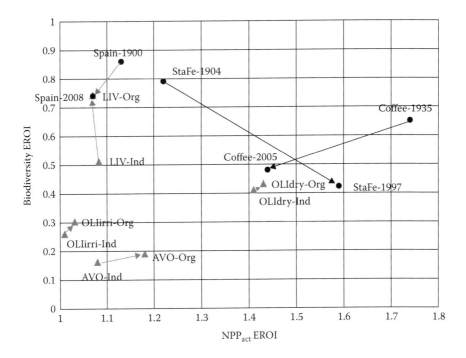

Figure 10.3 Evolution of biodiversity EROI and actual net primary productivity (NPP_{act}) EROI during the transition from traditional to industrial farming and from industrial to certified organic farming. *Notes*: NPP_{act} EROI = NPP_{act}/TIC; biodiversity EROI = unharvested biomass/TIC, where TIC = external inputs + reused biomass + unharvested biomass. (●) Transition from traditional to industrialized management. (▲) Transition from industrialized to certified organic. The cases of Liv-org and Liv-ind correspond, respectively, to Groups 1 and 2 from Chapter 9. Group 2 encompassed all conventional farms and a few more intensively managed organic ones. Group 1 was made up of organic farms only. The arrow point indicates the direction of the transition process. (Author data.)

root:shoot ratio) were applied to traditional and modern cereal varieties (Appendix I), which have led to a lower proportion of *UhB* in these crops in the present day, after deducting hay consumed by livestock. Second, chemical inputs have broken the equilibrium in land uses required by traditional farming, in favor of expanding cultivated areas, which allow for greater human appropriation of biomass. This phenomenon has affected four of the case studies. In the cases of Santa Fe and Spain (Chapters 5 and 6), even livestock currently feed mainly on crops, following a major decline in silvopastoral uses in these lands. In the case of coffee growing (Chapter 7), the use of chemical fertilizers has enabled the internal equilibrium between leguminous shade trees and crops to be broken, in favor of the latter (coffee, nonleguminous timber-yielding species), decreasing the percentage of NPP_{act} that is returned in the form of *UhB*. In the case of livestock farming, the Group Liv-Ind has intensified its holdings, increasing the dedication to crops as opposed to silvopastoral uses. This has increased the appropriation of biomass by society, in contrast to the Group Liv-Org. Finally, industrialized farming uses herbicides to prevent the growth of phytomass that farmers are not interested in appropriating. This consciously eliminates *UhB* that could feed other trophic chains in the agroecosystems. This phenomenon affects all the case studies examined here.

10.3.2 NPP$_{act}$ EROI

NPP$_{act}$ EROI is also greater in traditional or modern organic agroecosystems, with the exception of two cases. As we saw in Chapter 2, NPP$_{act}$ EROI explains the real productive capacity of the agroecosystem, whatever the origin of the energy it receives (solar for biomass or fossil for an important proportion of the *EI*). We use the term "real productivity" because it considers all of the vegetable biomass produced, not just that which is socialized, and because it is independent of the transformation of biomass through livestock farming and other factors that, as we have seen, can influence FEROI. The degradation processes affecting natural resources, such as soil salinization or erosion, genetic erosion, and so on, must be compensated by the incorporation of increasing amounts of energy to palliate the loss of productive capacity in the agroecosystems. Falling NPP$_{act}$ EROI values in an agroecosystem over time indicates degradation of productive capacity. The cases analyzed show, therefore, that this degradation has taken place with the transition to industrialized farming. However, the case of Santa Fe would seem to be a notable exception that belies that trend. As we saw in Chapter 5, however, it is not actually an exception. The growth of NPP$_{act}$ in the case of Santa Fe with industrialized farming has been possible not only through the increase in *EI*, but also through the continued increase in water consumption. Water is a peculiar input in terms of energy, because its gross energy content is "0" and, therefore, it does not have a direct repercussion on *EI*. It only has an indirect impact, as *EI* encompasses the energy costs of the impulsion and infrastructure required for irrigation. However, in semiarid climates, the availability of water is essential to produce biomass. Therefore, the conversion from dry-farmed to irrigated lands could be masking the deterioration in the agroecosystem in 1997. To demonstrate this hypothesis, in Chapter 5, we divided the Santa Fe agroecosystem according to when the modernization of different land uses took place. The results showed a decline in NPP$_{act}$ in all spaces, except in the last one, which had been recently modernized. This space, dedicated to a crop that produces large amounts of biomass (black poplar), was compensating in 1997 for the decline in NPP$_{act}$ EROI in the rest of the territory. In the case of livestock farming, the slight increase in this EROI in the Group Liv-Ind is also due to this same cause: the introduction of spaces cultivated with EI and partially irrigated.

In short, we could conclude that the industrialization of farming has led to the deterioration of agroecosystems, and that conversion to organic farming could help them to recover. We could also conclude that these processes can be detected by NPP$_{act}$ EROI. However, certain factors, such as the conversion from dry-farmed to irrigated lands in semiarid areas, could temporarily mask this deterioration. Under these circumstances, it might be necessary to apply complementary tools to evaluate the degradation of the fund elements.

10.3.3 Agroecological Final EROI

Figure 10.4 shows that agroecological final EROI (AE-FEROI) has increased in the modernization of traditional farming in two of the three cases, and has declined following conversion to organic farming in three of the four cases analyzed. This EROI reflects better than any other ultimate objective of industrialized farming: it does not seek to increase the total productivity of agroecosystems, but rather just the proportion of NPP_{act} intended for human consumption.

In Section 10.2, we saw that industrialized farming had not improved the efficiency of *EI* conversion, and, only occasionally, the conversion of *RuB* into *SB* (Figure 10.2). However, it has improved efficiency with regard to *UhB*. This improvement was due to the fact that industrialized farming has significantly reduced the unharvested part of NPP_{act} (Section 10.3.1). The improvement in efficiency with regard to *UhB* is shifted onto the whole AE-FEROI. Only in the case of the two tropical crops (avocado and coffee)—although *UhB* decreases in industrialized comparative

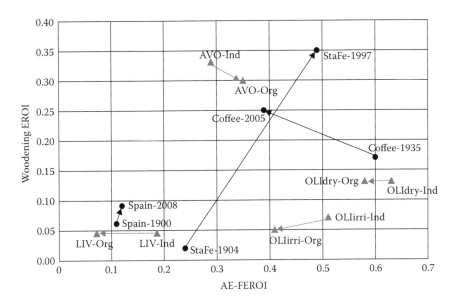

Figure 10.4 Evolution of agroecological FEROI (AE-FEROI) and woodening EROI during the transition from traditional to industrial farming, and from industrial to certified organic farming. *Notes*: AE-FEROI = socialized biomass/total inputs consumed (TIC); woodening EROI = accumulated biomass/TIC, where TIC = external inputs + reused biomass + unharvested biomass. (●) Transition from traditional to industrialized management. (▲) Transition from industrialized to certified organic. The cases of Liv-org and Liv-ind correspond, respectively, to Groups 1 and 2 from Chapter 9. Group 2 encompassed all conventional farms and a few more intensively managed organic ones. Group 1 was made up of organic farms only. The arrow point indicates the direction of the transition process. (Author data.)

to organic farming—it does not manage to compensate for the losses in efficiency recorded with regard to *RuB* and *EI* (Figure 10.2).

10.3.4 Woodening EROI

Figure 10.4 shows the evolution of woodening EROI during the transition from traditional to industrial farming, and from industrial to certified organic farming. This indicator increased during the modernization process in the three cases studied, and declined (two cases) or remained the same (two cases) during the transition from industrialized to certified organic production. Therefore, the process of industrialization fostered the increase in accumulated biomass (*AB*) with regard to total inputs consumed (*TIC*).

This indicator can increase for several reasons: (1) changes in usage from agricultural or pasture land to forestry uses; (2) the lower intensity of usage in wooded spaces (e.g., lower extraction of firewood, timber, or grazing fodder); (3) the replacement of herbaceous with woody crops; (4) transformation from monocropping to polycropping, adding woody species (e.g., "shadeless coffee or cocoa" to "shade-grown coffee or cocoa"; from herbaceous crops to integrated agroforestry systems; from treeless pastureland to wooded pastureland, etc.); (5) the shortening of the useful lifespan of woody species, accelerating their rate of growth; and (6) decrease in the denominator energy invested (*TIC*) without an alteration in the numerator (*AB*).

Only if causes 1 and 2 occur simultaneously in the growth of this EROI, there could be the freeing up of land predicted by the *land sparing* hypothesis (Phalan et al., 2011) as a consequence of the transition to industrialized farming. This situation could generate advantages for the sustenance of biodiversity, albeit limited by the relative decrease in *UhB* available for wildlife during the transition to industrialized farming. This situation is only observed in the case of Spain. Causes 3 and 4 would be linked with environmental services associated with agroforestry and/or silvopastoral systems. Cause 3 justifies the increase in woodening EROI in the case of Santa Fe, since black poplar trees have become just another crop. Finally, causes 5 and 6 would have environmental benefits in terms of mitigation of greenhouse gas emissions. This environmental service could extend to all cases, although it would have to be verified in each of them. Cause 5 explains the increase in woodening EROI in the case of the coffee plantations in Costa Rica, where the lifespan of the plantations has been shortened with the industrialization of the agroecosystem. In the case of avocado and irrigated olive groves, on the other hand, it has increased on account of the lower *TIC* in industrialized farming (cause 6).

The growth of woodening EROI in the case of Spain is particularly interesting. Accumulated biomass (*AB*) increased from 229,401 TJ in 1900 to 466,480 TJ in 2008. Of this increase (237,079 TJ), approximately 10% corresponded to the extension of woody crops, principally olive trees (23,225 TJ). A further 30% was due to the decline in the extraction of *SB*. In other words, it is not a collateral effect of industrialized farming, but principally a consequence of replacing firewood with fossil fuels. The rest (60%) of the growth observed in *AB* corresponds strictly to the growth of the wooded land area freed up from agricultural uses. However, this growth is due to the abandonment of spaces (pastureland and dryland) dedicated to animal feed, which have largely been replaced by commercial feed imported chiefly from Latin America (Soto et al., 2016). In all likelihood, this partial outsourcing of the land costs of Spanish livestock farming to third countries has led to major deforestation in these latter countries, a phenomenon that is not covered here. In fact, scientists using NASA satellite data have found that the size of the clearings used for crops has averaged twice the size of clearings used for pasture in the Amazon (Bettwy, 2006). If we discount biomass not accumulated in other agroecosystems, we would probably be in negative figures, but this issue would have to be examined in subsequent research.

In summary, by applying the proposed methodology, we have been able to improve our understanding, from an energy perspective, of the processes of transition from traditional to industrialized farming, and from industrialized to certified organic farming, on several scales: crop, farm, municipal, and national. We have also been able to identify the processes in which society has focused its energy investments: the growth of agriculture, the growth of livestock farming, forestation, the sustenance of wild biodiversity, and/or the sustenance of human society. This makes it a particularly useful tool when analyzing complex agroecosystems. Finally, it has allowed us to explore whether the foundations of agroecosystem renewability have been put at risk.

10.4 APPLICATION TO THE DESIGN OF SUSTAINABLE AGROECOSYSTEMS

The challenges faced by farming are tremendous in the twenty-first century. We expect it to feed over 9000 million inhabitants, the predicted global population for the year 2050, which would entail a huge increase in *socialized biomass*. It also has to become more sustainable and contribute to reduce greenhouse gas emissions and promote carbon sequestration, which in energy terms involves reducing dependence on *external inputs* based on fossil fuels and, in parallel, increasing investments of internal biomass flows to restore the fund elements of agroecosystems. And all this within a context of climate change that is increasing uncertainty about the stability of harvests and the production of biomass within our agroecosystems. All of the aforementioned puts us in a scenario of increasing competition between alternative uses of biomass and demands so that we design agroecosystems that maximize not only energy efficiency in obtaining SB (FEROI, AE-FEROI), but also the energy efficiency of total productivity (NPP$_{act}$ EROI), and fund elements (biodiversity EROI and woodening EROI).

The results obtained by applying our theoretical and methodological proposal to diachronic and synchronic case studies, at different scales (crop, farm, municipal, and national) allow us to draw a few conclusions that are useful for the design of sustainable agroecosystems:

1. The magnitude reached by *SB* is strongly influenced by its composition. Agroecosystem specialization in products of little calorific value (Chapter 7) and above all specialization in livestock products (increasing livestock farming) greatly

depress the returns of energy flowing back to society (Chapters 6 and 9). In Chapter 9, we tackled the need to decrease the livestock component of the average human diet, and basing livestock feed on resources (pasture, crop by-products, etc.) that do not compete with human food. In contrast, agroecosystems that diversify the composition of *SB*, including products that offer greater energy content such as firewood, substantially increase energy returns for society. For example, using this strategy, the FEROI of a Mediterranean livestock system (*dehesa* pastureland) is much higher than in other livestock agroecosystems (see Chapter 9), putting it on a par with the FEROI obtained from a tropical agroecosystem that specializes in a relatively high calorie vegetable product, the avocado (see Figure 10.1: LIV versus AVO).

2. Given that traditional agroecosystems displayed a better energy performance than their industrialized counterparts in the majority of the EROIs examined, we must learn from them to design sustainable agroecosystems today. One of the keys is that they base their functioning on internal loops of biomass with practically no dependence on the external incorporation of energy, with high-entropic costs. The theoretical foundations that justify these results were set out in Chapters 1 and 2. Therefore, reducing investments in *EI* would allow us to improve the energy performance of agroecosystems.

However, to strengthen the internal functioning of agroecosystems by fostering internal flows of energy, we must invest in rebuilding internal loops as efficiently as possible, using low-entropy loops. For example, replacing external fossil energy dedicated to mechanization with animal labor cannot be implemented on a large scale, since this is a high-entropy internal loop. In other words, this internal loop is mediated by a relatively inefficient converter that needs to consume large quantities of biomass (and, therefore, land) per unit of labor provided. However, the fuel used to power farm machinery could be replaced with agrofuel, reducing the investment of biomass to obtain the same service (Guzmán et al., 2011), or with another type of solar-based renewable energy (e.g., photovoltaic), which has no biomass cost whatsoever.

The case studies have illustrated several low-entropy internal loops. Of particular note is the use of leguminous plants as the basis for nitrogenous fertilization in crop rotations and polycrops. Chapter 7 showed how the traditional polycropping strategy of coffee/leguminous trees was the cornerstone of the superb energy performance of this agroecosystem in the year 1935 (see Figures 10.1 through 10.4). Polycrops are an outstanding strategy for the reducing land cost (land equivalent ratio [LER]) of biomass production owing to the synergies produced between the crops that they integrate. In many of them, leguminous crops are an essential component. In traditional Mediterranean farming (Chapters 5 and 6), rotation of leguminous crops was also key to optimizing EROIs in these agroecosystems.

In the cases of Santa Fe and Spain (Chapters 5 and 6), the mobilization of nutrients through livestock from spaces with a low intensity of usage (silvopastoral areas) to crop areas was one of the keys to the energy functioning of the agroecosystem. However, due to the fact that this loop has a high cost in terms of biomass (and, therefore, a high land cost as well), it would be difficult to implement today, on account of the growing scarcity of farmland and increasing pressure from potential uses (food, livestock, and energy). Other proposals (e.g., composting of agroindustrial and urban waste) must be examined to close as far as possible the cycle of nutrients, particularly phosphorus, given the increasing concern about the scarcity of this nutrient (Cordell et al., 2009; Zhu et al., 2016) and the broad consensus about the need to develop recycling strategies to avoid losses (Koppelaar et al., 2013).

In the case of olive groves (Chapter 8), we have seen that groundcover is a key element since it generates a low-entropy internal loop capable of improving the fund elements of the agroecosystem.

Finally, in Chapter 9, we concluded that feeding livestock with pasture fodder (herbaceous or woody) would provide a lower entropy internal loop than feeding them with cultivated crops.

Therefore, by applying the proposed methodology to different case studies, we have been able to detect the design and management of agroecosystems that could be the key to achieving greater sustainability in farming from the perspective of energy.

3. Given that biomass constituted the main energy basis of society, traditional farming could never make major investments in accumulated biomass (woodening EROI). In this respect, the necessary replacement of *EI* with internal flows of biomass should be achieved through the development of integrated crop–livestock–forestry systems to avoid deforestation.

4. The application of the proposed EROIs shows that there is still a great deal of room to improve organic management from an agroecological perspective, moving the agroecosystems analyzed further away from conventional management and making them even more sustainable. This margin for improvement is directly linked with an increase in the density and connectivity of low-entropy internal loops. Some of them could be achieved by changing the structure and management of farms, for example, by including a large proportion of leguminous plants. However, others such as agrolivestock integration might need intervention at larger territorial scales. This is true when it comes to using pastureland through transhumance, making use of silvopastoral spaces, for example. Furthermore, as we indicated before, organic farmers need to invest huge amounts of biomass to restore the fund elements and repair internal loops.

In both matters—intervention at larger scales than the individual farm and the restoration of fund elements—organic farmers must be supported by public policies if organic conversion is to become more than a mere replacement of inputs. Especially in the case of smaller farmers (lower income, less land) and those located in environments that present a higher degree of degradation.

5. Ultimately, EROIs are a very useful tool not only to ascertain the distance between a certain agroecosystem and conventional management when it is attempting to switch back to organic management, but also to indicate the path that should be taken in its management to make it more sustainable.

In this regard, we also believe that EROIs, particularly agroecological ones, offer powerful tools for organic or agroecological certification.

Appendix I

Table Al.1 Residues

Scientific Name	Crops Cereals (Modern Varieties)	Harvest Index — kg Product/kg Aerial Biomass — Mean	Residue Indices — kg Residue/ kg Aerial Biomass (Fresh Matter) — Mean	kg Residue/kg Product (Fresh Matter) — Mean	kg Residue/ kg Product (Fresh Matter) — Standard Deviation	No. of References	References	Comments
Avena sativa	Oat	0.41	0.59	1.43	1.06	7	Di Blasi et al. (1997), González de Molina and Guzmán (2006), Hernández Díaz-Ambrona (1999), CAP (2008), Prince et al. (2001), Bolinder et al. (2007), Unkovich et al. (2010)	
Hordeum vulgare	Barley	0.45	0.55	1.20	0.31	11	Di Blasi et al. (1997), Apel (1984), Hernández Díaz-Ambrona (1999), CAP (2008), MAGRAMA (2012), Kyle et al. (2011), Siddique et al. (1990), Bolinder et al. (2007), Unkovich et al. (2010), Prince et al. (2001), Petr et al. (2002)	
Oryza sativa	Rice	0.45	0.55	1.20	0.44	3	Di Blasi et al. (1997), MAGRAMA (2012), Kyle et al. (2011)	
Panicum miliaceum	Millet	0.45	0.55	1.22		1	Kyle et al. (2011)	
Secale cereale	Rye	0.43	0.57	1.30	0.42	1	MAGRAMA (2012), Kyle et al. (2011)	
Sorghum spp.	Sorghum	0.37	0.63	1.69	0.88	4	MAGRAMA (2012), Kyle et al. (2011), Bolinder et al. (2007), Unkovich et al. (2010)	
Triticum aestivum, T. turgidum	Wheat	0.42	0.58	1.36	0.33	12	Di Blasi et al. (1997), López Bellido (1991), Hernández Díaz-Ambrona (1999), CAP (2008), MAGRAMA (2012), Kemanian et al. (2007), Siddique et al. (1990), Prince et al. (2001), Bolinder et al. (2007), Unkovich et al. (2010), Hay (1995), Sánchez-García et al. (2013)	

							References	Notes
Zea mays	Maize	0.52	0.48	0.94	0.05	5	MAGRAMA (2012), Kyle et al. (2011), Bolinder et al. (2007), Unkovich et al. (2010), Prince et al. (2001)	
	Winter cereals, other	0.45	0.55	1.22		0		It is equated with rye
	Summer cereals, other	0.45	0.55	1.22		0		Mean (millet and sorghum) Prior to the 1940s
Cereals (Old Varieties)								
Avena sativa	Oat	0.33	0.67	2.03		1	González de Molina and Guzmán (2006)	
Hordeum vulgare	Barley	0.35	0.65	1.88	0.22	4	Apel (1984), Hay (1995), Petr et al. (2002), González de Molina and Guzmán (2006)	
Oryza sativa	Rice	0.30	0.70	2.33		1	Hay (1995)	
Triticum aestivum, T. turgidum	Wheat	0.28	0.72	2.53	0.05	4	Siddique et al. (1989), Siddique et al. (1990), Hay (1995), Sánchez-García et al. (2013)	Calculated on the dry matter
Zea mays	Maize	0.45	0.55	1.22		1	Hay (1995)	
Legumes								
Arachis hypogaea	Peanuts	0.33	0.67	2.03		1	Unkovich et al. (2010)	
Cicer arietinum	Chickpea	0.37	0.63	1.70		1	Unkovich et al. (2010)	
Glycine max	Soybeans	0.35	0.65	1.86	0.87	4	Kyle et al. (2011), Prince et al. (2001), Bolinder et al. (2007), Unkovich et al. (2010)	
Lens culinaris	Lentils	0.32	0.68	2.08	0.07	2	Hanlan et al. (2006), Unkovich et al. (2010)	
Lupinus albus	Lupin	0.30	0.70	2.33		1	Unkovich et al. (2010)	
Pisum sativus	Pea, green, with pod	0.39	0.61	1.57	0.63	5	Di Blasi et al. (1997), González de Molina and Guzmán (2006), Hernández Díaz-Ambrona (1999), Kyle et al. (2011), Unkovich et al. (2010)	

(Continued)

Table AI.1 (Continued)   Residues

Scientific Name	Crops Cereals (Modern Varieties)	Harvest Index: kg Product/kg Aerial Biomass Mean	kg Residue/kg Aerial Biomass (Fresh Matter) Mean	kg Aerial Residue/kg Product (Fresh Matter) Mean	kg Residue/kg Product (Fresh Matter) Mean	Standard Deviation	No. of References	References	Comments
Vicia faba	Faba bean/ Broad bean	0.39	0.61	1.56		0.97	6	Di Blasi et al. (1997), González de Molina and Guzmán (2006), Hernández Díaz-Ambrona (1999), Kyle et al. (2011), Múñoz-Romero et al. (2011), Unkovich et al. (2010)	
Vicia sativa	Vetch	0.45	0.55	1.24		0.56	3	Di Blasi et al. (1997), González de Molina and Guzmán (2006), Unkovich et al. (2010)	
	Legumes, other	0.37	0.63	1.72			0		Mean (pea, faba bean, and vetch)
Root Crops									
Cyperus esculentus	Tigernuts	0.50	0.50	1.00			0		Estimated
Ipomoea batatas	Sweet potato	0.53	0.47	0.89			1	Kyle et al. (2011)	
Solanum tuberosum	Potato	0.59	0.41	0.70		0.42	2	Di Blasi et al. (1997), Kyle et al. (2011)	
Vegetables									
Allium ampeloprasum	Leek	1.00	0.00	0.00			1	Kyle et al. (2011)	
Allium cepa	Onion	0.62	0.38	0.61		0.24	2	Kyle et al. (2011), Salo (1999)	
Allium fistulosum	Welsh onion	1.00	0.00	0.00			1	Kyle et al. (2011)	
Allium sativum	Garlic	1.00	0.00	0.00			1	Kyle et al. (2011)	

Species	Common name						References	Notes
Apium graveolens	Celery	0.86	0.14	0.16		1	Aranguiz (2006)	
Asparagus officinalis	Asparagus	0.22	0.78	3.59	3.61	2	Kyle et al. (2011), Fuertes (2009)	
Beta vulgaris	Beet	0.51	0.49	0.95		0		It is equated with sugar beet
Beta vulgaris var. *Cicla*	Chard	0.91	0.09	0.10	0.07	2	Kyle et al. (2011), Aranguiz (2006)	
Borago officinalis	Borage	0.86	0.14	0.16		0		It is equated with celery
Brassica oleracea	Cabbage, broccoli	0.45	0.55	1.23	1.15	3	Di Blasi et al. (1997), Salo (1999), Kyle et al. (2011)	
Brassica oleracea var. *botrytis*	Cauliflower	0.65	0.35	0.54	0.40	2	Kyle et al. (2011), Erley et al. (2010)	
Brassica oleracea var. *viridis*	Collard	0.80	0.20	0.25		1	Kyle et al. (2011)	
Brassica rapa var. *rapa*	Turnip	0.53	0.47	0.89		1	Kyle et al. (2011)	
Capsicum annuum	Chili pepper	0.30	0.70	2.33		1	Kyle et al. (2011)	
Capsicum annuum	Pepper	0.58	0.42	0.73	0.14	3	Kyle et al. (2011), Fernández et al. (2005), Marcelis et al. (2005)	
Cichorium endivia	Endive	0.67	0.33	0.50		1	Aranguiz (2006)	
Cichorium intybus	Chicory	0.78	0.22	0.28		1		It is equated with lettuce
Cichorium intybus	Belgian endive	0.78	0.22	0.28		0		It is equated with lettuce
Citrullus lanatus	Watermelon	0.91	0.09	0.10		1	Kyle et al. (2011)	
Cucumis melo	Melon	0.75	0.25	0.33	0.33	2	Kyle et al. (2011), Huang et al. (2012)	

(Continued)

Table AI.1 (Continued) Residues

Scientific Name	Crops Cereals (Modern Varieties)	Harvest Index kg Product/kg Aerial Biomass Mean	Residue Indices				References	Comments
			kg Residue/kg Aerial Biomass (Fresh Matter) Mean	kg Residue/kg Product (Fresh Matter) Mean	kg Residue/kg Product (Fresh Matter) Standard Deviation	No. of References		
Cucumis sativus	Cultivar for pickled cucumber, gherkins	0.80	0.20	0.25		1	Kyle et al. (2011)	
Cucumis sativus	Cucumber	0.80	0.20	0.25		1	Kyle et al. (2011)	
Cucurbita pepo	Squash/pumpkin	0.88	0.12	0.14		1	Kyle et al. (2011)	
Cucurbita pepo var. cylindrica	Zucchini	0.80	0.20	0.25		0		It is equated with cucumber
Cynara cardunculus	Artichoke thistle	0.78	0.22	0.28		0		It is equated with lettuce
Cynara scolimus	Artichoke	0.42	0.58	1.40	1.56	2	Di Blasi et al. (1997), Kyle et al. (2011)	
Daucus carota	Carrot	0.53	0.47	0.87	0.02	2	Kyle et al. (2011), Salo (1999)	
Fragaria x ananassa	Strawberry	0.50	0.50	1.00		0		Estimated
Lactuca sativa	Lettuce	0.78	0.22	0.28	0.31	2	Kyle et al. (2011), Aranguiz (2006)	
Mentha spp.	Mint and peppermint	0.80	0.20	0.25		0		Estimated
Petroselinum crispum	Parsley	0.86	0.14	0.16		0		It is equated with celery
Phaseolus vulgaris	Beans, green	0.38	0.62	1.60	0.61	2	Kyle et al. (2011), González et al. (2002)	

Pisum sativum	Pea, green, with pod	0.30	0.70	2.33		1	Kyle et al. (2011)	
Raphanus sativus	Radish	0.53	0.47	0.89		0		It is equated with turnip
Solanum lycopersicum	Tomato	0.51	0.49	0.96	0.94	3	Kyle et al. (2011), Di Blasi et al. (1997), Zotarelli et al. (2009)	
Solanum melongena	Eggplant/aubergine	0.59	0.41	0.69		1	Kyle et al. (2011)	
Spinacia oleracea	Spinach	0.91	0.09	0.10	0.07	2	Kyle et al. (2011), Aranguiz (2006)	
Vicia faba	Faba bean/broad bean, green, without pod	0.34	0.66	1.97		0		Mean (pea and beans)
Industrial Crops								
Beta vulgaris	Sugar beet	0.51	0.49	0.95	0.78	2	Di Blasi et al. (1997), Kyle et al. (2011)	
Brassica napus	Rape	0.29	0.71	2.45	0.17	2	Kyle et al. (2011), Unkovich et al. (2010)	
Brassica sp.	Mustard (black mustard)	0.29	0.71	2.45		0		It is equated with rape
Cannabis sativa	Hemp	0.80	0.20	0.25		0		Estimated
Capparis spinosa	Caper	0.33	0.67	2.00		0		Estimated
Carthamus tinctorius	Safflower	0.22	0.78	3.54	0.65	2	Koutroubas et al. (2004), Dordas et al. (2009)	
Crocus sativus	Saffron	0.17	0.83	5.00		0		Estimated
Cuminum cyminum	Cumin	0.33	0.67	2.00		0		Estimated
Glycyrrhiza glabra	Liquorice	0.50	0.50	1.00		0		Estimated
Gossypium spp.	Cotton fiber	0.39	0.62	1.60		2	Mondino and Peterlin (2003), Cáceres Díaz (2012)	
Gossypium spp.	Cotton seed	0.40	0.60	1.50		1	Kyle et al. (2011)	

(Continued)

Table A1.1 (Continued) Residues

Scientific Name	Crops Cereals (Modern Varieties)	Harvest Index — kg Product/kg Aerial Biomass — Mean	Residue Indices — kg Residue/kg Aerial Biomass (Fresh Matter) — Mean	kg Aerial Residue/kg Product (Fresh Matter) — Mean	kg Residue/kg Product (Fresh Matter) — Standard Deviation	No. of References	References	Comments
Helianthus annuus	Sunflower	0.30	0.70	2.30	0.69	3	Kyle et al. (2011), Unkovich et al. (2010), Prince et al. (2001)	
Humulus lupulus	Hop	0.33	0.67	2.00		0		Estimated
Linum usitatissimum	Linseed	0.26	0.74	2.85		1	Kyle et al. (2011)	
Nicotiana tabacum	Tobacco	0.67	0.33	0.50	0.71	2	Kyle et al. (2011), Di Blasi et al. (1997)	
Pimpinella anisum	Anise	0.20	0.80	4.00		1	Kyle et al. (2011)	
Ricinus communis	Castor oil plant	0.33	0.67	2.00		0		Estimated
Rus coriaria	Sumac	0.40	0.60	1.50		1	Kyle et al. (2011)	It is equated with pistachio
Saccharum officinarum	Sugarcane	0.70	0.30	0.43		1	Kyle et al. (2011)	
Fruit Trees								
Actinidia deliciosa	Kiwifruit	0.68	0.32	0.46	0.09	2	Kyle et al. (2011), Smith et al. (1988)	
Carica papaya	Papaya	0.99	0.01	0.01	–	1	Kyle et al. (2011)	
Citrus reticulata	Mandarin	0.78	0.22	0.28	0.32	3	Di Blasi et al. (1997), Eubionet (2003), Kyle et al. (2011)	

Species	Common name						References
Citrus x limon	Lemon	0.83	0.17	0.20	0.22	3	Di Blasi et al. (1997), Eubionet (2003), Kyle et al. (2011)
Citrus x sinensis	Orange	0.86	0.14	0.17	0.12	4	Di Blasi et al. (1997), Eubionet (2003), Kyle et al. (2011), Roccuzzo et al. (2012)
Corylus avellana	Hazelnuts growing	0.37	0.63	1.70	0.28	2	Eubionet (2003), Kyle et al. (2011)
Elaeis guineensis/ E. oleifera	Oil palm	0.19	0.81	4.26	–	1	Kyle et al. (2011)
Ficus carica	Figs	0.62	0.38	0.61	–	1	Kyle et al. (2011)
Juglans regia	Walnut tree	0.40	0.60	1.50	–	1	Kyle et al. (2011)
Malus domestica	Apple	0.73	0.27	0.37	0.40	3	Di Blasi et al. (1997), Eubionet (2003), Kyle et al. (2011)
Musa x paradisiaca	Bananas, plantains	0.40	0.60	1.50	–	1	Kyle et al. (2011)
Olea europaea	Olive tree	0.51	0.49	0.95	0.46	4	Di Blasi et al. (1997), Eubionet (2003), Kyle et al. (2011), Civantos and Olid (1982)
Persea americana	Avocado	0.71	0.29	0.41	–	1	Kyle et al. (2011)
Pistacia vera	Pistachio	0.40	0.60	1.50	–	1	Kyle et al. (2011)
Prunus armeniaca	Apricot	0.71	0.29	0.41	0.08	2	Eubionet (2003), Kyle et al. (2011)
Prunus avium	Cherry	0.66	0.34	0.50	0.46	2	Eubionet (2003), Kyle et al. (2011)
Prunus dulcis	Almonds	0.30	0.70	2.28	1.14	3	Di Blasi et al. (1997), Eubionet (2003), Kyle et al. (2011)
Prunus persica	Peach	0.80	0.20	0.25	0.13	3	Di Blasi et al. (1997), Eubionet (2003), Kyle et al. (2011)
Punica granatum	Pomegranate	0.78	0.22	0.28	–	1	García-Gómez (2011)
Pyrus communis	Pear	0.74	0.26	0.34	0.39	3	Di Blasi et al. (1997), Eubionet (2003), Kyle et al. (2011)
Quercus ilex	Holm oak	0.23	0.77	3.33	–	1	Almoguera Millán (2007)
Vitis vinifera	Grapevine	0.65	0.35	0.53	0.29	3	Di Blasi et al. (1997), Eubionet (2003), Kyle et al. (2011)

Table AI.2   Roots

Scientific Name	Crops Cereals (Modern Varieties)	Aboveground Biomass (kg Dry Matter/ha) Mean	Belowground Biomass (kg Dry Matter/ha) Mean	Root:Shoot Ratio Mean	Root:Shoot Ratio Deviation Standard	No. of Data	Comments	References
Avena sativa	Oat	8,470	3,393	0.40	0.03	3		Bolinder et al. (1997)
Hordeum vulgare	Barley	9,537	1,544	0.21	0.18	25		Izaurralde et al. (1993) in Bolinder et al. (2007), Rutherford and Juma (1989), Xu and Juma (1992), Xu and Juma (1993), Haugen-Kozyra et al. (1993), Soon (1988), Bolinder et al. (1997), Chirinda et al. (2012)
Phalaris sp.	Canary grass	1,022	1,487	1.50	0.71	2		Bolinder et al. (2002)
Sorghum spp.	Sorghum			0.09		1		Piper and Kulakow (1994) en Bolinder et al. (2007)
Triticum aestivum, T. turgidum	Wheat	10,193	2,029	0.20	0.15	47		Izaurralde et al. (1992) en Bolinder et al. (2007), Buyanovsky and Wagner (1986), Bolinder et al. (1997), Campbell and de Jong (2001), López Bellido, Jong (2001), Chirinda et al. (2012)
x *Triticosecale*	Triticale	16,260	**3,115**	0.19	0.02	2		Bolinder et al. (1997)
Zea mays	Maize	13,585	4,616	0.24	0.15	20		Allmaras et al. (1975), Eghball and Maranville (1993) in Bolinder et al. (2007), Anderson (1988), Buyanovsky and Wagner (1986), Kiselle et al. (2001), Tran and Giroux (1998), Zan et al. (2001)

Cereals (Old Varieties)								
Triticum aestivum, T. turgidum	Wheat	6,224	3,972	0.64		1	Prior to the 1940s	Siddique et al., 1990
Legumes								
Glycine max	Soybeans	4,686	3,331	0.39	0.38	12	At anthesis phenological stage	Marvel et al. (1992), Allmaras et al. (1975), Mayaki et al. (1976) en Bolinder et al. (2007), Buyanovsky and Wagner (1986), House et al. (1984)
Pisum sativum	Pea, green, with pod	6,220	340	0.06	0.02	2		Williams et al. (2013)
Vicia faba	Faba bean/ broad bean			0.60	0.24	3	It is calculated on fresh matter	Muñoz-Romero et al. (2011)
Industrial Crops								
Beta vulgaris	Sugar beet			14.29		1	The harvested part is the root	European Commission (2011)
Cannabis sativa	Hemp fiber	2,810		0.18	0.06	2		Amaducci et al. (2008)
Green Fodder								
Bromus sp.	Brome	3,013	9,595	4.11	3.09	9		Walley et al. (1996) in Bolinder et al. (2007), Leyshon (1991), Bolinder et al. (2002)
Bromus spp.	Brome grasses			2.44	0.77	3		Walley et al. (1996) in Bolinder et al. (2007)

(Continued)

Table AI.2 (Continued) Roots

Scientific Name	Crops Cereals (Modern Varieties)	Aboveground Biomass (kg Dry Matter/ha) Mean	Belowground Biomass (kg Dry Matter/ha) Mean	Root:Shoot Ratio Mean	Root:Shoot Ratio Standard Deviation	No. of Data	Comments	References
Dactylis sp.	Cocksfoot	722	836	1.10	0.47	2		Bolinder et al. (2002)
Festuca sp.	Fescue grass	972	1,301	1.13	0.62	4		Izaurralde et al. (1993) in Bolinder et al. (2007), Bolinder et al. (2002)
Lolium multiflorum	Ryegrasss	5,684	3,340	0.51	0.31	22		Kunelius et al. (1992), Carter et al. (2003), Andrews et al. (2001)
Lolium perenne	Perennial ryegrass	719	1,101	1.58	1.30	2		Bolinder et al. (2002)
Medicago sativa	Alfalfa	2,051	2,011	1.20	0.58	8		Walley et al. (1996) in Bolinder et al. (2007), Bowren et al. (1969), Griffin et al. (2000)
Medicago sativa	Alfalfa (mixed cropping)			1.14	0.25	3		Walley et al. (1996) in Bolinder et al. (2007)
Panicum virgatum	Switchgrass	656	886	0.99	0.70	4		Zan et al. (2001), Bolinder et al. (2002)
Phleum sp.	Cat's-tail	863	1,351	1.42	0.82	2		Bolinder et al. (2002)
Secale cereale	Rye	2,750	2,386	0.85	0.19	3		Griffin et al. (2000)
Secale cereale+Vicia villosa	Rye+Hairy vetch	3,522	2,094	0.61	0.08	3		Griffin et al. (2000)

Species	Common name						Notes	References
Trifolium sp.	Clover	3,528	1,442	0.56	0.37	20		Sheaffer et al. (1991), in Bolinder et al. (2007), Andrews et al. (2001), Carter et al. (2003), Bolinder et al. (2002), Bowren et al. (1969), Kunelius et al. (1992)
Trifolium subterraneum	Subterranean clover			0.25		1		Almoguera Millán (2007)
Zea mays	Corn (silage)	12,830	1,280	0.10		1		Tran and Giroux (1998)
	Grass			0.80		1		European Commission (2011)
Fruits								
Actinidia deliciosa	Kiwifruit			0.67		1		Smith et al. (1988)
Coffea sp.	Coffee tree			0.20		1		Siles et al. (2010)
Olea europaea	Olive tree	17,298	3,604	0.21		1		Almagro et al. (2010)
Persea americana	Avocado			0.25		1		Cairns et al. (1997)
Forest Trees								
Eucalyptus sp.	Eucalyptus			0.28				Mokany et al. (2006)
Populus spp.	Poplar			0.50		2		Cannell and Willett (1976), Wullschleger et al. (2005)
Quercus ilex	Holm oak	40,700	30,300	0.84	0.43	5	It is calculated on fresh matter. 32 trees/ha	Almoguera Millán (2007)
Salix sp.	Willow, sallow	4,919		0.45	0.14	2		Zan et al. (2001)
	Mediterranean scrub		13,280	1.60	0.84	4		Martínez et al. (1998)

Table AI.3 Dry Matter

Scientific Name	Product Name	Product			Residues (Straw, Pruning, etc.)				
		kg Dry Matter/kg Fresh Matter	References	Comments	kg Dry Matter/kg Fresh Matter	Standard Deviation	No. of References	References	Comments
Cereals									
Avena sativa	Oat	0.867	Piat (1989), FEDNA (2010), Moreiras et al. (2011)		0.907		1	CIHEAM, 1990	
Hordeum vulgare	Barley	0.885	Piat (1989), FEDNA (2010)		0.864		1	CIHEAM, 1990	
Oryza sativa	Brown rice	0.864	Moreiras et al. (2011)		0.910		1	Kyle et al. (2011)	
Panicum miliaceum	Millet	0.881	Piat (1989), Moreiras et al. (2011)		0.900		1	Kyle et al. (2011)	
Secale cereale	Rye	0.876	Piat (1989), FEDNA (2010)		0.924		1	CIHEAM, 1990	
Sorghum spp.	Sorghum	0.865	Piat (1989)		0.870		1	Kyle et al. (2011)	
Triticum aestivum/ T. durum	Wheat	0.879	Piat (1989), FEDNA (2010)		0.867	0.03	2	CIHEAM (1990), Kyle et al. (2011)	
x *Triticosecale*	Triticale	0.878	Piat (1989), FEDNA (2010)		0.922		1	CIHEAM (1990)	
Zea mays	Maize	0.862	FEDNA (2010)		0.881		1	CIHEAM (1990)	

Species	Common name	Value	Source	Note	Value	Number	Source	Note
	Winter cereals, other	0.873	Estimated	Mean (barley, wheat and rye)	0.890	0		Mean (barley, wheat, oat and rye)
	Spring cereals, other	0.872	Estimated	Mean (rice, maize, sorghum)	0.881	0		It is equated with maize
Legumes								
Cicer arietinum	Chickpea	0.944	Moreiras et al. (2011)		0.893	1	CIHEAM (1990)	
Glycine max	Soybeans	0.860	Moreiras et al. (2011)		0.886	3	CIHEAM (1990), Aranguiz (2006), Kyle et al. (2011)	
Lathyrus sativus	Grass pea, chickling vetch				0.915	1	CIHEAM (1990)	
Lens culinaris	Lentils	0.897	FEDNA (2010), Moreiras et al. (2011)		0.928	1	CIHEAM (1990)	
Lupinus albus	White lupin	0.894	Piat (1989), FEDNA (2010)			0		
Phaseolus vulgaris	Beans, white	0.983	Moreiras et al. (2011)		0.871	1	CIHEAM (1990)	
Pisum sativus	Pea, dry	0.902	Piat (1989), FEDNA (2010), Moreiras et al. (2011)		0.907	1	CIHEAM (1990)	
Vicia articulata	Bard vetch/oneflower vetch	0.967	Moreiras et al. (2011)			0		

(Continued)

Table AI.3 (Continued) Dry Matter

		Product			Residues (Straw, Pruning, etc.)				
Scientific Name	Product Name	kg Dry Matter/kg Fresh Matter	References	Comments	kg Dry Matter/kg Fresh Matter	Standard Deviation	No. of References	References	Comments
Vicia ervilia	Bitter vetch	0.900	FEDNA (2010)		0.914		1	CIHEAM (1990)	
Vicia faba	Faba bean/broad bean, dry	0.915	Piat (1989), FEDNA (2010), Moreiras et al. (2011)		0.886		1	CIHEAM (1990)	
Vicia monantha	Hard vetch	0.900	FEDNA (2010)		0.886		1	CIHEAM (1990)	
Vicia sativa	Vetch	0.916	Estimated	Mean of all other	0.911		1	CIHEAM (1990)	
	Legumes, other				0.900		0		Mean of all other
Vegetables									
Agaricus bisporus	Common mushroom	0.086	Moreiras et al. (2011)				0		
Allium ampeloprasum	Leek	0.129	Moreiras et al. (2011)		0.21	0.13	2	Rahn and Lillywhite (2002), Kyle et al. (2011)	
Allium ascalonicum	Shallot	0.207	Moreiras et al. (2011)				0		
Allium cepa	Onion	0.061	Moreiras et al. (2011)		0.20	0.14	2	Rahn and Lillywhite (2002), Kyle et al. (2011)	
Allium fistulosum	Welsh onion	0.078	Moreiras et al. (2011)				0		
Allium sativum	Garlic	0.297	Moreiras et al. (2011)	Peeled fruit	0.30		1	Kyle et al. (2011)	
Allium schoenoprasum	Chives	0.077	Moreiras et al. (2011)				0		

Apium graveolens	Celery	0.046	Moreiras et al. (2011)	0.10			1	Kyle et al. (2011)
Asparagus officinalis	Asparagus	0.053	Moreiras et al. (2011)	0.30			1	Kyle et al. (2011)
Beta vulgaris	Beet	0.108	Moreiras et al. (2011)	0.12			1	Rahn and Lillywhite (2002)
Beta vulgaris var. Cicla	Chard	0.125	Moreiras et al. (2011)	0.19	0.16		2	Aranguiz (2006), Kyle et al. (2011)
Borago officinalis	Borage	0.065	Moreiras et al. (2011)	0.17			0	It is equated with lettuce
Brassica oleracea	Cabbage	0.103	Moreiras et al. (2011)	0.18	0.12		3	Aranguiz (2006), Rahn and Lillywhite (2002), Kyle et al. (2011)
Brassica oleracea var. botrytis	Cauliflower	0.076	Moreiras et al. (2011)	0.21	0.13		2	Aranguiz (2006), Kyle et al. (2011)
Brassica oleracea var. italica	Broccoli	0.097	Moreiras et al. (2011)	0.18	0.12		3	Aranguiz (2006), Rahn and Lillywhite (2002), Kyle et al. (2011)
Brassica rapa var. rapa	Turnip	0.089	Moreiras et al. (2011)	0.20			1	Kyle et al. (2011)
Capsicum annuum	Chili pepper	0.105	Moreiras et al. (2011)	0.30			1	Kyle et al. (2011)
Capsicum annuum	Green pepper	0.087	Author data, unpublished	0.30		Mean of three varieties, Whole fruit	1	Kyle et al. (2011)
Cichorium endivia	Endive	0.064	Moreiras et al. (2011)	0.17			0	It is equated with lettuce
Cichorium intybus	Chicory			0.17			1	Kyle et al. (2011)

(Continued)

Table AI.3 (Continued) Dry Matter

Scientific Name	Product Name	Product			Residues (Straw, Pruning, etc.)				
		kg Dry Matter/kg Fresh Matter	References	Comments	kg Dry Matter/kg Fresh Matter	Standard Deviation	No. of References	References	Comments
Cichorium intybus	Belgian endive	0.066	Moreiras et al. (2011)		0.17		0		It is equated with lettuce
Citrullus lanatus	Watermelon	0.057	Author data, unpublished	Mean of three varieties, Whole fruit	0.20		1	Kyle et al. (2011)	
Coriandum sativum	Coriander	0.148	Moreiras et al. (2011)				0		
Cucumis melo	Melon	0.076	Moreiras et al. (2011)	Peeled	0.20		1	Kyle et al. (2011)	
Cucumis melo var. *cantalupensis*	Cantaloupe	0.091	Moreiras et al. (2011)	Peeled			0		
Cucumis sativus	Cultivar for pickled cucumber, gherkins				0.20		1	Kyle et al. (2011)	
Cucumis sativus	Cucumber	0.033	Moreiras et al. (2011)	Peeled	0.18	0.03	2	Aranguiz (2006), Kyle et al. (2011)	
Cucurbita pepo	Squash/pumpkin	0.108	Author data, unpublished	Mean of two varieties, Whole fruit	0.30		1	Kyle et al. (2011)	
Cucurbita pepo var. *cylindrica*	Zucchini	0.035	Moreiras et al. (2011)	Peeled fruit	0.18		0		It is equated with cucumber
Cynara cardunculus	Artichoke thistle	0.061	Moreiras et al. (2011)		0.17		0		It is equated with lettuce
Cynara scolimus	Artichoke	0.119	Moreiras et al. (2011)		0.20		1	Kyle et al. (2011)	

Cyperus esculentus	Yellow nutsedge, tigernuts	0.897	Moreiras et al. (2011)				0	
Daucus carota	Carrot	0.081	Moreiras et al. (2011)	Whole fruit	0.21	0.01	2	Rahn and Lillywhite (2002), Kyle et al. (2011)
Foeniculum vulgare	Fennel	0.067	Moreiras et al. (2011)				0	
Fragaria x ananassa	Strawberry	0.104	Moreiras et al. (2011)		0.30		0	Estimated
Ipomoea batatas	Sweet potato	0.258	Moreiras et al. (2011)	Peeled fruit			0	
Lactuca sativa	Lettuce	0.047	Moreiras et al. (2011)		0.17	0.12	3	Aranguiz (2006), Rahn and Lillywhite (2002), Kyle et al. (2011)
Lactuca sativa	Miniature lettuce	0.047	Moreiras et al. (2011)				0	
Lactuca sativa	Iceberg lettuce	0.042	Moreiras et al. (2011)				0	
Manihot sculenta	Cassava	0.416	Moreiras et al. (2011)		0.30		1	Kyle et al. (2011)
Mentha spp.	Mint and peppermint	0.098	Moreiras et al. (2011)				0	
Nasturtium officinale	Watercress	0.074	Moreiras et al. (2011)				0	
Ocimum basilicum	Basil, fresh	0.075	Moreiras et al. (2011)				0	
Pastinaca sativa	Parsnip	0.183	Moreiras et al. (2011)				0	
Petroselinum crispum	Parsley	0.120	Moreiras et al. (2011)				0	
Phaseolus vulgaris	Beans, green	0.104	Moreiras et al. (2011)		0.30		1	Kyle et al. (2011)

(Continued)

Table AI.3 (Continued) Dry Matter

		Product			Residues (Straw, Pruning, etc.)				
Scientific Name	Product Name	kg Dry Matter/kg Fresh Matter	References	Comments	kg Dry Matter/kg Fresh Matter	Standard Deviation	No. of References	References	Comments
Pisum sativus	Pea, green, with pod	0.248	Moreiras et al. (2011)		0.30		1	Kyle et al. (2011)	
Raphanus sativus	Radish	0.047	Moreiras et al. (2011)		0.19	0.03	2	Rahn and Lillywhite (2002), Kyle et al. (2011)	
Rumex acetosa	Sorrel	0.070	Moreiras et al. (2011)				0		
Solanum lycopersicum	Tomato	0.062	Author data, unpublished; Monreal (2012)	Mean of 28 varieties, Whole fruit	0.13	0.12	2	Aranguiz (2006), Kyle et al. (2011)	
Solanum lycopersicum var.cerasiforme	Cherry tomato	0.108	Monreal Carsi (2012)	Mean of 10 varieties, Whole fruit			0		
Solanum melongena	Eggplant/ aubergine	0.090	Author data, unpublished	Mean of two varieties, Whole fruit	0.20		1	Kyle et al. (2011)	
Solanum tuberosum	Potato	0.227	Moreiras et al. (2011)	Peeled	0.20	0.00	2	Aranguiz (2006), Kyle et al. (2011)	
Spinacia oleracea	Spinach	0.104	Moreiras et al. (2011)		0.19	0.16	2	Aranguiz (2006), Kyle et al. (2011)	
Valerianella locusta	Corn salad/ mâche	0.044	Moreiras et al. (2011)				0		
Vicia faba	Faba bean/broad bean, green, without pod	0.178	Moreiras et al. (2011)		0.24	0.06	3	Aranguiz (2006), Rahn and Lillywhite (2002), Kyle et al. (2011)	
Zingiber officinale	Ginger, fresh	0.139	Moreiras et al. (2011)				0		

Industrial Crops (before Processing)

Beta vulgaris	Sugar beet	0.250	Baraja Rodriguez (1994)	0.16	0.04	3	Rahn and Lillywhite (2002), FEDNA (2010), Kyle et al. (2011)	
Brassica napus	Rape	0.912	FEDNA (2010)	1.00		1	Kyle et al. (2011)	
Brassica spp.	Mustard (black mustard)	0.912	It is equated with rape	1.00		0		It is equated with rape
Cannabis sativa	Hemp seed and fiber	0.911	González Vázquez (1944), Soroa (1953)	0.91		0		
Capparis spinosa	Caper	0.114	Moreiras et al. (2011)			0		
Carthamus tinctorius	Safflower	0.912	It is equated with rape	1.00		0		It is equated with rape
Gossypium spp.	Cotton fiber	0.900	González Vázquez (1944), Soroa (1953), Centro de Comercio Internacional (2007)			0		
Gossypium spp.	Cotton seed	0.920	FEDNA (2010)	0.92		1	Kyle et al. (2011)	
Helianthus annuus	Sunflower	0.936	FEDNA (2010), Moreiras et al. (2011)	0.93		1	Kyle et al. (2011)	
Linum usitatissimum	Linseed/flax	0.929	González Vázquez (1944), Soroa (1953)	0.85		1	Kyle et al. (2011)	

(Continued)

Table AI.3 (Continued) Dry Matter

		Product			Residues (Straw, Pruning, etc.)				
Scientific Name	Product Name	kg Dry Matter/kg Fresh Matter	References	Comments	kg Dry Matter/kg Fresh Matter	Standard Deviation	No. of References	References	Comments
Nicotiana tabacum	Tobacco	0.150	Guerrero (1987), Lázaro García (2000)		0.80		1	Kyle et al. (2011)	
Saccharum officinarum	Sugarcane	0.295	Guerra (2013), SAGARPA (2013)		0.48		2	Kyle et al. (2011), SAGARPA (2013)	
Stipa tenacissima	Esparto grass	0.721	CIHEAM (1990)				0		
Fruits									
Actinidia deliciosa	Kiwifruit	0.186	Smith et al. (1988)	Whole fruit	—	—	0		
Castanea sativa	Chestnut	0.500	Kader (2013)	Whole nuts with shells			0		
Citrus reticulata	Mandarin				0.63	0.05	4	Fundación Abertis (2005), Eubionet (2003), Voivontas et al. (2001), Di Blasi et al. (1997)	
Citrus x limon	Lemon				0.63	0.05	4	Fundación Abertis (2005), Eubionet (2003), Voivontas et al. (2001), Di Blasi et al. (1997)	
Citrus x sinensis	Orange	0.121	It is calculated from Cerón-Salazar and Cardona-Alzate (2011) and Moreiras et al. (2011)	Whole fruit	0.63	0.05	4	Fundación Abertis (2005), Eubionet (2003), Voivontas et al. (2001), Di Blasi et al. (1997)	

Corylus avellana	Hazelnuts growing	0.930	Infoagro (2013)	Whole nuts with shells	0.75	0.18	2	Bilandzija et al. (2012), Fundación Abertis (2005)
Ficus carica	Figs	0.197	Moreiras et al. (2011)	Peeled fruit	0.81	0.15	2	Bilandzija et al. (2012), Fundación Abertis (2005)
Juglans regia	Walnut tree	0.753	Vásquez Panizza (2013), Parra (2013)	Whole nuts with shells, after air drying	0.83	0.19	2	Bilandzija et al. (2012), Fundación Abertis (2005)
Malus domestica	Apple	0.160	García-Gómez (2011)	Whole fruit	0.69	0.14	5	Bilandzija et al. (2012), Fundación Abertis (2005), Eubionet (2003), Voivontas et al. (2001), Di Blasi et al. (1997)
Musa x paradisiaca	Bananas, plantains	0.249	Moreiras et al. (2011)	Peeled fruit			0	
Olea europaea	Olive tree	0.539	Ferreira et al. (1986)	Whole fruit	0.70	0.13	6	Bilandzija et al. (2012), Fundación Abertis (2005), Eubionet (2003), Voivontas et al. (2001), Di Blasi et al. (1997), Civantos and Olid, 1985
Prunus armeniaca	Apricot	0.186	Kernels: Lazos (1991), flesh and flesh/kernels ratio: Moreiras et al. (2011)	Whole fruit	0.71	0.16	4	Bilandzija et al. (2012), Fundación Abertis (2005), Eubionet (2003), Voivontas et al. (2001)
Prunus avium	Cherry	0.262	Kernels: Lazos (1991), flesh and flesh/kernels ratio: Moreiras et al. (2011)	Whole fruit	0.71	0.16	4	Bilandzija et al. (2012), Fundación Abertis (2005), Eubionet (2003), Voivontas et al. (2001)

(Continued)

Table AI.3 (Continued)   Dry Matter

		Product			Residues (Straw, Pruning, etc.)				
Scientific Name	Product Name	kg Dry Matter/kg Fresh Matter	References	Comments	kg Dry Matter/kg Fresh Matter	Standard Deviation	No. of References	References	Comments
Prunus cerasus	Sour cherry, wild cherry	0.262		Whole fruit. It is equated with cherry	0.82	0.16	2	Bilandzija et al. (2012), Fundación Abertis (2005)	
Prunus dulcis	Almonds	0.689	Espada Carbó (2011), Moreiras et al. (2011), IDAE (2007)	Whole nuts with shells, after air drying	0.69	0.14	5	Bilandzija et al. (2012), Fundación Abertis (2005), Eubionet (2003), Voivontas et al. (2001), Di Blasi et al. (1997)	
Prunus persica	Peach	0.208	Kernels: Lazos (1991), flesh and flesh/kernels ratio: Moreiras et al. (2011)	Whole fruit	0.69	0.14	5	Bilandzija et al. (2012), Fundación Abertis (2005), Eubionet (2003), Voivontas et al. (2001), Di Blasi et al. (1997)	
Prunus subg. *Prunus*	Plum				0.82	0.17	2	Bilandzija et al. (2012), Fundación Abertis (2005)	
Punica granatum	Pomegranate	0.200	García-Gómez (2011)	Whole fruit	–	–	0		
Pyrus communis	Pear	0.180	García-Gómez (2011)	Whole fruit	0.69	0.15	5	Bilandzija et al. (2012), Fundación Abertis (2005), Eubionet (2003), Voivontas et al. (2001), Di Blasi et al. (1997)	

Species	Common name	Value	Source	Type				Reference
Quercus ilex	Holm oak	0.625	FEDNA (2010)	Whole fruit			0	
Ribes nigrum	Blackcurrant	0.144	Moreiras et al. (2011)	Whole fruit			0	
Ribes rubrum	Redcurrant	0.096	Moreiras et al. (2011)	Whole fruit			0	
Rubus idaeus	Raspberry	0.130	Moreiras et al. (2011)	Whole fruit			0	
Rubus ulmifolius	Blackberry	0.128	Moreiras et al. (2011)	Whole fruit			0	
Vaccinium spp.	Blueberry	0.122	Moreiras et al. (2011)	Whole fruit			0	
Vitis vinifera	Grapevine	0.291	Centeno (2009)	Whole fruit	0.65	0.16	6	Bilandzija et al. (2012), Fundación Abertis (2005), Eubionet (2003), Voivontas et al. (2001), Di Blasi et al. (1997), AVEBIOM (2009)
Forest Trees								
Fagus sylvatica	European beech/commom beech	0.75		Estimated				
Picea spp.	Spruce	0.75		Estimated				
Populus spp.	Poplar	0.75		Estimated				
Salix spp.	Willow, sallow	0.75		Estimated				
	Conifers	0.75		Estimated				
	Bark (conifers)	0.75		Estimated				
	Bark (broad-leaved tree)	0.75		Estimated				
	Broad-leaved tree	0.75		Estimated				

(Continued)

Table AI.3 (Continued) Dry Matter

Scientific Name	Product Name	kg Dry Matter/kg Fresh Matter	References	Comments	kg Dry Matter/kg Fresh Matter	Standard Deviation	No. of References	References	Comments
			Product		Residues (Straw, Pruning, etc.)				
Green Fodder									
Avena sativa	Oat, green	0.303	CIHEAM (1990)						
Beta vulgaris	Fodder beet	0.184	CIHEAM (1990)						
Beta vulgaris	Sugar beet, neck	0.203	FEDNA (2010)						
Beta vulgaris	Fodder	0.050	Guerrero (1987)						
Brassica napus	Turnip, for fodder	0.126	CIHEAM (1990)						
Brassica oleracea	Fodder cabbage	0.165		It is equated with human food varieties					
Cucurbita pepo	Squash, for fodder	0.108		It is equated with human food varieties					
Cynara cardunculus	Artichoke thistle, for fodder	0.118		It is equated with human food varieties					
Daucus carota	Carrot, for fodder	0.126		It is equated with human food varieties					
Hedysarum coronarium	French honeysuckle	0.153	CIHEAM (1990)						
Helianthus tuberosus	Jerusalem artichoke	0.234	Agrobit (2013)						

(Continued)

Species	Common name	Value	Reference	Notes
Hordeum vulgare	Barley, green	0.248	CIHEAM (1990)	
Lolium perenne	Perennial ryegrass	0.239	CIHEAM (1990)	
Lolium spp.	Ryegrass	0.227	CIHEAM (1990) and FEDNA (2010)	
Medicago arborea	Tree medick	0.280		It is equated with alfalfa
Medicago sativa	Alfalfa	0.280	CIHEAM (1990)	
Onobrychis viciifolia	Common sainfoin	0.202	CIHEAM (1990)	
Pisum sativus	Pea, green, for fodder	0.182	CIHEAM (1990)	
Secale cereale	Rye, green	0.194	CIHEAM (1990)	
Sorghum spp.	Sorghum, green	0.202	CIHEAM (1990)	
Trifolium incarnatum	Crimson clover, in bloom	0.215	CIHEAM (1990)	
Trifolium spp.	Other clovers (white, hybrid, subterranean, etc.)	0.215		It is calculated as the average of several species
Trifolium subterraneum	Subterranean clover	0.158	CIHEAM (1990)	
Trigonella foenum-graecum	Fenugreek, green	0.308	CIHEAM (1990)	
Triticum aestivum	Wheat, green	0.158	CIHEAM (1990)	
Vicia articulata	Bard vetch/oneflower vetch, green	0.249	CIHEAM (1990)	It is equated with Vicia villosa

Table AI.3 (Continued) Dry Matter

		Product			Residues (Straw, Pruning, etc.)				
Scientific Name	Product Name	kg Dry Matter/kg Fresh Matter	References	Comments	kg Dry Matter/kg Fresh Matter	Standard Deviation	No. of References	References	Comments
Vicia ervilia	Bitter vetch, green	0.193	CIHEAM (1990)						
Vicia faba	Faba bean/broad bean, green, for fodder	0.168	CIHEAM (1990)						
Vicia sativa	Vetch, green	0.376	CIHEAM (1990)						
Zea mays	Maize, green	0.216	CIHEAM (1990)						
	Other true grasses for fodder	0.194		It is equated with rye					
	Other legumes for green fodder	0.180		Mean (faba bean, vetch, pea, common sainfoin, and french honeysuckle)					
	Other monospecific swards	0.200		It is calculated as the average of several species					
	Other roots and tubers for fodder	0.200		It is calculated as the average of several species					

(Continued)

	Other fodders (Lupin, thistle, parsnip, tree medick)	0.194		It is calculated as the average of several species
	Artificial swards	0.200		It is calculated as the average of several species
	Mixed swards	0.200		It is calculated as the average of several species
	Cereal-legume mixture	0.194		It is equated with rye
Transformed Products				
Sugars				
Beta vulgaris and Saccharum officinarum	White sugar	0.995	Moreiras et al. (2011)	
Saccharum officinarum	Brown sugar	0.965	Moreiras et al. (2011)	
Oils				
Arachis hypogaea	Peanut oil	0.999	Moreiras et al. (2011)	

Table AI.3 (Continued)   Dry Matter

| | | Product | | | Residues (Straw, Pruning, etc.) | | | | |
Scientific Name	Product Name	kg Dry Matter/kg Fresh Matter	References	Comments	kg Dry Matter/kg Fresh Matter	Standard Deviation	No. of References	References	Comments
Cocos nucifera	Coconut oil	0.999	Moreiras et al. (2011)						
Elaeis guineensis	Palm oil	0.999	Moreiras et al. (2011)						
Glycine max	Soybean oil	0.999	Moreiras et al. (2011)						
Helianthus annuus	Sunflower seed oil	0.999	Moreiras et al. (2011)						
Olea europaea	Olive oil	0.999	Moreiras et al. (2011)						
Zea mays	Maize germ oil	0.999	Moreiras et al. (2011)						
Products from Grape									
Vitis vinifera	Vinegar	0.010	Moreiras et al. (2011)						
Vitis vinifera	Wine	0.012	Moreiras et al. (2011)						
Vitis vinifera	Sweet fortified wine	0.132	Moreiras et al. (2011)						
Vitis vinifera	Fine wine	0.031	Moreiras et al. (2011)						
Vitis vinifera	Grape juice	0.165	Moreiras et al. (2011)						

Dried Fruits			
Ficus carica	Fig, dry	0.77	Moreiras et al. (2011)
Phoenix datilifera	Date, dry, stoneless	0.823	Moreiras et al. (2011)
Prunus armeniaca	Apricot, dry	0.705	Moreiras et al. (2011)
Prunus subg. *Prunus*	Plum, dry, stoneless	0.584	Moreiras et al. (2011)
Vitis vinifera	Raisins	0.745	Moreiras et al. (2011)
Agroindustry By-Products			
Arachis hypogaea	Peanut cake	0.912	FEDNA (2010)
Beta vulgaris	Sugar beet molasses	0.753	Piat (1989), FEDNA (2010)
Brassica napus	Rapeseed hulls	0.870	Piat (1989)
Brassica napus	Rapeseed cake	0.892	FEDNA (2010)
Cocos nucifera	Copra cake	0.909	FEDNA (2010)
Elaeis guineensis	Palmkernel cake	0.914	FEDNA (2010)
Fabaceae	Legume brans	0.890	FEDNA (2010)
Glycine max	Soy hulls	0.920	Piat (1989)
Glycine max	Soybean cake	0.880	FEDNA (2010)
Gossypium spp.	Cottonseed hulls	0.904	Piat (1989), CIHEAM (1990)
Gossypium spp.	Cottonseed cake	0.893	FEDNA (2010)

(Continued)

Table AI.3 (Continued) Dry Matter

		Product			Residues (Straw, Pruning, etc.)				
Scientific Name	Product Name	kg Dry Matter/kg Fresh Matter	References	Comments	kg Dry Matter/kg Fresh Matter	Standard Deviation	No. of References	References	Comments
Helianthus annus	Sunflower seed hulls	0.891	Piat (1989), CIHEAM (1990)						
Helianthus annuus	Sunflower seed cake	0.910	FEDNA (2010)						
Linum usitatissimum	Linseed cake	0.910	FEDNA (2010)						
Medicago sativa	Alfalfa meal and pellets	0.912	FEDNA (2010)						
Poaceae	Cereal brans	0.888	FEDNA (2010)						
Saccharum officinarum	Sugarcane molasses	0.737	FEDNA (2010)						
Zea mays	Maize gluten meal	0.890	FEDNA (2010)						
Zea mays	Corn cob	0.940	CIHEAM (1990)						
	Brewers grains	0.243	FEDNA (2010)						
	Blood meal	0.952	FEDNA (2010)						
	Citrus pulp	0.175	FEDNA (2010)						
	Whey	0.956	FEDNA (2010)						
Animal Products									
Milk Products									
Bos taurus	Cow milk	0.119	Moreiras et al. (2011)						
Capra aegagrus hircus	Goat milk	0.118	Moreiras et al. (2011)						

(Continued)

Eggs

Anas platyrhynchos domesticus	Duck eggs	0.281	Moreiras et al. (2011)
Coturnix coturnix japonica	Quail eggs	0.247	Moreiras et al. (2011)
Gallus gallus domesticus	Chicken eggs	0.236	Moreiras et al. (2011)

Honey

Apis mellifera	Honey	0.785	Moreiras et al. (2011)

Meat

Alectoris rufa	Red-legged partridge	0.246	Moreiras et al. (2011)
Anas platyrhynchos domesticus	Duck	0.360	Moreiras et al. (2011)
Bos taurus	Steer sirloin	0.217	Moreiras et al. (2011)
Bos taurus	Lean beef meat	0.261	Moreiras et al. (2011)
Bos taurus	Beef meat	0.377	Moreiras et al. (2011)
Bos taurus	Beef chop	0.375	Moreiras et al. (2011)
Capra aegagrus hircus	Goat meat	0.233	Moreiras et al. (2011)
Capra aegagrus hircus	Lamb chop	0.350	Moreiras et al. (2011)
Capra aegagrus hircus	Lamb, other cuts	0.483	Moreiras et al. (2011)

Table A1.3 (Continued) Dry Matter

| Scientific Name | Product Name | Product | | | | Residues (Straw, Pruning, etc.) | | | | |
		kg Dry Matter/kg Fresh Matter	References	Comments	kg Dry Matter/kg Fresh Matter	Standard Deviation	No. of References	References	Comments
Capra aegagrus hircus	Lamb, leg and chuck	0.366	Moreiras et al. (2011)						
Coturnix coturnix	Quay	0.246	Moreiras et al. (2011)						
Equus ferus caballus	Horse meat	0.220	Moreiras et al. (2011)						
Gallus gallus domesticus	Hen	0.297	Moreiras et al. (2011)						
Gallus gallus domesticus	Chicken	0.297	Moreiras et al. (2011)						
Gallus gallus domesticus	Chicken breast	0.246	Moreiras et al. (2011)						
Meleagris gallopavo mexicana	Turkey, boneless, skinless	0.241	Moreiras et al. (2011)						
Meleagris gallopavo mexicana	Turkey drumstick	0.273	Moreiras et al. (2011)						
Meleagris gallopavo mexicana	Turkey breast, skinless	0.233	Moreiras et al. (2011)						
Meleagris gallopavo mexicana	Turkey, skinless	0.243	Moreiras et al. (2011)						

Oryctolagus cuniculus and Lepus spp.	Rabbit and hare	0.276	Moreiras et al. (2011)
Sus scrofa	Wild boar meat	0.229	Moreiras et al. (2011)
Sus scrofa domestica	Lean pork meat	0.283	Moreiras et al. (2011)
Sus scrofa domestica	Pork meat	0.396	Moreiras et al. (2011)
Sus scrofa domestica	Pork chop	0.449	Moreiras et al. (2011)
Sus scrofa domestica	Pork loin (3% fat)	0.226	Moreiras et al. (2011)
Sus scrofa domestica	Pork loin (9% fat)	0.268	Moreiras et al. (2011)
Sus scrofa domestica	Pork lard	0.950	Moreiras et al. (2011)
Sus scrofa domestica	Pork chuck	0.507	Moreiras et al. (2011)
Sus scrofa domestica	Pork bacon	0.591	Moreiras et al. (2011)
Sus scrofa domestica	Pork sirloin	0.261	Moreiras et al. (2011)
Sus scrofa domestica	Pork fat	0.794	Moreiras et al. (2011)

Table AI.4   Weeds

Scientific Name	Crops	Method of Crop Production	Weed (kg Dry Matter/ha)	References	Comments
	Vegetables				
Allium cepa	Onion	Organic	4257	Vecina and Guzmán (1997)	This weed biomass does not decrease the onion yield
Brassica oleracea	Cabbage	Organic	2087	Gliesmann (1997)	
Brassica oleracea	Cabbage	Organic	615	Díaz del Cañizo et al. (1998)	
Brassica oleracea	Cabbage	Organic	491	Gliesmann (1997)	
Cucurbita pepo var. cylindrica	Zucchini	Organic	475	Díaz del Cañizo et al. (1998)	
Solanum lycopersicum	Tomato	Organic	4036	Vecina and Guzmán (1997)	This weed biomass did not decrease the tomato yield
Solanum lycopersicum	Tomato	Organic	587	Poudel et al. (2002)	
Solanum lycopersicum	Tomato	Low inputs	527	Poudel et al. (2002)	
Solanum lycopersicum	Tomato	Conventional	212	Poudel et al. (2002)	
	Mean	Organic/low inputs	1634		
	Mean	Conventional	212		
	Arable Crops				
Hordeum vulgare	Barley	Conventional	130	Chao et al. (2002)	Measured in April, 2 months before the end of cycle
Hordeum vulgare	Barley	Low inputs	669	Robledo et al. (2007)	No herbicides were used. Was measured at the end of cycle
Hordeum vulgare	Barley	Organic	225	Chao et al. (2002)	Measured in April, 2 months before the end of cycle

Linum usitatissimum	Flax	Organic	2385	Sánchez-Vallduví and Sarandón (2011)	
Linum usitatissimum	Flax	Organic	1650	Sánchez-Vallduví and Sarandón (2011)	Management was different to the previous example
Oryza sativa	Rice	Organic	300	Camacho et al. (2011)	This weed biomass did not decrease the rice yield
Oryza sativa	Rice	Organic	640	Camacho et al. (2011)	Fall of 30% rice yield
Triticum aestivum, T. turgidum	Wheat	Conventional	61	Rios and Carriquiry (2007)	
Triticum turgidum	Durum wheat	Conventional	60	García-Martín et al. (2007)	
Zea mays	Corn	Organic	1310	Poudel et al. (2002)	
Zea mays	Corn	Low inputs	717	Poudel et al. (2002)	
Zea mays	Corn	Conventional	678	Poudel et al. (2002)	
Zea mays	Corn	Organic	754	Fujiyoshi et al. (2007)	
Zea mays	Corn	Organic	84	Fujiyoshi et al. (2007)	Polyculture
	Mean	Organic/low inputs	873		
	Mean	Conventional	232		
Fruit Trees					
Citrus sp.	Citrus	Organic	3.800–4.500	Ingelmo et al. (1994)	Cover crop (*Vicia sativa*+*Avena sativa*)
Citrus sp.	Citrus	Conventional	700	Ingelmo et al. (1994)	Mechanical weed control, without herbicides. Resident ground vegetation

(Continued)

Table AI.4 (Continued) Weeds

Scientific Name	Crops	Method of Crop Production	Weed (kg Dry Matter/ha)	References	Comments
Olea europaea	Olive tree	Organic	3000	Guzmán and Foraster (2011)	Resident ground vegetation. After several years without using herbicides
Olea europaea	Olive tree	Organic	2248	Repullo et al. (2012)	Resident ground vegetation. After several years without using herbicides
Olea europaea	Olive tree		800	Pajarón et al. (1996)	Cover crop (*Vicia sativa*), growing in soils with high slope, heavily eroded
Olea europaea	Olive tree		6243	Alcántara et al. (2009)	Cover crop (*Sinapis alba*)
Vitis vinifera	Grapevine	Conventional	983	Pou et al. (2011)	Resident ground vegetation

Table AI.5   Gross Energy

Scientific Name	Product Name	Gross Energy of the Edible Portion (MJ/kg Fresh Matter)	Gross Energy of the Residue (MJ/kg Fresh Matter)	Gross Energy (MJ/ kg Fresh Matter)	References
	Cereals				
Avena sativa	Oat			15.18	Calculated from ⇒ Food composition: Moreiras et al. (2011), Gross energy of food components: Flores Mengual and Rodríguez Ventura (2013)
Fagopyrum esculentum (*Fam.* Polygonaceae)	Buckwheat			18.19	Calculated from ⇒ Food composition: Costa-Batllori (1978), Gross energy of food components: EWAN (1989)
Hordeum vulgare	Barley			15.63	Calculated from ⇒ Food composition: Mataix and Mañas (1998), Gross energy of food components: Flores Mengual and Rodríguez Ventura (2013)
Oryza sativa	Brown rice			15.18	Calculated from ⇒ Food composition: Moreiras et al. (2011), Gross energy of food components: Flores Mengual and Rodríguez Ventura (2013)
Oryza sativa	White rice			16.58	Calculated from ⇒ Food composition: Moreiras et al. (2011), Gross energy of food components: Flores Mengual and Rodríguez Ventura (2013)
Panicum miliaceum	Millet			15.12	Calculated from ⇒ Food composition: Moreiras et al. (2011), Gross energy of food components: Flores Mengual and Rodríguez Ventura (2013)
Phalaris canariensis	Canary grass			15.18	It is equated with oat
Secale cereale	Rye			14.14	Calculated from ⇒ Food composition: Mataix and Mañas (1998), Gross energy of food components: Flores Mengual and Rodríguez Ventura (2013)
Setaria italica	Foxtail millet			14.44	It is equated with maize
Sorghum bicolor	Sorghum			15.98	It is equated with sorghum

(*Continued*)

Table AI.5 (Continued) Gross Energy

Scientific Name	Product Name	Gross Energy of the Edible Portion (MJ/kg Fresh Matter)	Gross Energy of the Residue (MJ/kg Fresh Matter)	Gross Energy (MJ/kg Fresh Matter)	References
Sorghum spp.	Sorghum			15.98	Gross energy: Piat (1989)
Triticum aestivum	Triticale			15.77	Gross energy: Piat (1989)
Triticum aestivum	Wheat			13.84	Calculated from ⇒ Food composition: Mataix and Mañas (1998), Gross energy of food components: Flores Mengual and Rodríguez Ventura (2013)
Triticum monococcum	Einkorn			13.02	Calculated from ⇒ Food composition: Moreiras et al. (2011), Gross energy of food components: Flores Mengual and Rodríguez Ventura (2013)
Zea mays	Maize			14.44	Calculated from ⇒ Food composition: Mataix and Mañas (1998), Gross energy of food components: Flores Mengual and Rodríguez Ventura (2013)
	Winter cereals, other			14.91	Mean (oat, barley, rye, wheat, and triticale)
	Spring cereals, other			15.18	Mean (maize, sorghum, and millet)
Legumes					
Cicer arietinum	Chickpea			15.76	Calculated from ⇒ Food composition: Moreiras et al. (2011), Gross energy of food components: Flores Mengual and Rodríguez Ventura (2013)
Glycine max	Soybeans			18.20	Calculated from ⇒ Food composition: Moreiras et al. (2011), Gross energy of food components: Flores Mengual and Rodríguez Ventura (2013)
Lathyrus sativus	Grass pea, chickling vetch			18.35	It is equated with bitter vetch
Lens culinaris	Lentils			15.36	Calculated from ⇒ Food composition: Moreiras et al. (2011), Gross energy of food components: Flores Mengual and Rodríguez Ventura (2013)

Lupinus albus	White lupin	20.03		Calculated from ⇒ Food composition: FEDNA (2010), Gross energy of food components: EWAN (1989)
Phaseolus vulgaris	Beans, white	13.84		Calculated from ⇒ Food composition: Moreiras et al. (2011), Gross energy of food components: Flores Mengual and Rodríguez Ventura (2013)
Phaseolus vulgaris	Beans, black	15.30		Calculated from ⇒ Food composition: Moreiras et al. (2011), Gross energy of food components: Flores Mengual and Rodríguez Ventura (2013)
Phaseolus vulgaris	Beans, red	15.72		Calculated from ⇒ Food composition: Moreiras et al. (2011), Gross energy of food components: Flores Mengual and Rodríguez Ventura (2013)
Phaseolus vulgaris	Beans, white and red	13.79		Calculated from ⇒ Food composition: Moreiras et al. (2011), Gross energy of food components: Flores Mengual and Rodríguez Ventura (2013)
Pisum sativus	Pea, dry	15.39		Calculated from ⇒ Food composition: Moreiras et al. (2011), Gross energy of food components: Flores Mengual and Rodríguez Ventura (2013)
Pisum sativus	Pea, dry, fodder varieties	18.44	3.80	Calculated from ⇒ Food composition: FEDNA (2010), Gross energy of food components: EWAN (1989)
Pisum sativus	Pea, green, with pod	10.14	15.11	Calculated from ⇒ Food composition: Moreiras et al. (2011), Gross energy of food components: Flores Mengual and Rodríguez Ventura (2013), Residue moisture: González Vázquez (1944)
Trigonella foenum-graecum	Fenugreek	18.35		It is equated with bitter vetch
Vicia articulata	Bard vetch/ Oneflower vetch	13.84		Calculated from ⇒ Food composition: Moreiras et al. (2011), Gross energy of food components: Flores Mengual and Rodríguez Ventura (2013)
Vicia ervilia	Bitter vetch	18.35		Calculated from ⇒ Food composition: FEDNA (2010), Gross energy of food components: EWAN (1989)

(Continued)

Table AI.5 (Continued) Gross Energy

Scientific Name	Product Name	Gross Energy of the Edible Portion (MJ/kg Fresh Matter)	Gross Energy of the Residue (MJ/kg Fresh Matter)	Gross Energy (MJ/kg Fresh Matter)	References
Vicia faba	Faba bean/broad bean, green, with pod	2.68	15.22	11.46	Calculated from ⇒ Food composition: Moreiras et al. (2011), Gross energy of food components: Flores Mengual and Rodríguez Ventura (2013), Residue moisture: CIHEAM (1990)
Vicia faba	Faba bean/broad bean, green, without pod			2.68	Calculated from ⇒ Food composition: Moreiras et al. (2011), Gross energy of food components: Flores Mengual and Rodríguez Ventura (2013)
Vicia faba	Faba bean/broad bean, dry			15.59	Calculated from ⇒ Food composition: Moreiras et al. (2011), Gross energy of food components: Flores Mengual and Rodríguez Ventura (2013)
Vicia faba	Faba bean/broad bean, dry, fodder varieties			18.52	Calculated from ⇒ Food composition: FEDNA (2010), Gross energy of food components: EWAN (1989)
Vicia monantha	Hard vetch			18.35	It is equated with bitter vetch
Vicia sativa	Vetch			18.75	Calculated from ⇒ Food composition: FEDNA (2010), Gross energy of food components: EWAN (1989)
	Vegetables				
Agaricus bisporus	Common mushroom	1.21	1.51	1.27	Calculated from ⇒ Food composition: Moreiras et al. (2011), Gross energy of food components: Flores Mengual and Rodríguez Ventura (2013), Residue moisture: Moreiras et al. (2011)
Allium ampeloprasum	Leek	1.89	2.27	2.02	Calculated from ⇒ Food composition: Moreiras et al. (2011), Gross energy of food components: Flores Mengual and Rodríguez Ventura (2013), Residue moisture: Moreiras et al. (2011)

Species	Common name				References
Allium ascalonicum	Shallot	3.47	3.64	3.49	Calculated from ⇒ Food composition: Moreiras et al. (2011), Gross energy of food components: Flores Mengual and Rodríguez Ventura (2013), Residue moisture: Moreiras et al. (2011)
Allium cepa	Onion			1.00	Calculated from ⇒ Food composition: Moreiras et al. (2011), Gross energy of food components: Flores Mengual and Rodríguez Ventura (2013), Residue moisture: Moreiras et al. (2011)
Allium fistulosum	Welsh onion	1.19	1.37	1.22	Calculated from ⇒ Food composition: Moreiras et al. (2011), Gross energy of food components: Flores Mengual and Rodríguez Ventura (2013), Residue moisture: Moreiras et al. (2011)
Allium sativum	Garlic	5.25	5.22	5.24	Calculated from ⇒ Food composition: Moreiras et al. (2011), Gross energy of food components: Flores Mengual and Rodríguez Ventura (2013), Residue moisture: Moreiras et al. (2011)
Allium schoenoprasum	Chives			1.21	Calculated from ⇒ Food composition: Moreiras et al. (2011), Gross energy of food components: Flores Mengual and Rodríguez Ventura (2013), Residue moisture: Moreiras et al. (2011)
Apium graveolens	Celery	0.60	0.81	0.67	Calculated from ⇒ Food composition: Moreiras et al. (2011), Gross energy of food components: Flores Mengual and Rodríguez Ventura (2013), Residue moisture: Moreiras et al. (2011)
Asparagus officinalis	Asparagus	0.81	0.93	0.86	Calculated from ⇒ Food composition: Moreiras et al. (2011), Gross energy of food components: Flores Mengual and Rodríguez Ventura (2013), Residue moisture: Moreiras et al. (2011)
Beta vulgaris	Beet	1.39	1.90	1.48	Calculated from ⇒ Food composition: Moreiras et al. (2011), Gross energy of food components: Flores Mengual and Rodríguez Ventura (2013), Residue moisture: Moreiras et al. (2011)

(Continued)

Table A1.5 (Continued) Gross Energy

Scientific Name	Product Name	Gross Energy of the Edible Portion (MJ/kg Fresh Matter)	Gross Energy of the Residue (MJ/kg Fresh Matter)	Gross Energy (MJ/ kg Fresh Matter)	References
Beta vulgaris var. *Cicla*	Chard	1.38	2.20	1.63	Calculated from ⇒ Food composition: Moreiras et al. (2011), Gross energy of food components: Flores Mengual and Rodríguez Ventura (2013), Residue moisture: Moreiras et al. (2011)
Borago officinalis	Borage	1.21	1.14	1.20	Calculated from ⇒ Food composition: Moreiras et al. (2011), Gross energy of food components: Flores Mengual and Rodríguez Ventura (2013), Residue moisture: Moreiras et al. (2011)
Brassica oleracea	Cabbage	1.45	1.81	1.54	Calculated from ⇒ Food composition: Moreiras et al. (2011), Gross energy of food components: Flores Mengual and Rodríguez Ventura (2013), Residue moisture: Moreiras et al. (2011)
Brassica oleracea var. *botrytis*	Cauliflower	1.11	1.34	1.16	Calculated from ⇒ Food composition: Moreiras et al. (2011), Gross energy of food components: Flores Mengual and Rodríguez Ventura (2013), Residue moisture: Moreiras et al. (2011)
Brassica oleracea var. *capitata* f. *rubra*	Red cabbage	1.00	1.34	1.06	Calculated from ⇒ Food composition: Moreiras et al. (2011), Gross energy of food components: Flores Mengual and Rodríguez Ventura (2013), Residue moisture: Moreiras et al. (2011)
Brassica oleracea var. *gemmifera*	Brussels sprout	2.05	2.25	2.08	Calculated from ⇒ Food composition: Moreiras et al. (2011), Gross energy of food components: Flores Mengual and Rodríguez Ventura (2013), Residue moisture: Moreiras et al. (2011)
Brassica oleracea var. *italica*	Broccoli	1.67	1.70	1.68	Calculated from ⇒ Food composition: Moreiras et al. (2011), Gross energy of food components: Flores Mengual and Rodríguez Ventura (2013), Residue moisture: Moreiras et al. (2011)

Brassica oleracea var. *sabellica*	Cabbage (kale or borecole)	1.34	1.69	1.41	Calculated from ⇒ Food composition: Moreiras et al. (2011), Gross energy of food components: Flores Mengual and Rodríguez Ventura (2013), Residue moisture: Moreiras et al. (2011)
Brassica rapa var. *pekinensis*	Chinese cabbage	1.43	1.48	1.43	Calculated from ⇒ Food composition: Moreiras et al. (2011), Gross energy of food components: Flores Mengual and Rodríguez Ventura (2013), Residue moisture: Moreiras et al. (2011)
Brassica rapa var. *rapa*	Turnip greens/turnip tops	0.64	1.18	0.74	Calculated from ⇒ Food composition: Moreiras et al. (2011), Gross energy of food components: Flores Mengual and Rodríguez Ventura (2013), Residue moisture: Moreiras et al. (2011)
Brassica rapa var. *rapa*	Turnip	1.15	1.56	1.26	Calculated from ⇒ Food composition: Moreiras et al. (2011), Gross energy of food components: Flores Mengual and Rodríguez Ventura (2013), Residue moisture: Moreiras et al. (2011)
Capsicum annuum	Red pepper	1.47	1.70	1.52	Calculated from ⇒ Food composition: Moreiras et al. (2011), Gross energy of food components: Flores Mengual and Rodríguez Ventura (2013), Residue moisture: Moreiras et al. (2011)
Capsicum annuum	Green pepper	0.91	1.05	0.94	Calculated from ⇒ Food composition: Moreiras et al. (2011), Gross energy of food components: Flores Mengual and Rodríguez Ventura (2013), Residue moisture: Moreiras et al. (2011)
Cichorium endivia	Endive	0.97	1.12	1.03	Calculated from ⇒ Food composition: Moreiras et al. (2011), Gross energy of food components: Flores Mengual and Rodríguez Ventura (2013), Residue moisture: Moreiras et al. (2011)
Cichorium intybus	Chicory			0.83	Calculated from ⇒ Food composition: Mataix and Mañas (1998), Gross energy of food components: Flores Mengual and Rodríguez Ventura (2013)

(Continued)

Table A1.5 (Continued) Gross Energy

Scientific Name	Product Name	Gross Energy of the Edible Portion (MJ/kg Fresh Matter)	Gross Energy of the Residue (MJ/kg Fresh Matter)	Gross Energy (MJ/kg Fresh Matter)	References
Cichorium intybus	Belgian endive	1.05	1.16	1.07	Calculated from ⇒ Food composition: Moreiras et al. (2011), Gross energy of food components: Flores Mengual and Rodriguez Ventura (2013), Residue moisture: Moreiras et al. (2011)
Cucumis sativus	Cucumber	0.56	0.58	0.57	Calculated from ⇒ Food composition: Moreiras et al. (2011), Gross energy of food components: Flores Mengual and Rodriguez Ventura (2013), Residue moisture: Moreiras et al. (2011)
Cucurbita pepo	Squash/pumpkin	0.61	1.90	1.04	Calculated from ⇒ Food composition: Moreiras et al. (2011), Gross energy of food components: Flores Mengual and Rodriguez Ventura (2013), Residue moisture: Moreiras et al. (2011)
Cucurbita pepo var. cylindrica	Zucchini	0.59	0.62	0.60	Calculated from ⇒ Food composition: Moreiras et al. (2011), Gross energy of food components: Flores Mengual and Rodriguez Ventura (2013), Residue moisture: Author data
Cynara cardunculus	Artichoke thistle	1.00	1.07	1.01	Calculated from ⇒ Food composition: Moreiras et al. (2011), Gross energy of food components: Flores Mengual and Rodriguez Ventura (2013), Residue moisture: Moreiras et al. (2011)
Cynara scolimus	Artichoke	1.84	2.09	2.00	Calculated from ⇒ Food composition: Moreiras et al. (2011), Gross energy of food components: Flores Mengual and Rodriguez Ventura (2013), Residue moisture: Moreiras et al. (2011)
Daucus carota	Carrot	1.53	1.42	1.51	Calculated from ⇒ Food composition: Moreiras et al. (2011), Gross energy of food components: Flores Mengual and Rodriguez Ventura (2013), Residue moisture: Moreiras et al. (2011)

Ipomoea batatas	Sweet potato	4.17	4.53	4.24	Calculated from ⇒ Food composition: Moreiras et al. (2011), Gross energy of food components: Flores Mengual and Rodríguez Ventura (2013), Residue moisture: Moreiras et al. (2011)
Lactarius deliciosus	Saffron milk cap/red pine mushroom	0.71	1.28	0.94	Calculated from ⇒ Food composition: Moreiras et al. (2011), Gross energy of food components: Flores Mengual and Rodríguez Ventura (2013), Residue moisture: Moreiras et al. (2011)
Lactuca sativa	Lettuce	0.70	0.83	0.73	Calculated from ⇒ Food composition: Moreiras et al. (2011), Gross energy of food components: Flores Mengual and Rodríguez Ventura (2013), Residue moisture: Moreiras et al. (2011)
Lactuca sativa	Iceberg lettuce	0.60	0.74	0.62	Calculated from ⇒ Food composition: Moreiras et al. (2011), Gross energy of food components: Flores Mengual and Rodríguez Ventura (2013), Residue moisture: Moreiras et al. (2011)
Lactuca sativa	Miniature lettuce			0.70	Calculated from ⇒ Food composition: Moreiras et al. (2011), Gross energy of food components: Flores Mengual and Rodríguez Ventura (2013), Residue moisture: Moreiras et al. (2011)
Manihot sculenta	Cassava			6.92	Calculated from ⇒ Food composition: Moreiras et al. (2011), Gross energy of food components: Flores Mengual and Rodríguez Ventura (2013), Residue moisture: Moreiras et al. (2011)
Nasturtium officinale	Watercress	1.15	1.30	1.21	Calculated from ⇒ Food composition: Moreiras et al. (2011), Gross energy of food components: Flores Mengual and Rodríguez Ventura (2013), Residue moisture: Moreiras et al. (2011)
Pastinaca sativa	Parsnip	2.72	3.22	2.86	Calculated from ⇒ Food composition: Moreiras et al. (2011), Gross energy of food components: Flores Mengual and Rodríguez Ventura (2013), Residue moisture: Moreiras et al. (2011)

(Continued)

Table AI.5 (Continued)   Gross Energy

Scientific Name	Product Name	Gross Energy of the Edible Portion (MJ/kg Fresh Matter)	Gross Energy of the Residue (MJ/kg Fresh Matter)	Gross Energy (MJ/kg Fresh Matter)	References
Phaseolus vulgaris	Beans, green	1.46	1.83	1.49	Calculated from ⇒ Food composition: Moreiras et al. (2011), Gross energy of food components: Flores Mengual and Rodríguez Ventura (2013), Residue moisture: Moreiras et al. (2011)
Raphanus sativus	Radish	0.69	0.83	0.74	Calculated from ⇒ Food composition: Moreiras et al. (2011), Gross energy of food components: Flores Mengual and Rodríguez Ventura (2013), Residue moisture: Moreiras et al. (2011)
Rumex acetosa	Sorrel			1.28	Calculated from ⇒ Food composition: Moreiras et al. (2011), Gross energy of food components: Flores Mengual and Rodríguez Ventura (2013), Residue moisture: Moreiras et al. (2011)
Solanum lycopersicum	Tomato	0.87	1.05	0.88	Calculated from ⇒ Food composition: Moreiras et al. (2011), Gross energy of food components: Flores Mengual and Rodríguez Ventura (2013), Residue moisture: Moreiras et al. (2011)
Solanum melongena	Eggplant/ aubergine	1.10	1.23	1.12	Calculated from ⇒ Food composition: Moreiras et al. (2011), Gross energy of food components: Flores Mengual and Rodríguez Ventura (2013), Residue moisture: Moreiras et al. (2011)
Solanum tuberosum	Potato	3.71	3.99	3.74	Calculated from ⇒ Food composition: Moreiras et al. (2011), Gross energy of food components: Flores Mengual and Rodríguez Ventura (2013), Residue moisture: Moreiras et al. (2011)
Spinacia oleracea	Spinach	0.92	1.83	1.09	Calculated from ⇒ Food composition: Moreiras et al. (2011), Gross energy of food components: Flores Mengual and Rodríguez Ventura (2013), Residue moisture: Moreiras et al. (2011)

Valerianella locusta	Corn salad/mâche			0.69	Calculated from ⇒ Food composition: Moreiras et al. (2011), Gross energy of food components: Flores Mengual and Rodríguez Ventura (2013), Residue moisture: Moreiras et al. (2011)
	Mushrooms	1.21	1.51	1.27	Calculated from ⇒ Food composition: Moreiras et al. (2011), Gross energy of food components: Flores Mengual and Rodríguez Ventura (2013), Residue moisture: Moreiras et al. (2011)
Fruits					
Actinidia deliciosa	Kiwifruit	2.25	2.48	2.28	Calculated from ⇒ Food composition: Moreiras et al. (2011), Gross energy of food components: Flores Mengual and Rodríguez Ventura (2013), Residue moisture: Moreiras et al. (2011)
Ananas comosus	Pineapple	2.07	2.32	2.18	Calculated from ⇒ Food composition: Moreiras et al. (2011), Gross energy of food components: Flores Mengual and Rodríguez Ventura (2013), Residue moisture: Moreiras et al. (2011)
Annona cherimola	Cherimoya	3.71	4.06	3.85	Calculated from ⇒ Food composition: Moreiras et al. (2011), Gross energy of food components: Flores Mengual and Rodríguez Ventura (2013), Residue moisture: Moreiras et al. (2011)
Carica papaya	Papaya	1.65	2.06	1.75	Calculated from ⇒ Food composition: Moreiras et al. (2011), Gross energy of food components: Flores Mengual and Rodríguez Ventura (2013), Residue moisture: Moreiras et al. (2011)
Citrullus lanatus	Watermelon	0.86	0.95	0.90	Calculated from ⇒ Food composition: Moreiras et al. (2011), Gross energy of food components: Flores Mengual and Rodríguez Ventura (2013), Residue moisture: Moreiras et al. (2011)
Citrus reticulata	Mandarin	1.71	2.06	1.81	Calculated from ⇒ Food composition: Moreiras et al. (2011), Gross energy of food components: Flores Mengual and Rodríguez Ventura (2013), Residue moisture: Moreiras et al. (2011)

(Continued)

Table AI.5 (Continued) Gross Energy

Scientific Name	Product Name	Gross Energy of the Edible Portion (MJ/kg Fresh Matter)	Gross Energy of the Residue (MJ/kg Fresh Matter)	Gross Energy (MJ/kg Fresh Matter)	References
Citrus x aurantifolia	Lime	0.52	0.95	0.65	Calculated from ⇒ Food composition: Moreiras et al. (2011), Gross energy of food components: Flores Mengual and Rodríguez Ventura (2013), Residue moisture: Moreiras et al. (2011)
Citrus x limon	Lemon	1.85	1.95	1.88	Calculated from ⇒ Food composition: Moreiras et al. (2011), Gross energy of food components: Flores Mengual and Rodríguez Ventura (2013), Residue moisture: Moreiras et al. (2011)
Citrus x paradisi	Grapefruit	1.38	1.63	1.46	Calculated from ⇒ Food composition: Moreiras et al. (2011), Gross energy of food components: Flores Mengual and Rodríguez Ventura (2013), Residue moisture: Moreiras et al. (2011)
Citrus x sinensis	Orange	1.65	2.00	1.74	Calculated from ⇒ Food composition: Moreiras et al. (2011), Gross energy of food components: Flores Mengual and Rodríguez Ventura (2013), Residue moisture: Moreiras et al. (2011)
Cocos nucifera	Coconut, fresh	15.41	9.38	13.60	Calculated from ⇒ Food composition: Moreiras et al. (2011), Gross energy of food components: Flores Mengual and Rodríguez Ventura (2013), Residue moisture: Moreiras et al. (2011)
Cocos nucifera	Coconut, water			0.68	Calculated from ⇒ Food composition: Moreiras et al. (2011), Gross energy of food components: Flores Mengual and Rodríguez Ventura (2013)
Cocos nucifera	Coconut, milk			10.39	Calculated from ⇒ Food composition: Moreiras et al. (2011), Gross energy of food components: Flores Mengual and Rodríguez Ventura (2013)

Cucumis melo	Melon	1.16	1.34	1.23	Calculated from ⇒ Food composition: Moreiras et al. (2011), Gross energy of food components: Flores Mengual and Rodríguez Ventura (2013), Residue moisture: Moreiras et al. (2011)
Cucumis melo var. cantalupensis	Cantaloupe	1.46	1.60	1.52	Calculated from ⇒ Food composition: Moreiras et al. (2011), Gross energy of food components: Flores Mengual and Rodríguez Ventura (2013), Residue moisture: Moreiras et al. (2011)
Cydonia oblonga	Quince	1.25	2.39	1.69	Calculated from ⇒ Food composition: Moreiras et al. (2011), Gross energy of food components: Flores Mengual and Rodríguez Ventura (2013), Residue moisture: Moreiras et al. (2011)
Diospyros kaki	Persimmon	3.00	3.27	3.03	Calculated from ⇒ Food composition: Moreiras et al. (2011), Gross energy of food components: Flores Mengual and Rodríguez Ventura (2013), Residue moisture: Moreiras et al. (2011)
Eriobotrya japonica	Loquat	2.09	3.81	2.74	Calculated from ⇒ Food composition: Moreiras et al. (2011), Gross energy of food components: Flores Mengual and Rodríguez Ventura (2013), Residue moisture: Moreiras et al. (2011)
Ficus carica	Figs	3.00	3.46	3.07	Calculated from ⇒ Food composition: Moreiras et al. (2011), Gross energy of food components: Flores Mengual and Rodríguez Ventura (2013), Residue moisture: Moreiras et al. (2011)
Fragaria x ananassa	Strawberry	1.55	1.83	1.56	Calculated from ⇒ Food composition: Moreiras et al. (2011), Gross energy of food components: Flores Mengual and Rodríguez Ventura (2013), Residue moisture: Moreiras et al. (2011)
Litchi chinensis	Litchee	3.15	3.34	3.22	Calculated from ⇒ Food composition: Moreiras et al. (2011), Gross energy of food components: Flores Mengual and Rodríguez Ventura (2013), Residue moisture: Moreiras et al. (2011)

(Continued)

Table A1.5 (Continued) Gross Energy

Scientific Name	Product Name	Gross Energy of the Edible Portion (MJ/kg Fresh Matter)	Gross Energy of the Residue (MJ/kg Fresh Matter)	Gross Energy (MJ/ kg Fresh Matter)	References
Malpighia emarginata	Acerola	0.74	2.51	1.08	Calculated from ⇒ Food composition: Mataix and Mañas (1998), Gross energy of food components: Flores Mengual and Rodríguez Ventura (2013), Residue moisture: It is equated with apple
Malus domestica	Apple	2.11	2.51	2.17	Calculated from ⇒ Food composition: Moreiras et al. (2011), Gross energy of food components: Flores Mengual and Rodríguez Ventura (2013), Residue moisture: Moreiras et al. (2011)
Mangifera indica	Mango	2.64	3.15	2.80	Calculated from ⇒ Food composition: Moreiras et al. (2011), Gross energy of food components: Flores Mengual and Rodríguez Ventura (2013), Residue moisture: Moreiras et al. (2011)
Musa x paradisiaca	Bananas, plantains	3.79	4.38	3.99	Calculated from ⇒ Food composition: Moreiras et al. (2011), Gross energy of food components: Flores Mengual and Rodríguez Ventura (2013), Residue moisture: Moreiras et al. (2011)
Olea europaea	Table olives, with stone	8.15	7.26	7.98	Calculated from ⇒ Food composition: Moreiras et al. (2011), Gross energy of food components: Flores Mengual and Rodríguez Ventura (2013), Residue moisture and residue gross energy: IDAE (2007)
Olea europaea	Table olives, without stone	8.15		8.15	Calculated from ⇒ Food composition: Moreiras et al. (2011), Gross energy of food components: Flores Mengual and Rodríguez Ventura (2013)
Olea europaea	Oil olives, with stone	11.55		11.55	Calculated from ⇒ Food composition: Moreiras et al. (2011), Gross energy of food components: Flores Mengual and Rodríguez Ventura (2013)

Opuntia ficus-indica	Prickly pear	1.97	4.06	2.91	Calculated from ⇒ Food composition: Mataix and Mañas (1998), Gross energy of food components: Flores Mengual and Rodríguez Ventura (2013), Residue moisture: It is equated with cherimoya
P. persica var. *nucipersica*	Nectarine	1.89	2.23	1.93	Calculated from ⇒ Food composition: Moreiras et al. (2011), Gross energy of food components: Flores Mengual and Rodríguez Ventura (2013), Residue moisture: Moreiras et al. (2011)
Passiflora edulis	Passion fruit	2.32	2.43	2.36	Calculated from ⇒ Food composition: Moreiras et al. (2011), Gross energy of food components: Flores Mengual and Rodríguez Ventura (2013), Residue moisture: Moreiras et al. (2011)
Persea americana	Avocado	6.03	13.39	8.16	Calculated from ⇒ Food composition: Moreiras et al. (2011), Gross energy of food components: Flores Mengual and Rodríguez Ventura (2013), Residue moisture: Olaeta et al. (2007)
Prunus armeniaca	Apricot	1.80	15.87	2.92	Calculated from ⇒ Food composition: Moreiras et al. (2011), Gross energy of food components: Flores Mengual and Rodríguez Ventura (2013), Residue moisture: Lazos (1991)
Prunus avium	Cherry var. Picota	2.67	2.86	2.70	Calculated from ⇒ Food composition: Moreiras et al. (2011), Gross energy of food components: Flores Mengual and Rodríguez Ventura (2013), Residue moisture: Moreiras et al. (2011)
Prunus avium, Prunus cerasus	Cherry	2.67	16.20	4.43	Calculated from ⇒ Food composition: Moreiras et al. (2011), Gross energy of food components: Flores Mengual and Rodríguez Ventura (2013), Residue moisture: Lazos (1991)
Prunus persica	Peach	1.67	16.27	3.42	Calculated from ⇒ Food composition: Moreiras et al. (2011), Gross energy of food components: Flores Mengual and Rodríguez Ventura (2013), Residue moisture: Lazos (1991)

(Continued)

Table A1.5 (Continued) Gross Energy

Scientific Name	Product Name	Gross Energy of the Edible Portion (MJ/kg Fresh Matter)	Gross Energy of the Residue (MJ/kg Fresh Matter)	Gross Energy (MJ/ kg Fresh Matter)	References
Prunus subg. *Prunus*	Plum	2.01	2.41	2.07	Calculated from ⇒ Food composition: Moreiras et al. (2011), Gross energy of food components: Flores Mengual and Rodríguez Ventura (2013), Residue moisture: Moreiras et al. (2011)
Psidium guajava	Apple guava	1.39	2.18	1.48	Calculated from ⇒ Food composition: Moreiras et al. (2011), Gross energy of food components: Flores Mengual and Rodríguez Ventura (2013), Residue moisture: Moreiras et al. (2011)
Punica granatum	Pomegranate	1.48	1.49	1.49	Calculated from ⇒ Food composition: Moreiras et al. (2011), Gross energy of food components: Flores Mengual and Rodríguez Ventura (2013), Residue moisture: Moreiras et al. (2011)
Pyrus communis	Pear	1.89	2.34	1.95	Calculated from ⇒ Food composition: Moreiras et al. (2011), Gross energy of food components: Flores Mengual and Rodríguez Ventura (2013), Residue moisture: Moreiras et al. (2011)
Ribes nigrum	Blackcurrant			1.41	Calculated from ⇒ Food composition: Moreiras et al. (2011), Gross energy of food components: Flores Mengual and Rodríguez Ventura (2013)
Ribes rubrum	Redcurrant			1.15	Calculated from ⇒ Food composition: Moreiras et al. (2011), Gross energy of food components: Flores Mengual and Rodríguez Ventura (2013)
Rubus idaeus	Raspberry	1.22		1.22	Calculated from ⇒ Food composition: Moreiras et al. (2011), Gross energy of food components: Flores Mengual and Rodríguez Ventura (2013)
Rubus ulmifolius	Blackberry			1.15	Calculated from ⇒ Food composition: Moreiras et al. (2011), Gross energy of food components: Flores Mengual and Rodríguez Ventura (2013)

Tamarindus indica	Tamarind	11.50	12.48	12.15	Calculated from ⇒ Food composition: Moreiras et al. (2011), Gross energy of food components: Flores Mengual and Rodríguez Ventura (2013), Residue moisture: Moreiras et al. (2011)
Tamarindus indica	Tamarind, pulp			11.50	Calculated from ⇒ Food composition: Moreiras et al. (2011), Gross energy of food components: Flores Mengual and Rodríguez Ventura (2013)
Vaccinium spp.	Blueberry			1.41	Calculated from ⇒ Food composition: Moreiras et al. (2011), Gross energy of food components: Flores Mengual and Rodríguez Ventura (2013), Residue moisture: Moreiras et al. (2011)
Vitis vinifera	White grapes	2.88	3.09	2.90	Calculated from ⇒ Food composition: Moreiras et al. (2011), Gross energy of food components: Flores Mengual and Rodríguez Ventura (2013), Residue moisture: Moreiras et al. (2011)
Vitis vinifera	Black grapes	2.77	2.90	2.79	Calculated from ⇒ Food composition: Moreiras et al. (2011), Gross energy of food components: Flores Mengual and Rodríguez Ventura (2013), Residue moisture: Moreiras et al. (2011)
Nuts and Seeds					
Anacardium occidentale	Cashew, without shell			25.92	Calculated from ⇒ Food composition: Moreiras et al. (2011), Gross energy of food components: Flores Mengual and Rodríguez Ventura (2013)
Arachis hypogaea	Peanuts, without shell			26.77	Calculated from ⇒ Food composition: Moreiras et al. (2011), Gross energy of food components: Flores Mengual and Rodríguez Ventura (2013)
Brassica napus	Rapeseed			27.33	Calculated from ⇒ Food composition: FEDNA (2010), Gross energy of food components: EWAN (1989)
Cannabis sativa	Hemp seed			25.96	Calculated from ⇒ Food composition: FEDNA (2010), Gross energy of food components: EWAN (1989)

(Continued)

Table AI.5 (Continued) Gross Energy

Scientific Name	Product Name	Gross Energy of the Edible Portion (MJ/kg Fresh Matter)	Gross Energy of the Residue (MJ/kg Fresh Matter)	Gross Energy (MJ/ kg Fresh Matter)	References
Castanea sativa	Chestnut	8.50	14.94	9.66	Calculated from ⇒ Food composition: Moreiras et al. (2011), Gross energy of food components: Flores Mengual and Rodríguez Ventura (2013), Residue moisture (It is equated with hazelnuts and peanuts): IDAE (2007)
Ceratonia siliqua	Carobs			17.21	Calculated from ⇒ Food composition: FEDNA (2010), Gross energy of food components: EWAN (1989)
Corylus avellana	Hazelnuts, without shell			25.36	Calculated from ⇒ Food composition: Moreiras et al. (2011), Gross energy of food components: Flores Mengual and Rodríguez Ventura (2013)
Cyperus esculentus	Tigernuts			17.87	Calculated from ⇒ Food composition: Moreiras et al. (2011), Gross energy of food components: Flores Mengual and Rodríguez Ventura (2013)
Gossypium spp.	Cotton seed			22.23	Calculated from ⇒ Food composition: FEDNA (2010), Gross energy of food components: EWAN (1989)
Helianthus annuus	Sunflower seeds, with shell	26.38	15.65	23.38	Calculated from ⇒ Food composition: Moreiras et al. (2011), Gross energy of food components: Flores Mengual and Rodríguez Ventura (2013), Residue moisture: Piat (1989) and CIHEAM (1990)
Helianthus annuus	Sunflower seeds, without shell			26.38	Calculated from ⇒ Food composition: Moreiras et al. (2011), Gross energy of food components: Flores Mengual and Rodríguez Ventura (2013)
Juglans regia	Walnut, with shell	26.79	15.46	21.13	Calculated from ⇒ Food composition: Moreiras et al. (2011), Gross energy of food components: Flores Mengual and Rodríguez Ventura (2013)
Juglans regia	Walnut, without shell			26.79	Calculated from ⇒ Food composition: Moreiras et al. (2011), Gross energy of food components: Flores Mengual and Rodríguez Ventura (2013)

Macadamia integrifolia	Macadamia nut		32.20	Calculated from ⇒ Food composition: Moreiras et al. (2011), Gross energy of food components: Flores Mengual and Rodríguez Ventura (2013)
Pinus pinea	Pine nuts		30.65	Calculated from ⇒ Food composition: Moreiras et al. (2011), Gross energy of food components: Flores Mengual and Rodríguez Ventura (2013)
Pistacia vera	Pistachio	26.84	21.25	Calculated from ⇒ Food composition: Moreiras et al. (2011), Gross energy of food components: Flores Mengual and Rodríguez Ventura (2013), Residue moisture (It's equated with hazelnuts and peanuts): IDAE (2007)
Prunus dulcis	Almonds, with shell	26.06	19.67	Calculated from ⇒ Food composition: Moreiras et al. (2011), Gross energy of food components: Flores Mengual and Rodríguez Ventura (2013), Residue moisture and gross energy: IDAE (2007), Seed/shell ratio: Alonso et al. (2012)
Prunus dulcis	Almonds, without shell		26.06	Calculated from ⇒ Food composition: Moreiras et al. (2011), Gross energy of food components: Flores Mengual and Rodríguez Ventura (2013)
Quercus ilex	Acorn, with shell		18.33	Calculated from ⇒ Food composition: FEDNA (2010), Gross energy of food components: EWAN (1989)
Quercus ilex	Acorn, without shell		18.56	Calculated from ⇒ Food composition: FEDNA (2010), Gross energy of food components: EWAN (1989)
Sesamum indicum	Sesame		26.96	Calculated from ⇒ Food composition: Moreiras et al. (2011), Gross energy of food components: Flores Mengual and Rodríguez Ventura (2013)
Spices				
Anethum graveolens	Dill		13.47	Calculated from ⇒ Food composition: Moreiras et al. (2011), Gross energy of food components: Flores Mengual and Rodríguez Ventura (2013)
Capparis spinosa	Caper		1.74	Calculated from ⇒ Food composition: Moreiras et al. (2011), Gross energy of food components: Flores Mengual and Rodríguez Ventura (2013)

(*Continued*)

Table AI.5 (Continued) Gross Energy

Scientific Name	Product Name	Gross Energy of the Edible Portion (MJ/kg Fresh Matter)	Gross Energy of the Residue (MJ/kg Fresh Matter)	Gross Energy (MJ/ kg Fresh Matter)	References
Capsicum annuum	Jalapeño chili pepper	1.22	15.08	3.02	Calculated from ⇒ Food composition: Moreiras et al. (2011), Gross energy of food components: Flores Mengual and Rodríguez Ventura (2013), Residue moisture: Moreiras et al. (2011)
Capsicum annuum	Chili pepper	2.03	29.52	6.43	Calculated from ⇒ Food composition: Moreiras et al. (2011), Gross energy of food components: Flores Mengual and Rodríguez Ventura (2013), Residue moisture: Moreiras et al. (2011)
Capsicum annuum	Chili pepper, dry, milled			14.90	Calculated from ⇒ Food composition: Moreiras et al. (2011), Gross energy of food components: Flores Mengual and Rodríguez Ventura (2013)
Capsicum annuum	Red pepper, dry, milled			14.41	Calculated from ⇒ Food composition: Moreiras et al. (2011), Gross energy of food components: Flores Mengual and Rodríguez Ventura (2013)
Cinnamomum spp.	Cinnamon, milled			2.15	Calculated from ⇒ Food composition: Moreiras et al. (2011), Gross energy of food components: Flores Mengual and Rodríguez Ventura (2013)
Coriandum sativum	Coriander	2.39	26.01	4.75	Calculated from ⇒ Food composition: Moreiras et al. (2011), Gross energy of food components: Flores Mengual and Rodríguez Ventura (2013), Residue moisture: Moreiras et al. (2011)
Crocus sativus	Saffron			15.38	Calculated from ⇒ Food composition: Moreiras et al. (2011), Gross energy of food components: Flores Mengual and Rodríguez Ventura (2013)
Cuminum cyminum	Cumin			18.91	Calculated from ⇒ Food composition: Moreiras et al. (2011), Gross energy of food components: Flores Mengual and Rodríguez Ventura (2013)

Foeniculum vulgare	Fennel	0.64	23.55	5.22	Calculated from ⇒ Food composition: Moreiras et al. (2011), Gross energy of food components: Flores Mengual and Rodríguez Ventura (2013), Residue moisture: Moreiras et al. (2011)
Laurus nobilis	Bay laurel			13.29	Calculated from ⇒ Food composition: Moreiras et al. (2011), Gross energy of food components: Flores Mengual and Rodríguez Ventura (2013)
Mentha spp.	Mint and peppermint			2.05	Calculated from ⇒ Food composition: Moreiras et al. (2011), Gross energy of food components: Flores Mengual and Rodríguez Ventura (2013)
Ocimum basilicum	Basil, fresh			0.89	Calculated from ⇒ Food composition: Moreiras et al. (2011), Gross energy of food components: Flores Mengual and Rodríguez Ventura (2013)
Ocimum basilicum	Basil, dry			8.35	Calculated from ⇒ Food composition: Moreiras et al. (2011), Gross energy of food components: Flores Mengual and Rodríguez Ventura (2013)
Origanum vulgare	Oregano, fresh			2.94	Calculated from ⇒ Food composition: Moreiras et al. (2011), Gross energy of food components: Flores Mengual and Rodríguez Ventura (2013)
Origanum vulgare	Oregano, dry			14.96	Calculated from ⇒ Food composition: Moreiras et al. (2011), Gross energy of food components: Flores Mengual and Rodríguez Ventura (2013)
Petroselinum crispum	Parsley			1.66	Calculated from ⇒ Food composition: Moreiras et al. (2011), Gross energy of food components: Flores Mengual and Rodríguez Ventura (2013)
Piper nigrum	White pepper			3.21	Calculated from ⇒ Food composition: Moreiras et al. (2011), Gross energy of food components: Flores Mengual and Rodríguez Ventura (2013)
Piper nigrum	Black pepper			3.82	Calculated from ⇒ Food composition: Moreiras et al. (2011), Gross energy of food components: Flores Mengual and Rodríguez Ventura (2013)

(*Continued*)

Table AI.5 (Continued) Gross Energy

Scientific Name	Product Name	Gross Energy of the Edible Portion (MJ/kg Fresh Matter)	Gross Energy of the Residue (MJ/kg Fresh Matter)	Gross Energy (MJ/ kg Fresh Matter)	References
Rosmarinus officinalis	Rosemary			14.97	Calculated from ⇒ Food composition: Moreiras et al. (2011), Gross energy of food components: Flores Mengual and Rodríguez Ventura (2013)
Syzygium aromaticum	Cloves			18.98	Calculated from ⇒ Food composition: Moreiras et al. (2011), Gross energy of food components: Flores Mengual and Rodríguez Ventura (2013)
Thymus spp.	Thyme			14.69	Calculated from ⇒ Food composition: Moreiras et al. (2011), Gross energy of food components: Flores Mengual and Rodríguez Ventura (2013)
Vanilla planifolia	Vanilla, extract			2.21	Calculated from ⇒ Food composition: Moreiras et al. (2011), Gross energy of food components: Flores Mengual and Rodríguez Ventura (2013)
Zingiber officinale	Ginger, dry, milled	15.29		15.29	Calculated from ⇒ Food composition: Moreiras et al. (2011), Gross energy of food components: Flores Mengual and Rodríguez Ventura (2013)
Zingiber officinale	Ginger, fresh	2.28	14.66	3.02	Calculated from ⇒ Food composition: Moreiras et al. (2011), Gross energy of food components: Flores Mengual and Rodríguez Ventura (2013), Residue moisture: Moreiras et al. (2011)
	Green Fodder				
Avena sativa	Oat, green			5.32	Calculated from ⇒ Moisture: CIHEAM (1990)
Beta vulgaris	Fodder beet			3.23	Calculated from ⇒ Moisture: CIHEAM (1990)
Beta vulgaris	Sugar beet, necks			3.57	Calculated from ⇒ Moisture: FEDNA (2010)
Beta vulgaris	Beet pulp			4.38	Calculated from ⇒ Moisture: FEDNA (2010)

Brassica napus	Turnip, for fodder	2.21	Calculated from ⇒ Moisture: CIHEAM (1990)
Brassica oleracea	Fodder cabbage	1.45	It is equated with human food varieties
Cucurbita pepo	Squash, for fodder	1.90	It is equated with human food varieties
Cynara cardunculus	Artichoke thistle, for fodder	1.00	It is equated with human food varieties
Daucus carota	Carrot, for fodder	1.53	It is equated with human food varieties
Hedysarum coronarium	French honeysuckle	2.69	Calculated from ⇒ Moisture: CIHEAM (1990)
Helianthus tuberosus	Jerusalem artichoke	4.12	Agrobit (2013)
Hordeum vulgare	Barley, green	4.36	Calculated from ⇒ Moisture: CIHEAM (1990)
Lolium perenne	Perennial ryegrass	4.20	Calculated from ⇒ Moisture: CIHEAM (1990)
Lolium spp.	Ryegrass	3.99	Calculated from ⇒ Moisture: CIHEAM (1990) and FEDNA (2010)
Medicago arborea	Tree medick	4.92	It is equated with *Medicago sativa*
Medicago sativa	Alfalfa	4.92	Calculated from ⇒ Moisture: CIHEAM (1990)
Onobrychis viciifolia	Common sainfoin	3.55	Calculated from ⇒ Moisture: CIHEAM (1990)
Pastinaca sativa	Parsnip, for fodder	2.97	Calculated from ⇒ Food composition: Mataix and Mañas (1998), Gross energy of food components: Flores Mengual and Rodríguez Ventura (2013)
Pisum sativus	Pea, green, for fodder	3.20	Calculated from ⇒ Moisture: CIHEAM (1990)
Secale cereale	Rye, green	3.41	Calculated from ⇒ Moisture: CIHEAM (1990)

(Continued)

Table AI.5 (Continued)   Gross Energy

Scientific Name	Product Name	Gross Energy of the Edible Portion (MJ/kg Fresh Matter)	Gross Energy of the Residue (MJ/kg Fresh Matter)	Gross Energy (MJ/ kg Fresh Matter)	References
Sorghum spp.	Sorghum, green			3.55	Calculated from ⇒ Moisture: CIHEAM (1990)
Trifolium incarnatum	Crimson clover, in bloom			3.78	Calculated from ⇒ Moisture: CIHEAM (1990)
Trifolium spp.	Other clovers (white, hybrid, subterranean, etc.)			3.78	It is calculated as the average of several species
Trifolium subterraneum	Subterranean clover			2.78	Calculated from ⇒ Moisture: CIHEAM (1990)
Trigonella foenum-graecum	Fenugreek, green			5.41	Calculated from ⇒ Moisture: CIHEAM (1990)
Triticum aestivum	Wheat, green			6.61	Calculated from ⇒ Moisture: CIHEAM (1990)
Vicia ervilia	Bitter vetch, green			5.89	Calculated from ⇒ Moisture: CIHEAM (1990)
Vicia faba	Faba bean/ broad bean, green, for fodder			2.95	Calculated from ⇒ Moisture: CIHEAM (1990)
Vicia monanthos/ V. articulata	Bard vetch/ oneflower vetch, green			4.38	Calculated from ⇒ Moisture (it is equated with *Vicia villosa*): (CIHEAM, 1990)
Vicia sativa	Vetch, green			3.39	Calculated from ⇒ Moisture: CIHEAM (1990)
Zea mays	Maize, green			3.80	Calculated from ⇒ Moisture: CIHEAM (1990)
	Other true grasses for fodder			3.41	It is equated with rye

	Other legumes for green fodder	3.16	Mean (faba bean, vetch, pea, common sainfoin, and french honeysuckle)
	Other monospecific swards	3.51	It is calculated as the average of several species
	Other roots and tubers for fodder	3.51	It is calculated as the average of several species
	Other fodders (lupin, thistle, parsnip, medick, etc.)	3.41	It is calculated as the average of several species
	Artificial swards	3.51	It is calculated as the average of several species
	Mixed swards	3.51	It is calculated as the average of several species
	Fodder	3.41	It is equated with rye
Fiber			
Avena sativa	Oat	15.94	Calculated from $\Rightarrow$ Moisture: CIHEAM (1990)
Cannabis sativa	Hemp, fiber	16.01	Calculated from $\Rightarrow$ Moisture: González Vazquez (1944) and Soroa (1953)
Cicer arietinum	Chickpea	15.69	Calculated from $\Rightarrow$ Moisture: CIHEAM (1990)
Glycine max	Soybeans	15.56	Calculated from $\Rightarrow$ Moisture: CIHEAM (1990), Aranguiz (2006), Kyle et al. (2011)
Gossypium spp.	Cotton, fiber	15.82	Calculated from $\Rightarrow$ Moisture: Centro de Comercio Internacional (2013), González Vazquez (1944) and Soroa (1953)
Hordeum vulgare	Barley	15.18	Calculated from $\Rightarrow$ Moisture: CIHEAM (1990)
Lathyrus sativus	Grass pea, chickling vetch	16.08	Calculated from $\Rightarrow$ Moisture: CIHEAM (1990)

(Continued)

ENERGY IN AGROECOSYSTEMS

Table AI.5 (*Continued*) Gross Energy

Scientific Name	Product Name	Gross Energy of the Edible Portion (MJ/kg Fresh Matter)	Gross Energy of the Residue (MJ/kg Fresh Matter)	Gross Energy (MJ/ kg Fresh Matter)	References
Lens culinaris	Lentils			16.31	Calculated from ⇒ Moisture: CIHEAM (1990)
Linum usitatissimum	Flax			16.33	Calculated from ⇒ Moisture: González Vazquez (1944) and Soroa (1953)
Oryza sativa	Brown rice			15.99	Calculated from ⇒ Moisture: Kyle et al. (2011)
Panicum miliaceum	Millet			15.82	Calculated from ⇒ Moisture: Kyle et al. (2011)
Phaseolus vulgaris	Beans, white			15.31	Calculated from ⇒ Moisture: CIHEAM (1990)
Pisum sativus	Pea, dry			15.94	Calculated from ⇒ Moisture: CIHEAM (1990)
Secale cereale	Rye			16.24	Calculated from ⇒ Moisture: CIHEAM (1990)
Sorghum spp.	Sorghum			15.29	Calculated from ⇒ Moisture: Kyle et al. (2011)
Triticum aestivum/ T. durum	Wheat			15.23	Calculated from ⇒ Moisture: CIHEAM (1990), Kyle et al. (2011)
Vicia ervilia	Bitter vetch			16.06	Calculated from ⇒ Moisture: CIHEAM (1990)
Vicia faba	Faba bean/ broad bean, dry			15.57	Calculated from ⇒ Moisture: CIHEAM (1990)
Vicia monantha	Hard vetch			15.57	Calculated from ⇒ Moisture: CIHEAM (1990)
Vicia sativa	Vetch			16.01	Calculated from ⇒ Moisture: CIHEAM (1990)
x Triticosecale	Triticale			16.20	Calculated from ⇒ Moisture: CIHEAM (1990)
Zea mays	Maize			15.48	Calculated from ⇒ Moisture: CIHEAM (1990)
	Straw				
	Winter cereals, other			15.65	Mean (barley, wheat, oat, and rye)
	Spring cereals, other			15.48	It is equated with maize
	Legumes, other			15.81	Mean of all other

	Crops Residues		
Allium ampeloprasum	Leek	3.60	Calculated from ⇒ Moisture: Rahn and Lillywhite (2002), Kyle et al. (2011)
Allium cepa	Onion	3.49	Calculated from ⇒ Moisture: Rahn and Lillywhite (2002), Kyle et al. (2011)
Allium sativum	Garlic	5.27	Calculated from ⇒ Moisture: Kyle et al. (2011)
Apium graveolens	Celery	1.74	Calculated from ⇒ Moisture: Kyle et al. (2011)
Asparagus officinalis	Asparagus	5.27	Calculated from ⇒ Moisture: Kyle et al. (2011)
Beta vulgaris	Beet	2.13	Calculated from ⇒ Moisture: Rahn and Lillywhite (2002)
Beta vulgaris var. Cicla	Chard	3.25	Calculated from ⇒ Moisture: Aranguiz (2006), Kyle et al. (2011)
Borago officinalis	Borage	3.06	It is equated with lettuce
Brassica oleracea	Cabbage	3.10	Calculated from ⇒ Moisture: Aranguiz (2006), Rahn and Lillywhite (2002), Kyle et al. (2011)
Brassica oleracea var. botrytis	Cauliflower	3.69	Calculated from ⇒ Moisture: Aranguiz (2006), Kyle et al. (2011)
Brassica oleracea var. italica	Broccoli	3.10	Calculated from ⇒ Moisture: Aranguiz (2006), Rahn and Lillywhite (2002), Kyle et al. (2011)
Brassica rapa var. rapa	Turnip	3.51	Calculated from ⇒ Moisture: Kyle et al. (2011)
Capsicum annuum	Chili pepper	5.27	Calculated from ⇒ Moisture: Kyle et al. (2011)
Capsicum annuum	Green pepper	5.27	Calculated from ⇒ Moisture: Kyle et al. (2011)
Cichorium endivia	Endive	3.06	It is equated with lettuce
Cichorium intybus	Chicory	3.06	Calculated from ⇒ Moisture: Kyle et al. (2011)
Cichorium intybus	Belgian endive	3.06	It is equated with lettuce
Citrullus lanatus	Watermelon	3.51	Calculated from ⇒ Moisture: Kyle et al. (2011)

(Continued)

Table AI.5 (Continued) Gross Energy

Scientific Name	Product Name	Gross Energy of the Edible Portion (MJ/kg Fresh Matter)	Gross Energy of the Residue (MJ/kg Fresh Matter)	Gross Energy (MJ/ kg Fresh Matter)	References
Cucumis melo	Melon			3.51	Calculated from ⇒ Moisture: Kyle et al. (2011)
Cucumis sativus	Cultivar for pickled cucumber, gherkins			3.51	Calculated from ⇒ Moisture: Kyle et al. (2011)
Cucumis sativus	Cucumber			3.13	Calculated from ⇒ Moisture: Aranguiz (2006), Kyle et al. (2011)
Cucurbita pepo	Squash/pumpkin			5.27	Calculated from ⇒ Moisture: Kyle et al. (2011)
Cucurbita pepo var. *cylindrica*	Zucchini			3.13	It is equated with cucumber
Cynara cardunculus	Artichoke thistle			3.06	It is equated with lettuce
Cynara scolimus	Artichoke			3.51	Calculated from ⇒ Moisture: Kyle et al. (2011)
Daucus carota	Carrot			3.60	Calculated from ⇒ Moisture: Rahn and Lillywhite (2002), Kyle et al. (2011)
Fragaria x ananassa	Strawberry			5.27	Estimated
Lactuca sativa	Lettuce			3.06	Calculated from ⇒ Moisture: Aranguiz (2006), Rahn and Lillywhite (2002), Kyle et al. (2011)
Manihot sculenta	Cassava			5.27	Calculated from ⇒ Moisture: Kyle et al. (2011)
Phaseolus vulgaris	Beans, green			5.27	Calculated from ⇒ Moisture: Kyle et al. (2011)
Pisum sativus	Pea, green, with pod			5.27	Calculated from ⇒ Moisture: Kyle et al. (2011)
Raphanus sativus	Radish			3.29	Calculated from ⇒ Moisture: Rahn and Lillywhite (2002), Kyle et al. (2011)

Solanum lycopersicum	Tomato	2.33	Calculated from ⇒ Moisture: Aranguiz (2006), Kyle et al. (2011)
Solanum melongena	Eggplant/ aubergine	3.51	Calculated from ⇒ Moisture: Kyle et al. (2011)
Solanum tuberosum	Potato	3.51	Calculated from ⇒ Moisture: Aranguiz (2006), Kyle et al. (2011)
Spinacia oleracea	Spinach	3.25	Calculated from ⇒ Moisture: Aranguiz (2006), Kyle et al. (2011)
Vicia faba	Faba bean/ Broad bean, green, without pod	4.18	Calculated from ⇒ Moisture: Aranguiz (2006), Rahn and Lillywhite (2002), Kyle et al. (2011)
	Animal Products		
	Milk Products		
Bos taurus	Cow milk	3.01	Calculated from ⇒ Food composition: Moreiras et al. (2011), Gross energy of food components: Flores Mengual and Rodríguez Ventura (2013)
Capra aegagrus hircus	Goat milk	3.07	Calculated from ⇒ Food composition: Moreiras et al. (2011), Gross energy of food components: Flores Mengual and Rodríguez Ventura (2013)
Equus africuanus asinus	Donkey milk	1.89	Calculated from ⇒ Food composition: Mataix and Mañas (1998), Gross energy of food components: Flores Mengual and Rodríguez Ventura (2013)
Ovis orientais aries	Sheep milk	4.43	Calculated from ⇒ Food composition: Mataix and Mañas (1998), Gross energy of food components: Flores Mengual and Rodríguez Ventura (2013)
	Eggs		
Anas platyrhynchos domesticus	Duck eggs	3.15	Calculated from ⇒ Food composition: Moreiras et al. (2011), Gross energy of food components: Flores Mengual and Rodríguez Ventura (2013)

(Continued)

Table AI.5 (Continued) Gross Energy

Scientific Name	Product Name	Gross Energy of the Edible Portion (MJ/kg Fresh Matter)	Gross Energy of the Residue (MJ/kg Fresh Matter)	Gross Energy (MJ/ kg Fresh Matter)	References
Coturnix coturnix japonica	Quail eggs			2.41	Calculated from ⇒ Food composition: Moreiras et al. (2011), Gross energy of food components: Flores Mengual and Rodríguez Ventura (2013)
Gallus gallus domesticus	Chicken eggs			2.25	Calculated from ⇒ Food composition: Moreiras et al. (2011), Gross energy of food components: Flores Mengual and Rodríguez Ventura (2013)
Honey					
Apis mellifera	Honey			13.38	Calculated from ⇒ Food composition: Moreiras et al. (2011), Gross energy of food components: Flores Mengual and Rodríguez Ventura (2013)
Meat					
Alectoris rufa	Red-legged partridge			4.14	Calculated from ⇒ Food composition: Moreiras et al. (2011), Gross energy of food components: Flores Mengual and Rodríguez Ventura (2013)
Anas platyrhynchos domesticus	Duck			9.36	Calculated from ⇒ Food composition: Moreiras et al. (2011), Gross energy of food components: Flores Mengual and Rodríguez Ventura (2013)
Bos taurus	Steer sirloin			4.39	Calculated from ⇒ Food composition: Moreiras et al. (2011), Gross energy of food components: Flores Mengual and Rodríguez Ventura (2013)
Bos taurus	Lean beef meat			6.87	Calculated from ⇒ Food composition: Moreiras et al. (2011), Gross energy of food components: Flores Mengual and Rodríguez Ventura (2013)
Bos taurus	Beef meat			11.43	Calculated from ⇒ Food composition: Moreiras et al. (2011), Gross energy of food components: Flores Mengual and Rodríguez Ventura (2013)

Species	Common name	Value	Source
Bos taurus	Beef chop	9.29	Calculated from ⇒ Food composition: Moreiras et al. (2011), Gross energy of food components: Flores Mengual and Rodríguez Ventura (2013)
Bos taurus	Beef liver	6.86	Calculated from ⇒ Food composition: Moreiras et al. (2011), Gross energy of food components: Flores Mengual and Rodríguez Ventura (2013)
Bos taurus	Beef tongue	8.71	Calculated from ⇒ Food composition: Moreiras et al. (2011), Gross energy of food components: Flores Mengual and Rodríguez Ventura (2013)
Bos taurus	Beef kidney	4.69	Calculated from ⇒ Food composition: Mataix and Mañas (1998), Gross energy of food components: Flores Mengual and Rodríguez Ventura (2013)
Capra aegagrus hircus	Goat meat	4.20	Calculated from ⇒ Food composition: Moreiras et al. (2011), Gross energy of food components: Flores Mengual and Rodríguez Ventura (2013)
Capra aegagrus hircus	Lamb chop	5.39	Calculated from ⇒ Food composition: Moreiras et al. (2011), Gross energy of food components: Flores Mengual and Rodríguez Ventura (2013)
Capra aegagrus hircus	Lamb, sweetbreads	6.56	Calculated from ⇒ Food composition: Moreiras et al. (2011), Gross energy of food components: Flores Mengual and Rodríguez Ventura (2013)
Capra aegagrus hircus	Lamb, other cuts	9.97	Calculated from ⇒ Food composition: Moreiras et al. (2011), Gross energy of food components: Flores Mengual and Rodríguez Ventura (2013)
Capra aegagrus hircus	Lamb, leg and chuck	8.22	Calculated from ⇒ Food composition: Moreiras et al. (2011), Gross energy of food components: Flores Mengual and Rodríguez Ventura (2013)
Capra aegagrus hircus	Lamb, brain	5.49	Calculated from ⇒ Food composition: Moreiras et al. (2011), Gross energy of food components: Flores Mengual and Rodríguez Ventura (2013)
Coturnix coturnix	Quay	4.14	Calculated from ⇒ Food composition: Moreiras et al. (2011), Gross energy of food components: Flores Mengual and Rodríguez Ventura (2013)

(Continued)

Table AI.5 (Continued) Gross Energy

Scientific Name	Product Name	Gross Energy of the Edible Portion (MJ/kg Fresh Matter)	Gross Energy of the Residue (MJ/kg Fresh Matter)	Gross Energy (MJ/kg Fresh Matter)	References
Equus ferus caballus	Horse meat			5.22	Calculated from ⇒ Food composition: Moreiras et al. (2011), Gross energy of food components: Flores Mengual and Rodríguez Ventura (2013)
Gallus gallus domesticus	Hen			5.87	Calculated from ⇒ Food composition: Moreiras et al. (2011), Gross energy of food components: Flores Mengual and Rodríguez Ventura (2013)
Gallus gallus domesticus	Chicken			5.87	Calculated from ⇒ Food composition: Moreiras et al. (2011), Gross energy of food components: Flores Mengual and Rodríguez Ventura (2013)
Gallus gallus domesticus	Chicken breast			6.11	Calculated from ⇒ Food composition: Moreiras et al. (2011), Gross energy of food components: Flores Mengual and Rodríguez Ventura (2013)
Meleagris gallopavo mexicana	Turkey, boneless, skinless			5.90	Calculated from ⇒ Food composition: Moreiras et al. (2011), Gross energy of food components: Flores Mengual and Rodríguez Ventura (2013)
Meleagris gallopavo mexicana	Turkey drumstick			5.63	Calculated from ⇒ Food composition: Moreiras et al. (2011), Gross energy of food components: Flores Mengual and Rodríguez Ventura (2013)
Meleagris gallopavo mexicana	Turkey breast, skinless			5.58	Calculated from ⇒ Food composition: Moreiras et al. (2011), Gross energy of food components: Flores Mengual and Rodríguez Ventura (2013)
Meleagris gallopavo mexicana	Turkey, skinless			3.40	Calculated from ⇒ Food composition: Moreiras et al. (2011), Gross energy of food components: Flores Mengual and Rodríguez Ventura (2013)
Oryctolagus cuniculus and *Lepus* spp.	Rabbit and hare			4.60	Calculated from ⇒ Food composition: Moreiras et al. (2011), Gross energy of food components: Flores Mengual and Rodríguez Ventura (2013)

Sus scrofa	Wild boar meat	5.81	Calculated from ⇒ Food composition: Moreiras et al. (2011), Gross energy of food components: Flores Mengual and Rodríguez Ventura (2013)
Sus scrofa domestica	Lean pork meat	7.84	Calculated from ⇒ Food composition: Moreiras et al. (2011), Gross energy of food components: Flores Mengual and Rodríguez Ventura (2013)
Sus scrofa domestica	Pork meat	12.79	Calculated from ⇒ Food composition: Moreiras et al. (2011), Gross energy of food components: Flores Mengual and Rodríguez Ventura (2013)
Sus scrofa domestica	Pork chop	10.83	Calculated from ⇒ Food composition: Moreiras et al. (2011), Gross energy of food components: Flores Mengual and Rodríguez Ventura (2013)
Sus scrofa domestica	Pork liver	5.84	Calculated from ⇒ Food composition: Moreiras et al. (2011), Gross energy of food components: Flores Mengual and Rodríguez Ventura (2013)
Sus scrofa domestica	Pork loin (3% fat)	5.63	Calculated from ⇒ Food composition: Moreiras et al. (2011), Gross energy of food components: Flores Mengual and Rodríguez Ventura (2013)
Sus scrofa domestica	Pork loin (9% fat)	7.59	Calculated from ⇒ Food composition: Moreiras et al. (2011), Gross energy of food components: Flores Mengual and Rodríguez Ventura (2013)
Sus scrofa domestica	Pork lard	38.81	Calculated from ⇒ Food composition: Moreiras et al. (2011), Gross energy of food components: Flores Mengual and Rodríguez Ventura (2013)
Sus scrofa domestica	Pork chuck	16.33	Calculated from ⇒ Food composition: Moreiras et al. (2011), Gross energy of food components: Flores Mengual and Rodríguez Ventura (2013)
Sus scrofa domestica	Pork bacon	18.52	Calculated from ⇒ Food composition: Moreiras et al. (2011), Gross energy of food components: Flores Mengual and Rodríguez Ventura (2013)

(Continued)

Table AI.5 (Continued) Gross Energy

Scientific Name	Product Name	Gross Energy of the Edible Portion (MJ/kg Fresh Matter)	Gross Energy of the Residue (MJ/kg Fresh Matter)	Gross Energy (MJ/ kg Fresh Matter)	References
Sus scrofa domestica	Pork blood			4.53	Calculated from ⇒ Food composition: Moreiras et al. (2011), Gross energy of food components: Flores Mengual and Rodríguez Ventura (2013)
Sus scrofa domestica	Pork sirloin			6.82	Calculated from ⇒ Food composition: Moreiras et al. (2011), Gross energy of food components: Flores Mengual and Rodríguez Ventura (2013)
Sus scrofa domestica	Pork fat			29.62	Calculated from ⇒ Food composition: Moreiras et al. (2011), Gross energy of food components: Flores Mengual and Rodríguez Ventura (2013)
	Transformed Products				
	Sugars				
Beta vulgaris and *Saccharum officinarum*	White sugar			16.92	Calculated from ⇒ Food composition: Moreiras et al. (2011), Gross energy of food components: Flores Mengual and Rodríguez Ventura (2013)
Saccharum officinarum	Brown sugar			16.41	Calculated from ⇒ Food composition: Moreiras et al. (2011), Gross energy of food components: Flores Mengual and Rodríguez Ventura (2013)
	Oils				
Arachis hypogaea	Peanut oil			38.96	Calculated from ⇒ Food composition: Moreiras et al. (2011), Gross energy of food components: Flores Mengual and Rodríguez Ventura (2013)
Cocos nucifera	Coconut oil			38.96	Calculated from ⇒ Food composition: Moreiras et al. (2011), Gross energy of food components: Flores Mengual and Rodríguez Ventura (2013)

Species	Product	Value	Source
Elaeis guineensis	Palm oil	38.96	Calculated from ⇒ Food composition: Moreiras et al. (2011), Gross energy of food components: Flores Mengual and Rodríguez Ventura (2013)
Glycine max	Soybean oil	38.96	Calculated from ⇒ Food composition: Moreiras et al. (2011), Gross energy of food components: Flores Mengual and Rodríguez Ventura (2013)
Helianthus annuus	Sunflower seed oil	38.96	Calculated from ⇒ Food composition: Moreiras et al. (2011), Gross energy of food components: Flores Mengual and Rodríguez Ventura (2013)
Olea europaea	Olive oil	38.96	Calculated from ⇒ Food composition: Moreiras et al. (2011), Gross energy of food components: Flores Mengual and Rodríguez Ventura (2013)
Zea mays	Maize germ oil	38.96	Calculated from ⇒ Food composition: Moreiras et al. (2011), Gross energy of food components: Flores Mengual and Rodríguez Ventura (2013)
	Products from Grape		
Vitis vinifera	Vinegar	0.19	Calculated from ⇒ Food composition: Moreiras et al. (2011), Gross energy of food components: Flores Mengual and Rodríguez Ventura (2013)
Vitis vinifera	Wine	0.21	Calculated from ⇒ Food composition: Moreiras et al. (2011), Gross energy of food components: Flores Mengual and Rodríguez Ventura (2013)
Vitis vinifera	Sweet fortified wine	2.26	Calculated from ⇒ Food composition: Moreiras et al. (2011), Gross energy of food components: Flores Mengual and Rodríguez Ventura (2013)
Vitis vinifera	Fine wine	0.53	Calculated from ⇒ Food composition: Moreiras et al. (2011), Gross energy of food components: Flores Mengual and Rodríguez Ventura (2013)
Vitis vinifera	Grape juice	2.83	Calculated from ⇒ Food composition: Moreiras et al. (2011), Gross energy of food components: Flores Mengual and Rodríguez Ventura (2013)

(Continued)

Table AI.5 (Continued) Gross Energy

Scientific Name	Product Name	Gross Energy of the Edible Portion (MJ/kg Fresh Matter)	Gross Energy of the Residue (MJ/kg Fresh Matter)	Gross Energy (MJ/ kg Fresh Matter)	References
	Dried Fruits				
Ficus carica	Fig, dry			10.60	Calculated from ⇒ Food composition: Moreiras et al. (2011), Gross energy of food components: Flores Mengual and Rodríguez Ventura (2013)
Phoenix datilifera	Date, dry, with stone			11.97	Calculated from ⇒ Food composition: Moreiras et al. (2011), Gross energy of food components: Flores Mengual and Rodríguez Ventura (2013)
Phoenix datilifera	Date, dry, stoneless			12.73	Calculated from ⇒ Food composition: Moreiras et al. (2011), Gross energy of food components: Flores Mengual and Rodríguez Ventura (2013)
Prunus armeniaca	Apricot, dry			8.74	Calculated from ⇒ Food composition: Moreiras et al. (2011), Gross energy of food components: Flores Mengual and Rodríguez Ventura (2013)
Prunus subg. *Prunus*	Plum, dry, with stone			7.32	Calculated from ⇒ Food composition: Moreiras et al. (2011), Gross energy of food components: Flores Mengual and Rodríguez Ventura (2013)
Prunus subg. *Prunus*	Plum, dry, stoneless			7.33	Calculated from ⇒ Food composition: Moreiras et al. (2011), Gross energy of food components: Flores Mengual and Rodríguez Ventura (2013)
Vitis vinifera	Raisins			11.66	Calculated from ⇒ Food composition: Moreiras et al. (2011), Gross energy of food components: Flores Mengual and Rodríguez Ventura (2013)
	Agroindustry By-Products				
Arachis hypogaea	Peanut cake			19.81	Calculated from ⇒ Food composition: FEDNA (2010), Gross energy of food components: EWAN (1989)

Beta vulgaris	Sugar beet molasses	16.14	Calculated from ⇒ Food composition: FEDNA (2010), Gross energy of food components: EWAN (1989)
Brassica napus	Rapeseed cake	19.71	Calculated from ⇒ Food composition: FEDNA (2010), Gross energy of food components: EWAN (1989)
Cocos nucifera	Copra cake	19.35	Calculated from ⇒ Food composition: FEDNA (2010), Gross energy of food components: EWAN (1989)
Elaeis guineensis	Palmkernel cake	19.20	Calculated from ⇒ Food composition: FEDNA (2010), Gross energy of food components: EWAN (1989)
Fabaceae	Legume brans	17.80	Calculated from ⇒ Food composition: FEDNA (2010), Gross energy of food components: EWAN (1989)
Glycine max	Soybean cake	19.54	Calculated from ⇒ Food composition: FEDNA (2010), Gross energy of food components: EWAN (1989)
Gossypium spp.	Cotton seed cake	18.40	Calculated from ⇒ Food composition: FEDNA (2010), Gross energy of food components: EWAN (1989)
Helianthus annuus	Sunflower seed cake	17.70	Gross energy: Piat (1989)
Linum usitatissimum	Lin seed cake	20.06	Calculated from ⇒ Food composition: FEDNA (2010), Gross energy of food components: EWAN (1989)
Medicago sativa	Alfalfa meal and pellets	16.87	Calculated from ⇒ Food composition: FEDNA (2010), Gross energy of food components: EWAN (1989)
Poaceae	Cereal brans	19.40	Calculated from ⇒ Food composition: FEDNA (2010), Gross energy of food components: EWAN (1989)
Zea mays	Maize gluten meal	19.24	Calculated from ⇒ Food composition: FEDNA (2010), Gross energy of food components: EWAN (1989)
Zea mays	Maize meal	16.95	Gross energy: Piat (1989)
	Blood meal	19.15	Calculated from ⇒ Food composition: FEDNA (2010), Gross energy of food components: EWAN (1989)
	Whey	16.34	Calculated from ⇒ Food composition: FEDNA (2010), Gross energy of food components: EWAN (1989)

(Continued)

Table AI.5 (Continued) Gross Energy

Scientific Name	Product Name	Gross Energy of the Edible Portion (MJ/kg Fresh Matter)	Gross Energy of the Residue (MJ/kg Fresh Matter)	Gross Energy (MJ/kg Fresh Matter)	References
	Wood and Pruning				
Citrus reticulata	Mandarin, pruning			11.00	Calculated from ⇒ Moisture: Fundación Abertis (2005), Eubionet (2003), Blasi et al. (1997), Voivontas et al. (1991), Gross Energy: Voivontas et al. (1991)
Citrus x limon	Lemon, pruning			11.00	Calculated from ⇒ Moisture: Fundación Abertis (2005), Eubionet (2003), Blasi et al. (1997), Voivontas et al. (1991), Gross Energy: Voivontas et al. (1991)
Citrus x sinensis	Orange, pruning			11.59	Calculated from ⇒ Moisture: Fundación Abertis (2005), Eubionet (2003), Blasi et al. (1997), Voivontas et al. (1991), Gross Energy: Fundación Abertis (2005), Voivontas et al. (1991)
Fagus sylvatica	European beech/ commom beech, wood			13.80	Calculated from ⇒ Gross Energy: Francescato et al. (2008)
Malus domestica	Apple, pruning			12.65	Calculated from ⇒ Moisture: Fundación Abertis (2005), Eubionet (2003), Blasi et al. (1997), Voivontas et al. (1991), Bilandzija et al. (2012), Gross Energy: Fundación Abertis (2005), Voivontas et al. (1991)
Olea europaea	Olive tree, pruning			13.16	Calculated from ⇒ Moisture: Fundación Abertis (2005), Eubionet (2003), Blasi et al. (1997), Voivontas et al. (1991), Bilandzija et al. (2012), Civantos and Olid (1985), Gross Energy: Fundación Abertis (2005), Voivontas et al. (1991)
Picea spp.	Spruce, wood			14.10	Calculated from ⇒ Gross Energy: Francescato et al. (2008)
Populus spp.	Poplar, wood			13.88	Calculated from ⇒ Gross Energy: Francescato et al. (2008)

Prunus armeniaca	Apricot, pruning	13.70	Calculated from ⇒ Moisture: Fundación Abertis (2005), Eubionet (2003), Voivontas et al. (1991), Bilandzija et al. (2012), Gross Energy: Voivontas et al. (1991)
Prunus avium	Cherry, pruning	12.49	Calculated from ⇒ Moisture: Fundación Abertis (2005), Eubionet (2003), Voivontas et al. (1991), Bilandzija et al. (2012), Gross Energy: Voivontas et al. (1991)
Prunus dulcis	Almonds, pruning	12.81	Calculated from ⇒ Moisture: Fundación Abertis (2005), Eubionet (2003), Blasi et al. (1997), Voivontas et al. (1991), Bilandzija et al. (2012), Gross Energy: Fundación Abertis (2005), Voivontas et al. (1991)
Prunus persica	Peach, pruning	13.32	Calculated from ⇒ Moisture: Fundación Abertis (2005), Eubionet (2003), Blasi et al. (1997), Voivontas et al. (1991), Bilandzija et al. (2012), Gross Energy: Voivontas et al. (1991)
Pyrus communis	Pear, pruning	12.82	Calculated from ⇒ Moisture: Fundación Abertis (2005), Eubionet (2003), Blasi et al. (1997), Voivontas et al. (1991), Bilandzija et al. (2012), Gross Energy: Fundación Abertis (2005), Voivontas et al. (1991)
Salix spp.	Willow, sallow, wood	13.80	Calculated from ⇒ Gross Energy: Francescato et al. (2008)
Vitis vinifera	Grapevine, branches	12.61	Calculated from ⇒ Moisture: Fundación Abertis (2005), Eubionet (2003), Blasi et al. (1997), Voivontas et al. (1991), Bilandzija et al. (2012), AVEBIOM (2009), Gross Energy: Fundación Abertis (2005), Voivontas et al. (1991), IDAE (2007), Francescato et al. (2008)
	Conifers, wood	15.23	Calculated from ⇒ Gross Energy: Francescato et al. (2008), FAO (1991), IDAE (2007)
	Bark (conifers)	15.10	Calculated from ⇒ Gross Energy: Francescato et al. (2008), IDAE, 2007
	Bark (broad-leaved tree)	14.66	Calculated from ⇒ Gross Energy: IDAE, 2007
	Broad-leaved tree, wood	14.52	Calculated from ⇒ Gross Energy: Francescato et al. (2008), FAO (1991), IDAE (2007)

Methodology for the Estimation of Embodied Energy Coefficients

**Eduardo Aguilera, Gloria I. Guzmán, Juan Infante,
David Soto and Manuel González de Molina**

AII.1 FUELS

Direct energy values of main fuels (Tables 4.2 and AII.1.1) were preferentially taken from the energy statistics manual of the IEA (2004), as they represent the current international standard. In the case of natural gas and coal, we used specific values provided by ecoinvent (Frischknecht et al. 2007b) and Audsley et al. (2003) instead of IEA ranges. Distillates are estimated as the average of fuel oil and diesel.

Resource extraction energy of fossil fuels can be derived from published EROI data. Long-term evolution of oil and gas EROI data is only available for the United States (Guilford et al., 2011). World data is only available for the 1990–2010 period. To estimate a long-term world series, we used the world trend from Hall et al. (2014) for the 1990–2010 period and assumed that the relationship between world EROI and U.S. EROI is maintained for previous periods. We simulated the evolution of the EROI and energy requirements of world coal production based on Hall et al. (2014) relative share of United States and China in total production. The results of these calculations, expressed as MJ/kg fuel, are shown in Table AII.1.4.

Refining energy consumption data for oil-derived fuels in year 2006 were taken from Wang (2008). These values were modulated to take into account the increase in the share of unconventional oil from 1990 to 2010. We assumed that unconventional oil requires 2.5 more energy to refine than conventional oil, based on Karras (2010). The relative shares of the two types of oil were taken from IEA's World Energy Outlook (IEA, 2012, 2015). For coal processing energy, we assumed a fixed consumption of 6.5 kWh electricity/Mg coal during the whole study period based on data around year 2000 from ecoinvent (ecoinvent Centre, 2007), assuming that 90% of coal is hard coal and 10% is lignite. The results are shown in Table AII.1.6.

Section AII.1 Fuels

Energy

Table AII.1.1 Fuel Energy and Properties

	Density g/l	Gross Energy (GE) MJ/kg	MJ/l	Net Energy (NE) MJ/kg	MJ NE/l	MJ NE/MJ GE
Fuel oil, kerosene	802.6	46.2	37.1	43.9	35.3	0.95
Gasoline	740.7	47.1	34.9	44.8	33.2	0.95
Diesel	843.9	45.7	38.5	43.4	36.5	0.95
Naphtha	690.6	47.7	33.0	45.3	31.4	0.95
Distillates	823.3	45.9	37.8	43.6	35.9	0.95
LPG	522.2	50.1	26.2	46.2	24.1	0.92
Natural gas (m3)	799.6	50.4	40.0	45.4	36.0	0.90
Average liquids	795.7	45.7	36.9	43.4	35.1	0.95
Coal		22.4		21.7		0.97
Biomass		15.3				

Table AII.1.2 Share of International Transport over Total Production of Fossil Primary Energy Sources

	1900 (%)	1910 (%)	1920 (%)	1930 (%)	1940 (%)	1950 (%)	1960 (%)	1970 (%)	1980 (%)	1990 (%)	2000 (%)	2010 (%)
Oil	8	8	8	8	8	16	40	57	53	47	56	61
Coal	13	12	9	13	8	11	8	10	11	12	14	14
Natural gas	0	0	0	0	0	0	0	2	10	24	29	31

(Continued)

Section AII.1 (Continued)   Fuels

Table AII.1.3   World Average Distances Traveled by Crude Oil and Coal

	1900	1910	1920	1930	1940	1950	1960	1970	1980	1990	2000	2010
Crude Oil												
Pipe	68	68	68	67	67	142	353	506	468	418	500	540
Sea (tanker)	681	681	681	672	672	1420	3529	5061	4679	4175	5000	5403
Coal												
Sea (container)	1000	1000	1000	1000	1000	1000	1000	1000	1000	1000	1000	1000
Rail	200	200	200	200	200	200	200	200	200	200	200	200

Table AII.1.4   Resource Production Energy of Fossil Fuels (MJ/kg)

	1900	1910	1920	1930	1940	1950	1960	1970	1980	1990	2000	2010
Coal	0.4	0.4	0.4	0.4	0.4	0.4	0.5	0.5	0.5	0.6	0.6	0.8
Fuel oil	1.8	1.6	1.5	1.4	1.3	1.2	1.1	1.2	1.3	1.4	1.8	2.5
Gasoline	1.8	1.7	1.5	1.4	1.3	1.2	1.2	1.2	1.3	1.5	1.9	2.6
Diesel	1.8	1.6	1.5	1.4	1.3	1.2	1.1	1.2	1.3	1.4	1.8	2.5
Natural gas	1.9	1.8	1.6	1.5	1.4	1.3	1.3	1.3	1.4	1.6	2.0	2.7
Oil	1.8	1.6	1.5	1.4	1.3	1.2	1.1	1.2	1.3	1.4	1.8	2.5

Table AII.1.5   Raw Resource Transport Energy of Fossil Fuels (MJ/kg)

	1900	1910	1920	1930	1940	1950	1960	1970	1980	1990	2000	2010
Coal	1.7	1.7	1.7	1.7	1.7	1.7	0.8	0.7	0.5	0.4	0.4	0.3
Fuel oil	0.2	0.2	0.2	0.2	0.2	0.4	1.0	1.3	1.1	0.9	1.0	1.1
Gasoline	0.2	0.2	0.2	0.2	0.2	0.4	1.0	1.3	1.1	0.9	1.0	1.1
Diesel	0.2	0.2	0.2	0.2	0.2	0.4	1.0	1.3	1.1	0.9	1.0	1.1
Natural gas	0.3	0.3	0.3	0.3	0.3	0.3	0.3	0.5	0.8	1.2	1.4	1.4

(Continued)

Section AII.1 (Continued)   Fuels

Table AII.1.6   Refining or Processing Energy of Fossil Fuels (MJ/kg)

	1900	1910	1920	1930	1940	1950	1960	1970	1980	1990	2000	2010
Coal	0.1	0.1	0.1	0.1	0.1	0.1	0.1	0.1	0.1	0.1	0.1	0.1
Fuel oil	4.0	4.0	4.0	4.0	4.0	4.0	4.0	4.0	4.0	4.0	4.1	4.2
Gasoline	8.1	8.1	8.1	8.1	8.1	8.1	8.1	8.1	8.1	8.2	8.3	8.6
Diesel	6.3	6.3	6.3	6.3	6.3	6.3	6.3	6.3	6.3	6.4	6.5	6.7
Natural gas	0.0	0.0	0.0	0.0	0.0	0.0	0.0	0.0	0.0	0.0	0.0	0.0

Table AII.1.7   Embodied Energy of Transport of Refined Fossil Fuel Products up to the Farm (MJ/kg)

	1900	1910	1920	1930	1940	1950	1960	1970	1980	1990	2000	2010
Coal												
Fuel oil	1.2	1.2	1.2	1.2	1.2	1.2	1.2	1.2	0.9	0.9	0.9	0.9
Gasoline	1.2	1.2	1.2	1.2	1.2	1.2	1.2	1.2	0.9	0.9	0.9	0.9
Diesel	1.2	1.2	1.2	1.2	1.2	1.2	1.2	1.2	0.9	0.9	0.9	0.9
Natural gas												

Table AII.1.8   Total Energy Requirements of Fossil Fuels (MJ/kg)

	1900	1910	1920	1930	1940	1950	1960	1970	1980	1990	2000	2010
Coal	2.2	2.2	2.2	2.2	2.1	2.2	1.4	1.3	1.1	1.1	1.1	1.2
Fuel oil	7.2	7.1	6.9	6.8	6.7	6.8	7.3	7.7	7.3	7.3	7.8	8.7
Gasoline	11.3	11.2	11.1	10.9	10.8	11.0	11.4	11.8	11.5	11.4	12.1	13.1
Diesel	9.5	9.4	9.3	9.1	9.0	9.2	9.7	10.0	9.7	9.6	10.2	11.1
Natural gas	2.2	2.0	1.9	1.8	1.7	1.6	1.5	1.8	2.2	2.8	3.3	4.2

(Continued)

Section AII.1 (*Continued*) Fuels

Table AII.1.9 Total Embodied Energy of Fossil Fuels (MJ/kg)

	1900	1910	1920	1930	1940	1950	1960	1970	1980	1990	2000	2010
Coal	24.6	24.6	24.6	24.6	24.5	24.6	23.8	23.7	23.5	23.5	23.5	23.6
Fuel oil	53.4	53.3	53.1	53.0	52.9	53.0	53.5	53.9	53.5	53.5	54.0	54.9
Gasoline	58.4	58.3	58.2	58.0	57.9	58.1	58.5	58.9	58.6	58.5	59.2	60.2
Diesel	55.2	55.1	55.0	54.8	54.7	54.9	55.4	55.7	55.4	55.3	55.9	56.8
Natural gas	52.6	52.4	52.3	52.2	52.1	52.0	51.9	52.2	52.6	53.2	53.7	54.6

Table AII.1.10 Total Energy Requirements of Energy Carriers (MJ/MJ Direct)

	1900	1910	1920	1930	1940	1950	1960	1970	1980	1990	2000	2010
Fuel oil	0.16	0.15	0.15	0.15	0.15	0.15	0.16	0.17	0.16	0.16	0.17	0.19
Gasoline	0.24	0.24	0.23	0.23	0.23	0.23	0.24	0.25	0.24	0.24	0.26	0.28
Diesel	0.21	0.21	0.20	0.20	0.20	0.20	0.21	0.22	0.21	0.21	0.22	0.24
Oil fuels	0.20	0.20	0.20	0.19	0.19	0.19	0.20	0.21	0.20	0.20	0.22	0.24
Coal	0.10	0.10	0.10	0.10	0.10	0.10	0.06	0.06	0.05	0.05	0.05	0.05
Natural gas	0.04	0.04	0.04	0.04	0.03	0.03	0.03	0.03	0.04	0.05	0.07	0.08
Nuclear electricity							0.20	0.20	0.20	0.20	0.20	0.20
Hydro electricity	0.05	0.05	0.05	0.05	0.05	0.05	0.05	0.05	0.05	0.05	0.05	0.05
Biomass	0.10	0.10	0.10	0.10	0.10	0.10	0.10	0.10	0.10	0.10	0.10	0.10
Renewable electricity								0.06	0.06	0.06	0.06	0.06

For the estimation of raw fuel **transport** embodied energy, we assumed a conservative average distance of 5,000 km water transport and 500 km pipeline transport for oil, and 1,000 km water and 200 km rail for coal around year 2000, based on the average of the country-specific values in ecoinvent (ecoinvent Centre, 2007). For natural gas we used distribution energy European data (ecoinvent Centre, 2007) for around year 2000 as a reference value. For the construction of the long-term series of raw fuel transport distances (Table AII.1.3) we modified these baseline values using the share of international transport over total transport as a proxy of the changes (Table AII.1.2). In the case of oil, this series was based on UN (1952) data from 1929 to 1950 and BP (2014) data from 1970 to 2010. We estimated 1940 and 1960 data as the average of the values of the previous and following time steps. Furthermore, we assumed that the share of oil traded internationally in the period 1900–1920 was constant and similar to the value in 1930 (UN, 1952) (8%). In the case of coal, the available data (Podobnik, 2006) suggest that the changes in transport distances have been relatively small. Therefore, we assumed constant transport distances for coal (Table AII.1.3). We calculated raw fuel transport embodied energy (Table AII.1.5) by multiplying the transport distances (Table AII.1.3) by our own embodied energy coefficients of transport modes (Table AII.11.4). In the case of natural gas, we modulated present data using BP (2014) data from 1970 onward, and we assumed that in previous time steps the share of international transport was negligible, given the low values in 1970.

For the estimation of refined fuel transport energy (Table AII.1.7) we used our own assumptions described in Section AII.11. The sum of resource production energy (Table AII.1.4), raw resource transport energy (Table AII.1.5), refining energy (Table AII.1.6) and refined products transport energy (Table AII.1.7) results in the total energy requirements of fossil fuels (Tables AII.1.8 and AII.1.10). The sum of total energy requirements (Table AII.1.8) and gross energy (Table AII.1.1) results in the total embodied energy of fossil fuels (Table AII.1.9).

AII.2 ELECTRICITY

Direct energy requirements of electricity production with fossil fuels were obtained from Dahmus (2014), and EIA (2014) values were used for direct energy requirements of nuclear-based electricity (Table AII.2.1). **Indirect energy** consumption in electricity production (Table AII.2.2) includes fuel production energy and infrastructure. In the case of fossil fuels it was calculated using our own estimations of fuel production energy values. Indirect energy use in nuclear energy production was based on a meta-analysis of worldwide studies (Lenzen, 2008). The embodied energy values of renewable energy sources (hydro, wind, and solar) were taken from Asdrubali et al. (2015) data and they were assumed to be constant along the studied period.

The sum of direct and indirect energy results in the total embodied energy of electricity production with each type of technology (Table AII.2.3). We assumed a constant share of **grid losses** during the whole studied period, taking into account that grid electricity losses in the world averaged 8.3% from 1960 to 2010, with a low variability (IEA, 2015). The application of this percentage to electricity embodied

Section AII.2 Electricity

Table AII.2.1 Direct Energy Requirements in Power Generation for Various Sources of Electricity (MJ/MJ Electricity)

	1930	1940	1950	1960	1970	1980	1990	2000	2010
Coal	4.50	3.81	3.36	2.48	2.53	2.79	2.80	2.86	2.98
Oil	5.11	4.27	3.76	3.03	2.96	2.80	2.77	2.85	2.84
Natural gas	6.01	5.30	4.52	3.48	3.33	3.34	3.14	2.95	2.46
Nuclear				3.06	3.06	3.06	3.06	3.06	3.06

Table AII.2.2 Fuel Production and Infrastructure Energy Requirements in Power Generation (MJ/MJ Electricity)

	1930	1940	1950	1960	1970	1980	1990	2000	2010
Coal	0.45	0.36	0.33	0.15	0.14	0.14	0.14	0.14	0.16
Oil	0.63	0.51	0.46	0.40	0.42	0.39	0.39	0.43	0.48
Natural gas	0.21	0.18	0.14	0.11	0.12	0.14	0.17	0.20	0.20
Nuclear				0.20	0.20	0.20	0.20	0.20	0.20
Hydro	0.05	0.05	0.05	0.05	0.05	0.05	0.05	0.05	0.05
Solar					0.17	0.17	0.17	0.17	0.17
Wind					0.05	0.05	0.05	0.05	0.05
Wind + Solar					0.06	0.06	0.06	0.06	0.06
Weighted average	0.47	0.36	0.30	0.17	0.16	0.16	0.16	0.16	0.17

Table AII.2.3 Total Embodied Energy of Electricity at Power Plant Gate (MJ/MJ Electricity)

	1930	1940	1950	1960	1970	1980	1990	2000	2010
Coal	4.95	4.18	3.68	2.63	2.67	2.92	2.93	3.00	3.14
Oil	5.74	4.78	4.23	3.43	3.38	3.19	3.16	3.28	3.32

(Continued)

Section AII.2 (Continued) Electricity

Table AII.2.3 Total Embodied Energy of Electricity from Various Sources at Power Plant Gate (MJ/MJ Electricity)

	1930	1940	1950	1960	1970	1980	1990	2000	2010
Natural gas	6.22	5.47	4.67	3.58	3.45	3.49	3.31	3.15	2.66
Nuclear				3.26	3.26	3.26	3.26	3.26	3.26
Hydro	1.05	1.05	1.05	1.05	1.05	1.05	1.05	1.05	1.05
Solar					1.17	1.17	1.17	1.17	1.17
Wind					1.05	1.05	1.05	1.05	1.05
Wind + Solar					1.06	1.06	1.06	1.06	1.06

Table AII.2.4 Total Embodied Energy of Electricity Grid Losses for Various Sources of Electricity (MJ/MJ Electricity)

	1930	1940	1950	1960	1970	1980	1990	2000	2010
Coal	0.41	0.35	0.31	0.22	0.22	0.24	0.24	0.25	0.26
Oil	0.48	0.40	0.35	0.28	0.28	0.26	0.26	0.27	0.28
Natural gas	0.52	0.45	0.39	0.30	0.29	0.29	0.28	0.26	0.22
Nuclear				0.27	0.27	0.27	0.27	0.27	0.27
Hydro	0.09	0.09	0.09	0.09	0.09	0.09	0.09	0.09	0.09
Solar					0.10	0.10	0.10	0.10	0.10
Wind						0.09	0.09	0.09	0.09
Wind + Solar					0.09	0.09	0.09	0.09	0.09

Table AII.2.5 Total Embodied Energy of Electricity from Various Sources at Consumer (MJ/MJ Electricity)

	1930	1940	1950	1960	1970	1980	1990	2000	2010
Coal	5.36	4.52	3.99	2.85	2.90	3.16	3.18	3.25	3.41
Oil	6.22	5.18	4.58	3.72	3.66	3.45	3.42	3.56	3.60
Natural gas	6.74	5.93	5.05	3.88	3.73	3.78	3.59	3.41	2.89

(Continued)

Section AII.2 (*Continued*)   Electricity

Table AII.2.5   Total Embodied Energy of Electricity from Various Sources at Consumer (MJ/MJ Electricity)

	1930	1940	1950	1960	1970	1980	1990	2000	2010
Nuclear				3.53	3.53	3.53	3.53	3.53	3.53
Hydro	1.14	1.14	1.14	1.14	1.14	1.14	1.14	1.14	1.14
Solar					1.26	1.26	1.26	1.26	1.26
Wind					1.14	1.14	1.14	1.14	1.14
Wind + Solar					1.15	1.15	1.15	1.15	1.15

Table AII.2.6   World Electricity Mix (% of Electricity Production)

	1930	1940	1950	1960	1970	1980	1990	2000	2010
Coal	64	59	53	46	41	37	38	38	40
Oil	29	27	24	21	19	17	10	7	4
Natural gas	0	1	2	6	9	10	11	18	22
Nuclear	0	0	0	0	1	10	19	17	13
Hydro	4	8	14	20	26	25	21	18	17
Wind + Solar	0	0	0	0	0	0	1	2	4

Table AII.2.7   Total Embodied Energy of World Average Electricity Production at Power Plant Gate (MJ/MJ Electricity)

	1930	1940	1950	1960	1970	1980	1990	2000	2010
Coal	3.16	2.46	1.95	1.22	1.10	1.09	1.11	1.14	1.26
Oil	1.68	1.29	1.03	0.73	0.64	0.55	0.31	0.23	0.13
Natural gas	0.00	0.05	0.09	0.22	0.31	0.35	0.36	0.57	0.59

(Continued)

Section AII.2 (*Continued*) Electricity

Table AII.2.7 Total Embodied Energy of World Average Electricity Production at Power Plant Gate (MJ/MJ Electricity)

	1930	1940	1950	1960	1970	1980	1990	2000	2010
Nuclear				0.00	0.03	0.33	0.62	0.56	0.42
Hydro	0.04	0.08	0.15	0.21	0.27	0.26	0.22	0.19	0.18
Wind + Solar					0.00	0.00	0.01	0.02	0.04
Total	4.88	3.89	3.22	2.37	2.36	2.58	2.64	2.71	2.62

Table AII.2.8 Total Embodied Energy of World Average Electricity Grid Losses (MJ/MJ Electricity)

	1930	1940	1950	1960	1970	1980	1990	2000	2010
Coal	0.26	0.20	0.16	0.10	0.09	0.09	0.09	0.10	0.10
Oil	0.14	0.11	0.09	0.06	0.05	0.05	0.03	0.02	0.01
Natural gas	0.00	0.00	0.01	0.02	0.03	0.03	0.03	0.05	0.05
Nuclear				0.00	0.00	0.03	0.05	0.05	0.04
Hydro	0.00	0.01	0.01	0.02	0.02	0.02	0.02	0.02	0.01
Wind + Solar						0.00	0.00	0.00	0.00
Total	0.41	0.32	0.27	0.20	0.20	0.21	0.22	0.22	0.22

Table AII.2.9 Total Embodied Energy of World Average Electricity at Consumer (MJ/MJ Electricity)

	1930	1940	1950	1960	1970	1980	1990	2000	2010
Coal	3.42	2.66	2.11	1.32	1.19	1.18	1.20	1.24	1.36
Oil	1.82	1.40	1.11	0.79	0.69	0.59	0.34	0.25	0.14
Natural gas	0.00	0.06	0.10	0.23	0.34	0.38	0.39	0.61	0.64
Nuclear				0.00	0.04	0.35	0.67	0.60	0.46
Hydro	0.05	0.09	0.16	0.23	0.30	0.29	0.24	0.21	0.19
Wind + Solar						0.01	0.02	0.02	0.04
Total	5.29	4.21	3.48	2.57	2.55	2.79	2.86	2.93	2.84

energy results in the estimated grid losses values shown in Table AII.2.4, and the total embodied energy at the point of consumption shown in Table AII.2.5.

The evolution of the **electricity mix** (Table AII.2.6), i.e., the relative contribution of the major energy sources to world electricity production, is taken from IEA (2015) from 1980 onward. We estimated the contribution of the different energy sources before 1980 assuming that all hydro, renewables (wind and solar) and nuclear are electricity, and that the share of the other energy sources is similar to their relative share of primary energy demand, as reported by Koppelaar (2012). The relative contribution of oil and coal in electricity production was assumed to be constant. The embodied energy of the average world electricity mix and the contribution of each technology are shown in Tables AII.2.7 through AII.2.9.

AII.3 MATERIALS

Our estimations of the evolution of energy use in the production of **ferrous metals** (Table AII.3.1) are based on Smil (1999) data on direct energy use in pig iron production up to 1990 and IEA (2007) data on global trends from 1990 to 2010. This series is complemented with an estimation of indirect energy use–based ecoinvent (ecoinvent Centre, 2007) data on additional energy requirements of ferrous metals production (Table AII.3.2). A similar approach was used for **aluminum**, using Dahmus (2014) and IEA (2007) (Table AII.3.1). The data in both sources is shown KWh/kg aluminum. We converted them to primary energy consumption requirements (Table AII.3.3) by using our own estimation of world average electricity embodied energy (Table AII.2.9). The energy requirements of the **other metals**, based in the data reported by ecoinvent (ecoinvent Centre, 2007) are presented in two categories: lead and an aggregated category calculated as the weighted average of the remaining metals, copper, zinc, and brass. The decadal change in the energy efficiency of these categories is modeled as the mean of the change of pig iron and aluminum in each period (Table AII.3.1). The energy required for the production of iron-based irrigation and greenhouse infrastructure components ("Steel (irrig.)" category in Table AII.3.1) was modeled assuming that these materials are made by 15% chromium steel and 85% regular steel.

We estimated the evolution of the energy required for **plastic** pipes production (Table AII.3.6) assuming a constant rate of efficiency gain between the values of Batty et al. (1975) and recent values by Ambrose et al. (2002), for polyethylene and PVC-O, and the data calculated by Piratla et al. (2012) using original data from various PVC types from Ambrose et al (2002) for PVC.

In the case of **concrete**, we assumed 300 kg cement, 1890 kg gravel, 186 kg water, and 226 GJ direct energy consumption per m^3 of concrete, with a density of 2.38 Mg/m^3. In the case of reinforced concrete, we assumed 4.7% steel content. Additional energy for concrete manufacturing was also included, using data from ecoinvent (Kellenberger et al. 2007). Gravel energy requirements were considered to be fixed along the studied period and were taken from ecoinvent (Kellenberger et al., 2007). Cement energy requirements were calculated based on the average value of all country-specific data in Madlool (2011) as the reference value for thermal and

Section AII.3   Materials

Table AII.3.1   Direct Energy Requirements of Metallic Materials (MJ/kg)

	1910	1920	1930	1940	1950	1960	1970	1980	1990	2000	2010
Pig iron	48.0	41.0	35.8	32.5	31.0	23.1	20.0	19.0	16.0	16.0	16.0
Steel (machin.)	49.1	41.9	36.6	33.2	31.7	23.6	20.5	19.4	16.4	16.4	16.4
Steel (irrig.)			50.8	45.4	42.5	34.7	29.4	27.3	23.5	22.3	22.3
Chromium steel			131.2	114.6	104.0	97.5	80.1	71.6	63.8	55.9	55.9
Lead		29.3	25.0	21.8	19.8	18.6	15.3	13.7	12.2	10.7	10.7
Aluminium			110.5	100.4	92.4	82.9	76.6	63.6	58.0	54.7	54.7
Other metals		70.8	60.5	52.8	48.0	44.9	36.9	33.0	29.4	25.8	25.8

Table AII.3.2   Indirect Energy Requirements of Metallic Materials (MJ/kg)

	1910	1920	1930	1940	1950	1960	1970	1980	1990	2000	2010
Pig iron	21.3	18.2	15.9	14.4	13.8	10.3	8.9	8.4	7.1	7.1	7.1
Steel (machin.)	21.8	18.6	16.3	14.8	14.1	10.5	9.1	8.6	7.3	7.3	7.3
Steel (irrig.)			22.6	20.2	18.9	15.4	13.1	12.1	10.4	9.9	9.9
Chromium steel			58.3	50.9	46.2	43.3	35.5	31.8	28.3	24.8	24.8
Lead		13.0	11.1	9.7	8.8	8.2	6.8	6.1	5.4	4.7	4.7
Aluminium			429.4	289.7	204.8	113.9	103.8	99.1	92.9	91.5	87.5
Other metals		31.4	26.9	23.4	21.3	19.9	16.4	14.7	13.0	11.4	11.4

(Continued)

Section AII.3 (*Continued*) Materials

Table AII.3.3 Total Embodied Energy of Metallic Materials (MJ/kg)

	1910	1920	1930	1940	1950	1960	1970	1980	1990	2000	2010
Pig iron	69.3	59.2	51.7	46.9	44.8	33.4	28.9	27.4	23.1	23.1	23.1
Steel (machin.)	70.9	60.5	52.9	48.0	45.8	34.1	29.5	28.1	23.6	23.6	23.6
Steel (irrig.)			73.3	65.6	61.4	50.1	42.4	39.4	33.9	32.2	32.2
Chromium steel			189.5	165.4	150.2	140.7	115.6	103.5	92.1	80.7	80.7
Lead		42.3	36.1	31.5	28.6	26.8	22.0	19.7	17.5	15.4	15.4
Aluminium			540.0	390.1	297.2	196.9	180.4	162.7	150.8	146.2	142.2
Other metals		102.2	87.3	76.2	69.2	64.9	53.3	47.7	42.4	37.2	37.2

Table AII.3.4 Direct Energy Requirements of Nonmetallic Materials (MJ/kg)

	1930	1940	1950	1960	1970	1980	1990	2000	2010
Construction									
Cement	10.2	9.0	7.9	7.0	6.1	5.4	4.8	4.2	3.7
Concrete	1.5	1.4	1.2	1.1	1.0	0.9	0.8	0.8	0.7
Reinforced concrete	4.0	3.3	2.7	2.2	1.9	1.8	1.5	1.5	1.4
Other									
Glass			19.8	16.7	14.0	11.8	9.9	8.3	7.0

(Continued)

Section AII.3 (*Continued*) Materials

Table AII.3.5 Indirect Energy Requirements of Nonmetallic Materials (MJ/kg)

	1930	1940	1950	1960	1970	1980	1990	2000	2010
Construction									
Cement	3.6	3.2	2.8	2.5	2.2	1.9	1.7	1.5	1.3
Concrete	0.3	0.2	0.2	0.1	0.1	0.1	0.0	0.0	0.0
Reinforced concrete	1.0	1.0	1.1	1.2	1.2	0.7	0.6	0.5	0.3
Other									
Glass			6.2	4.3	3.9	3.8	3.6	3.5	3.3

Table AII.3.6 Total Embodied Energy of Nonmetallic Materials (MJ/kg)

	1930	1940	1950	1960	1970	1980	1990	2000	2010
Plastics									
Polyethylene (PE)			264.7	205.8	160.0	124.4	96.7	75.2	58.5
PVC		192.2	164.2	140.4	120.0	102.5	87.6	74.9	64.0
PVC-O							102.8	87.9	75.1
Plexiglass			314.6	260.5	215.8	178.7	148.0	122.6	101.5
Construction									
Cement	13.8	12.2	10.7	9.5	8.3	7.4	6.5	5.7	5.0
Concrete	1.8	1.6	1.4	1.2	1.1	1.0	0.9	0.8	0.7
Reinforced concrete	5.0	4.3	3.8	3.4	3.1	2.5	2.2	2.0	1.7
Other									
Glass			26.0	21.0	17.9	15.6	13.5	11.8	10.3
Rubber		110	110	110	110	110	110	110	110
Other materials	64	64	64	64	64	64	64	64	64

electricity energy consumption in 2000. To this value, we added the energy required to produce the fuels used in thermal energy production (average of coal, oil, and natural gas, Table AII.1.10) and the electricity, using our own estimation of world average electricity embodied energy (Table AII.2.9). We also added the extra energy (transport, buildings, raw materials) needed for producing cement as a percentage of direct energy use-related energy requirements, using data from ecoinvent (Kellenberger et al., 2007). Over this basis in year 2000, we modeled the changes during the studied period (1950–2010) assuming that the average rate of efficiency gain is constant and equal to the average of efficiency gains in the United States (Worrell and Galitsky, 2008) and China (Hu et al., 2014). Direct and indirect energy requirements are shown in Table AII.3.4 through AII.3.6.

In the case of **glass**, we took European glass production (flat glass, uncoated) data (Kellenberger et al., 2007) as the reference for partitioning direct energy use between electricity and other fuels and for including energy consumption not related to direct energy use (mainly transport and buildings). Changes in direct energy use (Table AII.3.4) were modeled based on Van Der Woude (2013). We estimated indirect energy related to the production of the energy carriers employed (Table AII.3.5) based on our own estimations of the energy requirements of electricity (Table AII.2.9) and fossil fuels (Table AII.1.10).

In our farm machinery model, the composition of machinery includes **rubber** used in wheels and an additional category named "**Other materials.**" This category includes alkyd paint, flat glass, polypropylene and paper, that jointly represent roughly 5% of tractor weight in ecoinvent inventory (ecoinvent Centre, 2007), being polypropylene the main contributor to total energy. The embodied energy of these two categories (rubber and other materials) is assumed to be constant during the studied period, given the lack of specific historical information. We took the energy requirements of rubber from Lawson and Rudder (1996). In the case of other materials, we calculated the weighted average of the cumulative energy demand of these materials in ecoinvent database (ecoinvent Centre, 2007). The estimated coefficients of rubber and other materials are shown in Table AII.3.6.

AII.4 MACHINERY

We attempted to reconstruct the energy requirements of machinery production by taking into account the changes in the specific weight of the machines, in the efficiency of the production of the raw materials, in the raw material composition, and in the useful life of the machines.

In Figure AII.1a, we plot the **specific weight** data from the Nebraska Tractor Tests, gathered from the compilation by Evans (2004) from various sources, including Wendel (1985), and Grisso (2007). This data suggest that the average specific weight of tractors decreased steadily during 1920–1980. However, from that year to 2006 no clear trend can be observed. An exponential trend was fitted to the data in the first period (Figure AII.1b), and the average of all data points was assumed for the second period. The resulting trend is shown in Table AII.4.1.

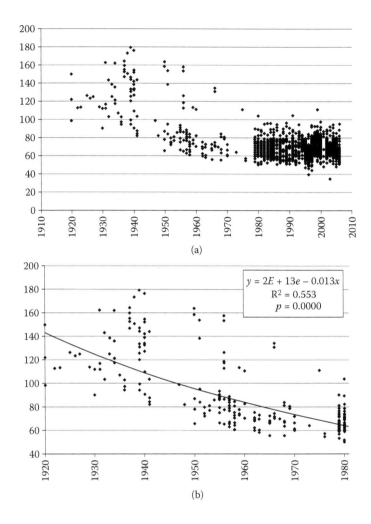

Figure AII.1 Estimated trend in the evolution of tractor specific weight, 1920–2006 (a) and 1920–1980 (b) (kg/kW rated power). (Data points from Evans, J., Nebraska tractor test data, 2004, accessed 02/15/2015; Wendel, C.H., *Nebraska Tractor Tests Since 1920,* Crestline Publishing Co., Sarasota, CA, 1985; and Grisso, R.D., Nebraska Tractor Test Data, 2007, available at http://filebox.vt.edu/users/rgrisso/Pres/Nebdata_07.xls.)

The **composition** of the actual machinery was modeled based on ecoinvent (ecoinvent Centre, 2007). In the case of rubber, the material requirements that can be attributed to maintenance are classified in that category. The same is done for lubricating oil. We estimated these changes based on the qualitative information reviewed (Tables AII.4.2 through AII.4.4).

The **embodied energy of raw materials** was studied in Section 4.5. The embodied energy of the materials used in farm machinery are shown again in Table AII.4.5. We multiplied the energy coefficients of the materials by their share in machinery

Section AII.4 Machinery

Table AII.4.1 Tractor Specific Weight (kg/kW Rated Power)

	1920	1930	1940	1950	1960	1970	1980	1990	2000	2010
Specific weight	142	125	109	95	83	73	64	64	64	64

Table AII.4.2 Tractor Composition (% of Weight)

	1920	1930	1940	1950	1960	1970	1980	1990	2000	2010
Steel, other	93.6	93.6	89.4	89.4	87.9	85.1	79.9	76.5	76.8	76.8
Chromium steel	0.0	0.0	0.0	0.0	0.5	1.0	2.0	3.0	3.1	3.1
Aluminium	0.0	0.0	0.0	0.0	1.0	2.3	3.6	5.0	4.6	4.6
Lead	3.4	3.4	3.4	3.4	3.4	3.4	3.4	3.4	3.4	3.4
Other metals	1.0	1.0	1.0	1.0	1.0	1.0	2.0	3.1	3.1	3.1
Rubber	0.0	0.0	4.2	4.2	4.2	4.2	4.2	4.2	4.2	4.2
Other materials	2.0	2.0	2.0	2.0	2.0	3.0	4.8	4.8	4.8	4.8
Total	100.0	100.0	100.0	100.0	100.0	100.0	100.0	100.0	100.0	100.0

Table AII.4.3 Tillage Machinery Composition (% of Weight)

	1920	1930	1940	1950	1960	1970	1980	1990	2000	2010
Steel, other	99.1	99.0	98.7	98.0	96.8	94.3	89.3	89.3	89.3	89.3
Chromium steel	0.2	0.3	0.6	1.2	2.5	5.0	10.0	10.0	10.0	10.0
Aluminium	0.0	0.0	0.0	0.0	0.0	0.0	0.0	0.0	0.0	0.0

(Continued)

Section AII.4 (Continued) Machinery

	1920	1930	1940	1950	1960	1970	1980	1990	2000	2010
Lead	0.0	0.0	0.0	0.0	0.0	0.0	0.0	0.0	0.0	0.0
Other metals	0.5	0.5	0.5	0.5	0.5	0.5	0.5	0.5	0.5	0.5
Rubber	0.2	0.2	0.2	0.2	0.2	0.2	0.2	0.2	0.2	0.2
Other materials	0.0	0.0	0.0	0.0	0.0	0.0	0.0	0.0	0.0	0.0
Total	100.0	100.0	100.0	100.0	100.0	100.0	100.0	100.0	100.0	100.0

Table AII.4.4 Other Machinery Composition (% of Weight)

	1920	1930	1940	1950	1960	1970	1980	1990	2000	2010
Steel, other	98.9	98.8	98.7	98.4	97.7	96.5	94.0	94.0	94.0	94.0
Chromium steel	0.1	0.2	0.3	0.6	1.2	2.5	5.0	5.0	5.0	5.0
Aluminium	0.0	0.0	0.0	0.0	0.0	0.0	0.0	0.0	0.0	0.0
Lead	0.0	0.0	0.0	0.0	0.0	0.0	0.0	0.0	0.0	0.0
Other metals	0.5	0.5	0.5	0.5	0.5	0.5	0.5	0.5	0.5	0.5
Rubber	0.5	0.5	0.5	0.5	0.5	0.5	0.5	0.5	0.5	0.5
Other materials	0.0	0.0	0.0	0.0	0.0	0.0	0.0	0.0	0.0	0.0
Total	100.0	100.0	100.0	100.0	100.0	100.0	100.0	100.0	100.0	100.0

(Continued)

Section AII.4 (Continued) Machinery

Table AII.4.5 Embodied Energy of Machinery Raw Materials (MJ/kg Material)

	1920	1930	1940	1950	1960	1970	1980	1990	2000	2010
Steel (machinery)	61	53	48	46	34	30	28	24	24	24
Chromium steel				150	141	116	103	92	81	81
Aluminium					197	181	164	153	148	144
Lead	42	36	32	29	27	22	20	18	15	15
Other metals	102	87	76	69	65	53	48	42	37	37
Rubber			110	110	110	110	110	110	110	110
Other materials	64	64	64	64	64	64	64	64	64	64

Table AII.4.6 Embodied Energy of Tractor Raw Materials (MJ/kg Machinery)

	1920	1930	1940	1950	1960	1970	1980	1990	2000	2010
Steel (machinery)	57	49	43	41	30	25	22	18	18	18
Chromium steel	0	0	0	0	1	1	2	3	2	2
Aluminium	0	0	0	0	2	4	6	8	7	7
Lead	1	1	1	1	1	1	1	1	1	1
Other metals	1	1	1	1	1	1	1	1	1	1
Rubber	0	0	5	5	5	5	5	5	5	5
Other materials	1	1	1	1	1	2	3	3	3	3
Total	60	53	51	49	40	38	40	38	37	37

(Continued)

Section AII.4 (Continued) Machinery

Table AII.4.7 Embodied Energy of Tillage Machinery Raw Materials (MJ/kg Machinery)

	1920	1930	1940	1950	1960	1970	1980	1990	2000	2010
Steel, other	60	52	47	45	33	28	25	21	21	21
Chromium steel	0	0	0	2	4	6	10	9	8	8
Aluminium	0	0	0	0	0	0	0	0	0	0
Lead	0	0	0	0	0	0	0	0	0	0
Other metals	0	0	0	0	0	0	0	0	0	0
Rubber	0	0	0	0	0	0	0	0	0	0
Other materials	0	0	0	0	0	0	0	0	0	0
Total	60	53	48	47	37	34	36	31	30	30

Table AII.4.8 Embodied Energy of Other Machinery Raw Materials (MJ/kg Machinery)

	1920	1930	1940	1950	1960	1970	1980	1990	2000	2010
Steel, other	60	52	47	45	33	28	26	22	22	22
Chromium steel	0	0	0	1	2	3	5	5	4	4
Aluminium	0	0	0	0	0	0	0	0	0	0
Lead	0	0	0	0	0	0	0	0	0	0
Other metals	1	0	1	1	1	1	1	1	1	1
Rubber	0	0	0	0	0	0	0	0	0	0
Other materials	0	0	0	0	0	0	0	0	0	0
Total	60	53	48	47	36	32	32	28	27	27

(Continued)

Section AII.4 (Continued) Machinery

Table AII.4.9 Embodied Energy Tractors Raw Materials, Related to Engine Power (GJ/kW)

	1920	1930	1940	1950	1960	1970	1980	1990	2000	2010
Steel (machinery)	8.1	6.2	4.7	3.9	2.5	1.8	1.4	1.2	1.2	1.2
Chromium steel	0.0	0.0	0.0	0.0	0.1	0.1	0.1	0.2	0.2	0.2
Aluminium	0.0	0.0	0.0	0.0	0.2	0.3	0.4	0.5	0.4	0.4
Lead	0.2	0.2	0.1	0.1	0.1	0.1	0.0	0.0	0.0	0.0
Other metals	0.1	0.1	0.1	0.1	0.1	0.0	0.1	0.1	0.1	0.1
Rubber	0.0	0.0	0.5	0.4	0.4	0.3	0.3	0.3	0.3	0.3
Other materials	0.2	0.2	0.1	0.1	0.1	0.1	0.2	0.2	0.2	0.2
Total	8.6	6.6	5.5	4.6	3.3	2.8	2.5	2.4	2.4	2.3

Table AII.4.10 Manufacture, Maintenance, and Useful Life Factors of Farm Machinery

Unit	Manufacture (MJ/kg)	Maintenance (% of Production Energy)
Tractors	14.6	45
Harvesters	12.9	23
Tillage machinery	8.6	30
Other machinery	7.4	26

Table AII.4.11 Useful Life of Self-Propelled Machinery (Hours)

	1920	1930	1940	1950	1960	1970	1980	1990	2000	2010
Tractors	16,000	16,000	16,000	16,000	16,000	14,400	12,800	11,200	9600	8000
Harvesters	2000	2000	2000	2000	2000	1860	1720	1580	1440	1300

(Continued)

Section AII.4 (*Continued*) Machinery

Table AII.4.12 Weight and Useful Life of Farm Implements

	Type	Weight (kg)	UL (h)
Tillage Machinery			
Plow: two-furrow plow	B	600	2000
Plow: four-furrow plow	B	1300	2000
Rotary cultivator (3 m)	B	1000	1500
Rotary cultivator (4 m)	B	1300	1500
Cultivator (2.2 m)	B	700	2000
Spring tine cultivator (6 m)	B	500	2000
Harrow with spring teeth (3 m)	B	650	2000
Clod-breaking rollers (3 m)	B	700	2000
Other Machinery			
Drill: 3 m	C	550	1500
Drill: 6 m	C	1200	1500
Disc broadcaster: Under 450R (12 m)	C	130	1200
Disc broadcaster: Over 450R (12 m)	C	280	1200
Mounted crop sprayer: 600R (12 m)	C	400	1750
Mounted crop sprayer: 1000R (12 m)	C	800	1750
Twin wheels C	C	160	1500
Four-wheel trailer (8t)	C	2500	3000
Round baler	C	1700	2000

(*Continued*)

Section AII.4 (Continued) Machinery

Table AII.4.12 Weight and Useful Life of Farm Implements

	Type	Weight (kg)	UL (h)
Frontloader	C	400	2000
Straw chopper	C	500	1200
Manure spreader (4.5t – 5.5t)	C	1400	1500
Hydraulic loader	C	1600	1500
Slurry pump	C	380	1500
Three-point reel (300 m)	C	450	1500
PVC hoses (100 m)	C	200	1500
Three-point spreader	C	110	1500
Round bale press	C	1700	1500

Table AII.4.13 Direct Manufacture Energy Requirements of Farm Machinery (MJ/kg Machinery)

	1920	1930	1940	1950	1960	1970	1980	1990	2000	2010
Tractors	14.6	14.6	14.6	14.6	14.6	14.6	14.6	14.6	14.6	14.6
Harvesters	12.9	12.9	12.9	12.9	12.9	12.9	12.9	12.9	12.9	12.9
Tillage machinery	8.6	8.6	8.6	8.6	8.6	8.6	8.6	8.6	8.6	8.6
Other machinery	7.4	7.4	7.4	7.4	7.4	7.4	7.4	7.4	7.4	7.4

(Continued)

Section AII.4 (*Continued*)   Machinery

Table AII.4.14   Indirect Manufacture Energy Requirements of Farm Machinery (MJ/kg Machinery)

	1920	1930	1940	1950	1960	1970	1980	1990	2000	2010
Tractors	62.6	62.6	46.8	36.3	22.9	22.7	26.2	27.1	28.2	26.9
Harvesters	55.3	55.3	41.4	32.0	20.3	20.0	23.1	24.0	24.9	23.8
Tillage machinery	36.9	36.9	27.6	21.4	13.5	13.4	15.4	16.0	16.6	15.8
Other machinery	31.7	31.7	23.7	18.4	11.6	11.5	13.3	13.8	14.3	13.6

Table AII.4.15   Total Manufacture Energy Requirements of Farm Machinery (MJ/kg Machinery)

	1920	1930	1940	1950	1960	1970	1980	1990	2000	2010
Tractors	77	77	61	51	38	37	41	42	43	41
Harvesters	68	68	54	45	33	33	36	37	38	37
Tillage machinery	45	45	36	30	22	22	24	25	25	24
Other machinery	39	39	31	26	19	19	21	21	22	21

Table AII.4.16   Total Embodied Energy of Farm Machinery Production (Raw Materials + Manufacture) (MJ/kg Machinery)

	1920	1930	1940	1950	1960	1970	1980	1990	2000	2010
Tractors	138	130	112	99	78	76	81	80	80	78
Harvesters	129	121	105	93	73	71	76	75	75	73
Tillage machinery	106	98	84	77	59	56	60	55	55	54
Other machinery	100	92	79	73	55	51	53	49	49	48

(Continued)

Section AII.4 (*Continued*) Machinery

Table AII.4.17 Total Embodied Energy of Farm Machinery (MJ/kg Machinery)

	1920	1930	1940	1950	1960	1970	1980	1990	2000	2010
Tractors										
Raw materials	60	53	51	49	40	38	40	38	37	37
Manufacture	77	77	61	51	38	37	41	42	43	41
Transport	4	4	4	4	2	2	2	2	2	2
Maintenance	62	59	50	45	35	34	36	36	36	35
Rubber (maintenance)			49	49	49	44	39	34	29	25
Lubricating oil	33	33	33	33	33	30	26	23	20	17
Total	237	226	249	231	197	185	184	175	167	156
Harvesters										
Raw materials	60	53	51	49	40	38	40	38	37	37
Manufacture	68	68	54	45	33	33	36	37	38	37
Transport	4	4	4	4	2	2	2	2	2	2
Maintenance	30	28	24	21	17	16	17	17	17	17
Rubber (maintenance)	0	0	14	14	14	13	12	11	10	9
Lubricating oil	2	2	2	2	2	2	2	1	1	1
Total	164	155	149	135	108	104	108	106	105	102
Tillage Machinery										
Raw materials	60	53	48	47	37	34	36	31	30	30
Manufacture	45	45	36	30	22	22	24	25	25	24
Transport	4	4	4	4	2	2	2	2	2	2
Maintenance	32	29	25	23	18	17	18	17	16	16
Total	142	132	114	105	79	75	79	74	73	72

(*Continued*)

Section AII.4 (*Continued*) Machinery

Table AII.4.17 Total Embodied Energy of Farm Machinery (MJ/kg Machinery)

	1920	1930	1940	1950	1960	1970	1980	1990	2000	2010
Other Machinery										
Raw materials	60	53	48	47	36	32	32	28	27	27
Manufacture	39	39	31	26	19	19	21	21	22	21
Transport	4	4	4	4	2	2	2	2	2	2
Maintenance	26	24	21	19	14	13	14	13	13	12
Total	130	120	104	96	72	66	68	63	63	62

Table AII.4.18 Total Embodied Energy of the Hourly Use of Self-Propelled Machinery (MJ/h)

	1920	1930	1940	1950	1960	1970	1980	1990	2000	2010
50 kW Tractor										
Raw materials	27	21	17	14	10	10	10	11	12	15
Manufacture	34	30	21	15	10	9	10	12	14	17
Transport	2	2	1	1	1	0	0	0	1	1
Maintenance	28	23	17	13	9	9	9	10	12	14
Rubber (maintenance)	0	0	17	15	13	11	10	10	10	10
Lubricating oil	15	13	11	10	9	8	7	7	7	7
Total	105	88	85	69	51	47	46	50	55	62
100 kW Combine										
Raw materials	430	329	276	231	167	150	148	154	163	180
Manufacture	486	425	296	214	138	129	134	149	168	180
Transport	29	26	22	21	9	8	6	7	8	10
Maintenance	211	173	132	102	70	64	65	70	76	83
Rubber (maintenance)	0	0	74	65	56	49	43	43	43	43

(*Continued*)

Section AII.4 (*Continued*) Machinery

Table AII.4.18 Total Embodied Energy of the Hourly Use of Self-Propelled Machinery Embodied Energy (MJ/h)

	1920	1930	1940	1950	1960	1970	1980	1990	2000	2010
Lubricating oil	15	13	11	10	9	7	6	6	6	5
Total	1170	966	811	643	451	408	402	429	464	501

Table AII.4.19 Total Embodied Energy of the Hourly Use of Farm Implements (MJ/h)

	1920	1930	1940	1950	1960	1970	1980	1990	2000	2010
Tillage Machinery										
Plow: Two-furrow plow	43	40	34	31	24	22	24	22	22	22
Plow: Four-furrow plow	92	86	74	68	52	49	52	48	48	47
Rotary cultivator (3 m)	95	88	76	70	53	50	53	49	49	48
Rotary cultivator (4 m)	123	114	98	91	69	65	69	64	63	63
Cultivator (2.2 m)	50	46	40	37	28	26	28	26	26	25
Spring tine cultivator (6 m)	35	33	28	26	20	19	20	18	18	18
Harrow with spring teeth (3m)	46	43	37	34	26	24	26	24	24	23
Clod-breaking rollers (3 m)	50	46	40	37	28	26	28	26	26	25
Other Machinery										
Drill: 3 m	48	44	38	35	26	24	25	23	23	23
Drill: 6 m	104	96	83	77	57	53	55	50	51	50
Disc broadcaster: Under 450R (12 m)	14	13	11	10	8	7	7	7	7	7
Disc broadcaster: Over 450R (12 m)	30	28	24	22	17	15	16	15	15	15
Mounted crop sprayer: 600R (12 m)	30	27	24	22	16	15	16	14	14	14

(*Continued*)

Section AII.4 (*Continued*) Machinery

Table AII.4.19 Total Embodied Energy of the Hourly Use of Farm Implements (MJ/h)

	1920	1930	1940	1950	1960	1970	1980	1990	2000	2010
Mounted crop sprayer: 1000R (12 m)	59	55	48	44	33	30	31	29	29	29
Twin wheels C	14	13	11	10	8	7	7	7	7	7
Four-wheel trailer (8t)	108	100	87	80	60	55	57	53	53	52
Round baler	110	102	89	82	61	56	58	54	54	53
Frontloader	26	24	21	19	14	13	14	13	13	12
Straw chopper	54	50	43	40	30	28	28	26	26	26
Manure spreader (4.5t – 5.5t)	121	112	97	90	67	62	64	59	59	58
Hydraulic loader	138	128	111	102	76	71	73	67	67	67
Slurry pump	33	30	26	24	18	17	17	16	16	16
Three-point reel (300 m)	39	36	31	29	21	20	21	19	19	19
PVC hoses (100 m)	17	16	14	13	10	9	9	8	8	8
Three-point spreader	10	9	8	7	5	5	5	5	5	5
Round bale press	147	136	118	109	81	75	77	72	72	71

(Continued)

Section AII.4 (Continued) Machinery

Table AII.4.20 Specific Fuel Consumption of Tractors (g Fuel/kWh)

	1920	1930	1940	1950	1960	1970	1980	1990	2000	2010
NTT year ave. (old)	312	251	178	181	169	161				
NTT year ave. (Grisso)							187	166	169	
Stout						247	223	198		
NTT (Grisso) trend						202	190	178	166	
Own estimation (test)	312	251	239	227	214	202	190	178	166	166
Own estimation (field)	497	400	381	362	342	323	303	284	264	264

Table AII.4.21 Specific Fuel Consumption of Tractors, Field (L/kWh)

	1920	1930	1940	1950	1960	1970	1980	1990	2000	2010
Specific fuel consumption	0.62	0.49	0.47	0.45	0.38	0.36	0.34	0.32	0.30	0.30

Table AII.4.22 Specific Fuel Energy Consumption of Tractors, Field (MJ/kWh)

	1920	1930	1940	1950	1960	1970	1980	1990	2000	2010
Specific fuel consumption	23.8	19.1	18.1	17.2	14.8	14.0	13.1	12.3	11.5	11.5

Table AII.4.23 Ratio of Used Power to Rated Power for Different Tasks (%)

	Load (R) (%)
Cultivating	66
Plowing	75
Rolling	25
Seeding	50

(Continued)

Section AII.4 (Continued) Machinery

Table AII.4.23 Ratio of Used Power to Rated Power for Different Tasks (%)

	Load (R) (%)
Fertilizing	25
Spraying	25
Harvest	85

Table AII.4.24 Total Embodied Energy per Rated Power per Hour Tillage Work (MJ/kW h) for A 50 kW Tractor at Full Load

	1920	1930	1940	1950	1960	1970	1980	1990	2000	2010
Tractor	2.1	1.8	1.7	1.4	1.0	0.9	0.9	1.0	1.1	1.2
Tillage implement	1.9	1.8	1.5	1.4	1.1	1.0	1.1	1.0	1.0	1.0
Fuel production	6.0	4.8	3.6	3.4	3.1	3.1	2.8	2.6	2.6	2.8
Fuel direct	29.8	23.8	18.1	17.2	14.8	14.0	13.1	12.3	11.5	11.5
Total	39.8	32.1	24.9	23.4	20.0	19.0	17.9	16.9	16.1	16.5

Fuel and Energy Consumption per Hectare

Table AII.4.25 Time Employed in Different Tasks (h) for Three Levels of Tractor Power

	Rated Tractor Power		
	20 kW	50 kW	100 Kw
Cultivating	3.02	1.10	0.60
Plowing	7.03	1.83	1.41
Rolling	0.84	0.36	0.17
Seeding	3.86	0.55	0.77
Fertilizing	1.24	0.68	0.25
Spraying	1.75	0.28	0.35
Harvest	6.16	1.40	1.23

(Continued)

Section AII.4 (Continued) Machinery

Table AII.4.26 Fuel Consumption for Different Tasks (L/ha)

	1920	1930	1940	1950	1960	1970	1980	1990	2000	2010
Cultivating	28.0	22.4	17.0	16.2	14.0	13.2	12.4	11.6	10.8	10.8
Ploughing	52.9	42.3	32.1	30.5	26.3	24.8	23.3	21.8	20.3	20.3
Rolling	3.5	2.8	2.1	2.0	1.7	1.6	1.5	1.4	1.3	1.3
Seeding	10.6	8.5	6.4	6.1	5.3	5.0	4.7	4.4	4.1	4.1
Fertilizing	6.6	5.3	4.0	3.8	3.3	3.1	2.9	2.7	2.5	2.5
Spraying	2.7	2.1	1.6	1.5	1.3	1.2	1.2	1.1	1.0	1.0
Harvest	45.9	36.8	27.9	26.5	22.9	21.6	20.3	19.0	17.7	17.7

Table AII.4.27 Total Embodied Energy for Different Tasks, with a 50 kWh Tractor (MJ/ha)

	1920	1930	1940	1950	1960	1970	1980	1990	2000	2010
Cultivating										
Tractor	116	97	93	76	57	52	51	55	61	69
Implement	55	51	44	40	31	29	31	28	28	28
Fuel production	219	173	130	125	114	111	101	94	93	101
Fuel direct	1081	865	657	623	538	508	477	447	416	416
Total	1470	1185	923	864	739	699	659	624	598	614
Plowing										
Tractor	192	160	155	126	94	86	84	91	101	114
Implement	78	72	62	57	43	41	44	40	40	40
Fuel production	413	326	245	236	214	209	190	178	175	191
Fuel direct	2039	1631	1239	1175	1015	958	900	842	785	785
Total	2722	2190	1700	1594	1367	1294	1218	1151	1101	1129

(Continued)

Section AII.4 (*Continued*) Machinery

Table AII.4.27 Total Embodied Energy for Different Tasks, with a 50 kWh Tractor (MJ/ha)

	1920	1930	1940	1950	1960	1970	1980	1990	2000	2010
Rolling										
Tractor	38	32	31	25	19	17	17	18	20	23
Implement	18	17	14	13	10	10	10	9	9	9
Fuel production	27	22	16	16	14	14	13	12	12	13
Fuel direct	135	108	82	78	67	63	60	56	52	52
Total	219	178	143	132	110	104	99	95	93	96
Seeding										
Tractor	58	48	47	38	28	26	25	27	30	34
Implement	42	38	33	31	23	21	22	20	20	20
Fuel production	83	66	49	47	43	42	38	36	35	38
Fuel direct	410	328	249	236	204	192	181	169	158	158
Total	592	480	378	352	298	281	266	253	244	250
Fertilizing										
Tractor	72	60	58	47	35	32	31	34	38	43
Implement	15	14	12	11	8	8	8	7	7	7
Fuel production	51	41	30	29	27	26	24	22	22	24
Fuel direct	254	203	154	146	126	119	112	105	98	98
Total	392	318	255	234	196	185	175	168	165	171
Spraying										
Tractor	29	24	23	19	14	13	13	14	15	17
Implement	4	4	3	3	2	2	2	2	2	2
Fuel production	21	16	12	12	11	10	10	9	9	10
Fuel direct	102	82	62	59	51	48	45	42	39	39
Total	156	126	101	93	78	73	69	67	65	68

(*Continued*)

Section AII.4 (*Continued*) Machinery

Table AII.4.27 Total Embodied Energy for Different Tasks, with a 50 kWh Tractor (MJ/ha)

	1920	1930	1940	1950	1960	1970	1980	1990	2000	2010
Harvest										
Machinery	147	123	119	96	72	66	64	70	78	87
Fuel production	359	284	213	205	186	182	165	154	152	166
Fuel direct	1772	1417	1076	1021	882	832	782	732	682	682
Total	2278	1824	1407	1322	1140	1080	1012	956	912	935

composition to obtain raw materials embodied energy values per kg of machinery shown in Tables AII.4.6 through AII.4.8. In the case of tractors, the embodied energy values are also expressed related to rated power (Table AII.4.9).

Machinery **manufacture** direct energy use (Tables AII.4.10 and AII.4.13) was taken from Doering (1980), who provides a value of electricity consumption that has been used in many other works (e.g., Audsley et al. 2003, Guzmán and Alonso, 2008). This value was assumed constant, but the energy requirements of electricity production and delivery (Table AII.4.14) were modeled using our own world average estimations described in Section 4.4 and shown in Table AII.2.9.

The sum of direct and indirect energy results in total manufacture energy requirements of farm machinery (Table AII.4.15), which is added to raw materials embodied energy (Tables AII.4.6 thorugh AII.4.8) to obtain the total embodied energy of farm machinery production (Table AII.4.16).

The energy in **repairs and maintenance** is usually expressed in the literature as percentage in total machinery production energy requirements. We took the values from Audsley et al. (2003) (Table AII.4.10), but we also included the extra rubber and lubricating oil required for machinery use in the case of self-propelled machinery (Table AII.4.17).

The **useful life** values published from the early 1960s to the early 2000s range between 10,000 and 16,000 hours for tractors and 2,000 hours for combine harvesters (Rotz, 1987; ASAE, 2000), while in Audsley et al. (2003) they range between 2,500 and 7,200 hours for tractors and 1,400 hours for combine harvesters, and ecoinvent Centre (2007) assumes 7,000 hours for tractors and 1,300 hours for combine harvesters. Therefore, we assumed that the average useful life of self-propelled machinery decreased from 1960 to 2010. Based on the information reviewed above, we assumed the useful life values shown in Table AII.4.11. In the case of tillage machinery and other we chose ASAE values (ASAE, 2000), which express useful life in hours. When a particular farm implement was missing in ASAE database, we took the value of a similar item. ASAE useful life values are usually higher than those reported by Audsley et al. (2003) and ecoinvent Centre (2007). Table AII.4.12 shows the selected weight and useful life values of relevant types of machinery implements. If we multiply the embodied energy values of Table AII.4.17 by the specific weight of the machines (Tables AII.4.1 and AII.4.12) and divde by the useful life (Table AII.4.11 and AII.4.12), we obtain the total embodied energy of the hourly use of self-propelled machinery (Table AII.4.18) and farm implements (Table AII.4.19).

We estimated the evolution of tractor **fuel consumption** (Table AII.4.20) using the reviewed Nebraska Tractor Test series (Figure AII.2a). We divided the data in two periods. The period between 1920 and 1970 is based mainly on the data compiled by Evans (2004), using different sources, mainly the extensive review of long-term Nebraska tractor tests by Wendel (1985). This dataset is not representative of the average trend, as it only covers a few companies. Hence, this data was complemented with some Ford models data from Wendel (2005). For the period between 1980 and 2010, the dataset of Nebraska tractor tests data from Grisso (2007) was used. This is a very comprehensive dataset covering about 1500 Nebraska tractor tests from 1972 to 2006. In this dataset, we can observe a trend toward decreased

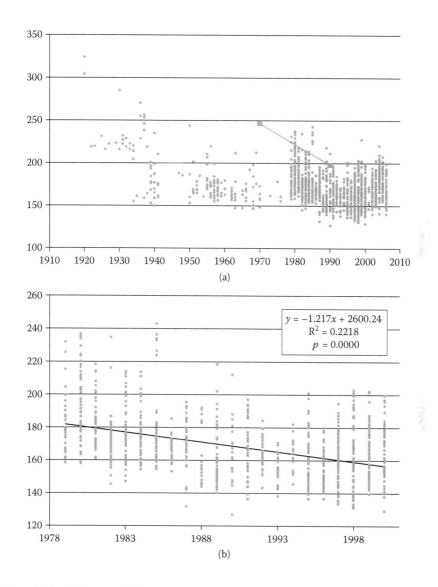

Figure AII.2 Brake-specific fuel consumption of tractors tested in the Nebraska Tractor Tests, 1920–2006 (a) and 1979–1998 (b) (g fuel/kWh). (Data points from Evans, J., Nebraska tractor test data, 2004, accessed 02/15/2015; Wendel, C.H., *Nebraska Tractor Tests Since 1920*, Crestline Publishing Co., Sarasota, CA, 1985; Wendel, C.H., *Standard Catalog of Farm Tractors 1890–1980*, 2nd Edition, Krause, Lola, WI, 2005; and Grisso, R.D., Nebraska Tractor Test Data, 2007, available at http://filebox.vt.edu/users/rgrisso/Pres/Nebdata_07.xls). The trend in Stout and McKiernan (1992) is shown in panel (a), and best fit is shown in panel (b).

fuel consumption from 1976 to about year 2000 (Figure AII.2b). From 2000 to 2006, there seems to be an increasing trend. We did not have comparable data for the years after 2006, so we could not confirm if this trend continues. Therefore, assumed that fuel efficiency during 2000–2010 remains constant. We extrapolated backward the 1980-2000 trend shown in Figure AII.2b to estimate specific diesel consumption up to 1940. Our estimated data points for 1920 and 1930 correspond to the average of our reviewed Nebraska tractor tests for these years.

Last, we used a multiplier for correcting NTT-based values for field operating conditions. Following ASAE standards, we added 15% to NTT fuel consumption data to simulate engine inefficiency under **field conditions**. We also added 39% to the vale obtained to take into account higher relative fuel consumption under lower than rated power output. This percentage is the average of 5 data points representing a range between 20% and 100% of the rated power output of the tractors in Grisso (2004). The "Field estimation" was converted to volume and energy units to provide a series of tractor fuel consumption over the 1920–2010 period (Tables AII.4.21 and AII.4.22). Table AII.4.24 provides a summary example of the total embodied energy of an agricultural operation performed at full load, expressed per rated tractor power and hour.

For the estimation of the embodied energy of each **agricultural task**, we took the engine load values (ratios of actually used power to rated power) from Leach (1976) (Table AII.4.23). We also took machinery working time data for specific tasks from Leach (1976), and converted them to three different levels of tractor power (Table AII.4.25). With this information and our estimations of the evolution of machinery energy requirements through the studied period, we calculated total fuel consumption (Table AII.4.26) and total energy consumption (Table AII.4.27) for each agricultural task. Twenty-five percent extra fuel consumption was added in 1920 and 1930 to account for the extra fuel consumed in the field by tractors with metallic wheels.

AII.5 NITROGEN

The basic composition of the major industrial fertilizers is shown in Table AII.5.1. We did not find any data to model the embodied energy of the early nitrogen fertilizers such as guano, saltpeter, or ammonium sulfate obtained from coke production. Therefore, we equated it to other processes reviewed. In the case of **ammonium sulfate** from coke production, we used the same energy consumption than in the processing of Haber-Bosch N fertilizers, excluding ammonia synthesis (Table AII.5.6). This assumption involves a relatively high energy consumption (despite much less than for ammonia synthesis at the time), which is in line with the fact that heat is needed to recover ammonia from coke oven gas. In the case of guano and saltpeter production, which are obtained by mining, we used phosphate rock mining as the reference. To these figures we added indirect energy consumption (Table AII.5.7), buildings and packaging energy (Table AII.5.8), and transport energy (Table AII.5.9), as explained below. Unlike the rest of agricultural inputs studied in this working paper, **guano and sodium nitrate** were assumed to be transported by water

Section AII.5 Nitrogen (N) Fertilizer Production

Table AII.5.1 Nutrient Content of Fertilizer (% Nutrient)

| Name | Abbreviation | Final Product Mass (%) | | | |
		N	P_2O_5	K_2O	SO_3
Ammonia		82			
Ammonium nitrate	AN	35			
Ammonium sulfate	AS	21			59
Calcium-ammonium nitrate	CAN	25			
Calcium nitrate	CN	16			
Urea	U	46			
Potassium nitrate	NK	13–25		15–46	
Complex NPK fertilizers	NPK	5–25	5–25	5–25	
Monoammonium phosphate	MAP	11	52		
Diammonium phosphate	DAP	18	46		
Ammonium phosphate[a]	AP	14.5	49		
Phosphate rock	P rock		32		
Triple superphosphate	TSP		48		
Single superphosphate	SSP		21		25
Slag	Slag		5–15		

[a]Average of monoammonium phosphate and diammonium phosphate

(Continued)

Section AII.5 (Continued) Nitrogen (N) Fertilizer Production

Table AII.5.1 Nutrient Content of Fertilizer (% Nutrient)

		Final Product Mass (%)	
Complex PK fertilizers	PK	22	22
Muriate of potash (potassium chloride)	MOP (KCl)	60	
Sulfate of potash	SOP (KS)	50	46

Table AII.5.2 Direct Energy Requirements of N Fixation (MJ/kg N)

	1900	1910	1920	1930	1940	1950	1960	1970	1980	1990	2000	2010
Haber–Bosch, China						105.8	89.1	76.3	70.1	64.5	61.0	57.7
Haber–Bosch, World excl. China		243.2	117.9	100.3	85.6	73.8	62.1	53.3	48.9	45.0	42.6	40.3
Haber–Bosch, World		243.2	117.9	100.3	85.6	73.9	62.9	56.8	53.2	49.3	48.8	46.5
Cyanamide	250.0	230.0	210.0	190.0	170.0	150.0	130.0	130.0	130.0	130.0	130.0	130.0

Table AII.5.3 Fuel Production Energy Requirements of N Fixation (MJ/kg N)

	1900	1910	1920	1930	1940	1950	1960	1970	1980	1990	2000	2010
Haber–Bosch, China						9.5	5.8	4.8	4.1	3.9	3.9	4.2
Haber–Bosch, World excl. China		24.2	11.7	4.6	3.8	3.1	2.6	2.5	2.7	3.0	3.3	3.8

(*Continued*)

Section AII.5 (*Continued*)   Nitrogen (N) Fertilizer Production

Table AII.5.3   Energy Requirements of the Production of Fuels Used in N Fixation (MJ/kg N)

	1900	1910	1920	1930	1940	1950	1960	1970	1980	1990	2000	2010
Haber–Bosch, World		24.2	11.7	4.6	3.8	3.2	2.7	2.8	3.0	3.2	3.5	4.0
Cyanamide	349.3	321.4	293.4	265.5	191.0	141.6	89.5	88.4	95.5	97.7	100.0	97.6

Table AII.5.4   Embodied Energy of Buildings Used in N Fixation (MJ/kg N)

	1900	1910	1920	1930	1940	1950	1960	1970	1980	1990	2000	2010
Haber–Bosch, China						1.3	1.3	1.3	1.3	1.3	1.3	1.3
Haber–Bosch, World excl. China		1.3	1.3	1.3	1.3	1.3	1.3	1.3	1.3	1.3	1.3	1.3
Haber–Bosch, World		1.3	1.3	1.3	1.3	1.3	1.3	1.3	1.3	1.3	1.3	1.3
Cyanamide	1.3	1.3	1.3	1.3	1.3	1.3	1.3	1.3	1.3	1.3	1.3	1.3

Table AII.5.5   Total Embodied Energy of N Fixation (MJ/kg N)

	1900	1910	1920	1930	1940	1950	1960	1970	1980	1990	2000	2010
Haber–Bosch, China						116.6	96.1	82.4	75.5	69.7	66.2	63.2
Haber–Bosch, World excl. China		268.7	131.0	106.3	90.7	78.3	66.1	57.1	52.9	49.3	47.2	45.4
Haber–Bosch, World		268.7	131.0	106.3	90.7	78.4	67.0	60.9	57.5	53.8	53.6	51.8
Cyanamide	600.6	552.7	504.7	456.8	362.3	292.9	220.9	219.7	226.8	229.0	231.3	229.0

(*Continued*)

Section AII.5 (Continued) Nitrogen (N) Fertilizer Production

Table AII.5.6 Process Direct Energy Requirements of N Fertilizer Production, Excluding NH_3 (MJ/kg N)

	1900	1910	1920	1930	1940	1950	1960	1970	1980	1990	2000	2010
AN		141.4	120.8	100.2	79.6	59.0	38.4	17.8	7.1	1.2	−1.8	−4.8
DAP, MAP						14.0	11.7	9.4	7.7	6.4	5.4	4.3
AS		13.1	10.3	7.5	4.8	2.0	−0.8	−3.6	−6.3	−9.1	−11.9	−14.7
CAN						74.6	50.9	27.2	13.6	6.6	1.5	0.3
NPK						66.7	57.9	49.1	51.2	52.6	49.1	45.8
U						25.8	21.8	17.8	13.6	11.2	8.0	5.7
Haber–Bosch N fertilizers average		73.6	65.0	56.5	47.9	39.3	30.8	22.2	16.8	14.2	10.3	8.0
Guano	3.9	3.9	3.9	3.9	3.9		3.9	3.9	3.9	3.9	3.9	3.9
NaN	3.9	3.9	3.9	3.9	3.9	3.9	3.9	3.9	3.9	3.9	3.9	3.9
AS from coke oven	73.6	73.6	65.0	56.5	47.9	39.3	30.8	22.2	16.8	14.2	10.3	8.0
Cyanamide	11.2	11.2	11.2	11.2	11.2	11.2	11.2	11.2	11.2	11.2	11.2	11.2
N fertilizers average	27.1	30.3	29.9	35.6	40.8	35.1	29.8	22.0	16.7	14.2	10.3	8.0

Table AII.5.7 Process Indirect Energy Requirements of N Fertilizer Production, Excluding NH_3 (MJ/kg N)

	1900	1910	1920	1930	1940	1950	1960	1970	1980	1990	2000	2010
AN		5.7	4.6	3.6	2.7	1.9	1.2	0.6	0.3	0.1	0.0	0.0
DAP, MAP						0.4	0.4	0.3	0.3	0.3	0.4	0.4
AS		0.5	0.4	0.3	0.2	0.1	0.0	0.0	0.0	0.0	0.0	0.0
CAN						2.4	1.5	0.9	0.6	0.4	0.1	0.0

(Continued)

Section AII.5 (*Continued*) Nitrogen (N) Fertilizer Production

Table AII.5.7 Process Indirect Energy Requirements of N Fertilizer Production, Excluding NH_3 (MJ/kg N)

	1900	1910	1920	1930	1940	1950	1960	1970	1980	1990	2000	2010
NPK						2.1	1.8	1.7	2.2	2.9	3.3	3.8
U						0.8	0.7	0.6	0.6	0.6	0.5	0.5
Haber–Bosch N fertilizers average		1.7	1.6	1.4	1.3	1.1	0.9	0.8	0.7	0.8	0.7	0.7
Guano	0.4	0.4	0.4	0.4	0.4							
NaN	0.4	0.4	0.4	0.4	0.4	0.4	0.4	0.5	0.5	0.5	0.5	0.5
AS from coke oven	1.7	1.7	1.6	1.4	1.3	1.1	0.9	0.8	0.7	0.8	0.7	0.7
Cyanamide	1.1	1.1	1.1	1.1	1.1	1.1	0.7	0.6	0.5	0.5	0.5	0.6
N fertilizers average	0.9	0.9	1.0	1.1	1.2	1.1	0.9	0.8	0.7	0.8	0.7	0.7

Table AII.5.8 Embodied Energy in Buildings and Packing Used in N Fertilizer Production, Excluding NH_3 (MJ/kg N)

	1900	1910	1920	1930	1940	1950	1960	1970	1980	1990	2000	2010
AN		9.4	9.4	9.4	9.4	9.4	9.4	9.4	9.4	9.4	9.4	9.4
DAP, MAP						7.1	7.1	7.1	7.1	7.1	7.1	7.1
AS		2.6	2.6	2.6	2.6	2.6	2.6	2.6	2.6	2.6	2.6	2.6
CAN						10.6	10.6	10.6	10.6	10.6	10.6	10.6
NPK						8.1	8.1	8.1	8.1	8.1	8.1	8.1
U						5.6	5.6	5.6	5.6	5.6	5.6	5.6
N fertilizers average	8.1	8.1	8.1	8.1	8.1	8.1	8.1	8.1	8.1	8.1	8.1	8.1

(*Continued*)

Section AII.5 (*Continued*) Nitrogen (N) Fertilizer Production

Table AII.5.9 Embodied Energy in Transport of N Fertilizers (MJ/kg N)

	1900	1910	1920	1930	1940	1950	1960	1970	1980	1990	2000	2010
AN		11.8	11.8	11.8	11.7	12.5	6.4	5.6	4.6	4.9	5.2	5.6
DAP, MAP						7.4	3.8	3.3	2.7	2.9	3.1	3.3
AS		19.6	19.6	19.6	19.5	20.8	10.7	9.3	7.6	8.1	8.7	9.3
CAN						17.5	9.0	7.8	6.4	6.8	7.3	7.8
NPK						9.7	5.0	4.4	3.5	3.8	4.1	4.3
U						9.5	4.9	4.3	3.5	3.7	4.0	4.2
Haber–Bosch N fertilizers average				13.4	13.3	14.2	7.3	5.5	4.2	4.3	4.4	4.6
Guano	88.9	79.5	55.2	45.4	44.6							
NaN	59.3	53.0	36.8	30.2	29.7	28.7	26.8	23.6	21.1	19.4	18.0	16.6
AS from coke oven	19.6	19.6	19.6	19.6	19.5	20.8	10.7	9.3	7.6	8.1	8.7	9.3
Cyanamide	20.6	20.6	20.6	20.6	20.5	21.9	11.3	9.8	8.0	8.5	9.2	9.8
N fertilizers average	47.7	40.3	27.7	20.5	16.4	16.4	8.0	5.8	4.3	4.4	4.4	4.6

Table AII.5.10 Total Embodied Energy of N Fertilizers, Excluding NH₃ Production (MJ/kg N)

	1900	1910	1920	1930	1940	1950	1960	1970	1980	1990	2000	2010
AN		168.3	146.5	124.9	103.4	82.7	55.4	33.4	21.3	15.5	12.8	10.2
DAP, MAP							23.0	20.1	17.8	16.7	15.9	15.1
AS		35.8	32.9	30.0	27.1	25.5	12.6	8.4	3.8	1.6	-0.6	-2.8
CAN							72.0	46.6	31.1	24.4	19.5	18.8
NPK							72.8	63.4	65.1	67.4	64.6	62.1
U							32.9	28.3	23.2	21.1	18.1	16.0
Haber–Bosch N fertilizers average		96.9	88.1	79.4	70.6	62.8	47.2	36.7	29.9	27.4	23.5	21.5

(*Continued*)

Section AII.5 (Continued) Nitrogen (N) Fertilizer Production

Table AII.5.10 Total Embodied Energy of N Fertilizers, Excluding NH₃ Production (MJ/kg N)

	1900	1910	1920	1930	1940	1950	1960	1970	1980	1990	2000	2010
Guano	101.4	92.0	67.7	57.8	57.0		39.2	36.1	33.6	31.9	30.5	29.1
NaN	71.7	65.5	49.3	42.7	42.2	41.2	50.6	40.5	33.2	31.2	27.9	26.1
AS from coke oven	103.1	103.1	94.3	85.6	76.8	69.4						
Cyanamide	41.0	41.0	41.0	41.0	40.9	42.3	31.3	29.8	27.9	28.4	29.1	29.7
N fertilizers average	83.8	79.7	66.8	65.3	66.5	60.7	46.9	36.7	29.9	27.4	23.5	21.5

Table AII.5.11 Total Embodied Energy of N Fertilizers, China (MJ/kg N)

	1900	1910	1920	1930	1940	1950	1960	1970	1980	1990	2000	2010
AN							151.5	115.8	96.8	85.2	79.1	73.4
DAP, MAP							119.1	102.6	93.3	86.4	82.2	78.3
AS							108.7	90.8	79.3	71.3	65.7	60.5
CAN							168.2	129.0	106.6	94.1	85.7	82.0
NPK							169.0	145.8	140.6	137.1	130.8	125.3
U							129.1	110.7	98.7	90.8	84.3	79.2
Haber–Bosch N fertilizers average						179.3	143.3	119.1	105.4	97.1	89.8	84.7

Table AII.5.12 Total Embodied Energy of N Fertilizers, World Excluding China (MJ/kg N)

	1900	1910	1920	1930	1940	1950	1960	1970	1980	1990	2000	2010
AN							121.4	90.5	74.2	64.8	60.0	55.6
DAP, MAP							89.0	77.2	70.6	66.0	63.1	60.5

(Continued)

Section AII.5 (*Continued*)   Nitrogen (N) Fertilizer Production

Table AII.5.12   Total Embodied Energy of N Fertilizers, World Excluding China (MJ/kg N)

	1900	1910	1920	1930	1940	1950	1960	1970	1980	1990	2000	2010
AS							78.6	65.4	56.7	50.9	46.6	42.6
CAN							138.1	103.6	84.0	73.6	66.7	64.2
NPK							138.9	120.4	118.0	116.7	111.8	107.5
U							99.0	85.3	76.1	70.4	65.3	61.4
Haber–Bosch N fertilizers average						141.0	113.2	93.7	82.8	76.7	70.7	66.9

Table AII.5.13   Total Embodied Energy of N Fertilizers, World (MJ/kg N)

	1900	1910	1920	1930	1940	1950	1960	1970	1980	1990	2000	2010
AN		437.0	277.5	231.2	194.1	161.1	122.3	94.3	78.8	69.3	66.4	62.0
DAP, MAP							89.9	81.1	75.3	70.5	69.6	66.9
AS		304.5	163.9	136.3	117.7	103.8	79.5	69.3	61.3	55.4	53.1	49.0
CAN							139.0	107.5	88.6	78.1	73.1	70.6
NPK							139.8	124.3	122.6	121.2	118.2	113.9
U							99.9	89.2	80.7	74.9	71.7	67.8
Haber–Bosch N fertilizers average		365.6	219.1	185.7	161.3	141.1	114.1	97.6	87.4	81.2	77.2	73.3
Guano	101.4											
NaN	71.7	65.5	49.3	42.7	42.2	41.2	39.2	36.1	33.6	31.9	30.5	29.1
AS from coke oven	103.1	103.1	94.3	85.6	76.8	69.4	50.6	40.5	33.2	31.2	27.9	26.1
Cyanamide	641.7	593.7	545.8	497.8	403.3	335.2	252.2	249.5	254.7	257.4	260.4	258.7
Average AS	103.1	103.1	107.9	116.9	109.7	99.2	77.1	68.5	60.9	55.2	52.9	49.0
N fertilizers average	83.8	88.7	129.5	169.5	163.9	140.5	111.3	97.1	87.1	81.0	77.0	73.2

Note: Nutrients are expressed following the standard conventions: elemental nitrogen (N), phosphate equivalents (P_2O_5), and potash equivalents (K_2O).

to Europe or North America, which meant a distance of ca. 16,000 km in the beginning of the century, which dropped to ca. 10.000 km after the opening of the Panama Canal in 1914 (Table AII.11.5). The resulting transport energy, expressed per kg N, is shown in Table AII.5.9. To simplify, we allocated all guano embodied energy to nitrogen, despite it also contains phosphorus and potassium (Table AII.5.10).

We modelled the evolution of **cyanamide** direct energy consumption based on the data in Jenssen and Kongshaug (2003) and Smil (2001b), assuming a linear efficiency gain from 1900 to 1960 (Table AII.5.2). To calculate indirect energy use (Table AII.5.3), we assumed that 75% of the energy used in cyanamide production was thermal energy from coal and 25% was electricity, obtaining the corresponding coefficients from Tables AII.1.10 and AII.2.9.

The energy consumed in **NH_3** production through Haber-Bosch process (Table AII.5.2) was estimated based on different data sources. The work by Ramirez and Worrell (2006) probably represents the most comprehensive review of the evolution of energy consumption in world ammonia synthesis along the twentieth century. Their estimated values are intermediate between other two long-term estimates available in the literature, Smil (2001b), Jenssen and Kongshaug (2003). However, the value offered for year 2000 is based on very few data points, and it is much lower than the 2005 average world value provided in an extensive study conducted by IEA (2007) (42.6 and 50.5 GJ/Mg NH_3–N, respectively). China is the first global producer of NH_3 in the twenty-first century, with a share ranging from 33% to 39%, and the largest urea exporter. Energy use in ammonia production is very high in this country, where coal is the main ammonia feedstock. However, China was usually omitted in previous assessments of world ammonia energy consumption, and it significantly raises the world average in IEA study. Therefore, we corrected the series by Ramirez and Worrell (2006) taking into account the energy use of ammonia production in China. To simplify, we divided the world in two regions: China and the rest of the world. The latter was modeled as in Ramirez and Worrell (2006), extrapolating the 1990–2000 efficiency trend up to 2010. For China we took IEA (2007) data for 2005 and assumed the same efficiency changes as in the rest of the world. We added the energy embodied in the raw materials used for ammonia production (Table AII.5.3) based on our own estimations of the evolution of the energy intensities of fuels (Table AII.1.10), and assuming that the fuel composition of ammonia production in China and the rest of the world were static along the studied period, with 70%, 20%, and 10% of coal, natural gas, and oil in China and 92% and 8% of natural gas and oil in the rest of the world. We also included the energy embodied in buildings (including equipment) (Table AII.5.4), gathered from ecoinvent database (ecoinvent Centre, 2007).

The total embodied energy of N fixation is shown in Table AII.5.5 and includes direct energy requirements (Table AII.5.2), fuel production energy (Table AII.5.3) and buildings energy (Table AII.5.4).

Last, the changes in the direct energy employed in the **manufacture of commercial N fertilizers** (Table AII.5.6) were estimated based mainly in Ramirez and Worrell (2006). We added the energy required for producing the primary fuels employed (Table AII.5.7) using the coefficients in Table AII.1.10. We also included the energy embedded in buildings (including equipment) (Nemecek et al., 2007) and packaging (Helsel, 1992)

Section AII.6 P Phosphorus (P) Fertilizer Production

Energy

Table AII.6.1 Direct Energy Requirements of Mining and Process of P Fertilizers (MJ/kg P$_2$O$_5$)

	1900	1910	1920	1930	1940	1950	1960	1970	1980	1990	2000	2010
PK 22-22						74.7	57.1	43.7	33.8	26.5	21.2	17.1
AP						30.0	23.2	18.0	14.2	11.3	9.0	7.2
TSP						36.6	27.2	20.2	15.5	11.7	9.3	7.4
SSP						25.5	19.3	14.6	11.3	8.5	6.6	5.1
MAP						34.8	26.9	20.8	16.4	13.0	10.5	8.4
DAP						25.3	19.6	15.1	11.9	9.5	7.6	6.1
NPK						53.6	40.8	31.0	25.4	20.0	15.7	12.3
Slag						7.4	7.4	7.4	7.4	7.4	7.4	7.4
Ground rock						3.9	3.9	3.9	3.9	3.9	3.9	3.9
P fertilizers average	26.17	26.17	26.17	26.17	26.17	26.17	20.09	17.61	15.28	12.74	9.78	7.76

Table AII.6.2 Energy Requirements of the Production of Fuels Used in P Fertilizer Mining and Process (MJ/kg P$_2$O$_5$)

	1900	1910	1920	1930	1940	1950	1960	1970	1980	1990	2000	2010
PK 22-22						8.6	6.2	5.4	4.2	3.2	2.7	2.2
AP						3.4	2.5	2.2	1.7	1.4	1.2	0.9
TSP						4.2	2.9	2.5	1.9	1.4	1.2	1.0
SSP						2.9	2.1	1.8	1.4	1.0	0.8	0.7
MAP						4.0	2.9	2.6	2.0	1.6	1.3	1.1
DAP						2.9	2.1	1.9	1.5	1.1	1.0	0.8
NPK						6.1	4.4	3.8	3.1	2.4	2.0	1.6

(Continued)

Section AII.6 (Continued) P Phosphorus (P) Fertilizer Production

Table AII.6.2 Energy Requirements of the Production of Fuels Used in P Fertilizer Mining and Process (MJ/kg P$_2$O$_5$)

	1900	1910	1920	1930	1940	1950	1960	1970	1980	1990	2000	2010
Slag						0.8	0.8	0.9	0.9	0.9	0.9	1.0
Ground rock						0.4	0.4	0.5	0.5	0.5	0.5	0.5
P fertilizers average	2.7	2.7	2.8	2.8	2.8	3.0	2.2	2.2	1.9	1.5	1.3	1.0

Table AII.6.3 Embodied Energy in Buildings and Packaging Used in P Fertilizer Production (MJ/kg P$_2$O$_5$)

	1900	1910	1920	1930	1940	1950	1960	1970	1980	1990	2000	2010
PK 22-22						4.0	4.0	4.0	4.0	4.0	4.0	4.0
AP						3.2	3.2	3.2	3.2	3.2	3.2	3.2
TSP						6.7	6.7	6.7	6.7	6.7	6.7	6.7
SSP						10.1	10.1	10.1	10.1	10.1	10.1	10.1
MAP						3.4	3.4	3.4	3.4	3.4	3.4	3.4
DAP						3.1	3.1	3.1	3.1	3.1	3.1	3.1
NPK						2.1	2.1	2.1	2.1	2.1	2.1	2.1
Slag						3.9	3.9	3.9	3.9	3.9	3.9	3.9
Ground rock						2.6	2.6	2.6	2.6	2.6	2.6	2.6
P fertilizers average	7.1	7.1	7.1	7.1	7.1	7.1	7.1	5.8	5.2	4.6	4.6	4.6

(Continued)

Section AII.6 (*Continued*)   P Phosphorus (P) Fertilizer Production

Table AII.6.4   Embodied Energy in Transport of P Fertilizers (MJ/kg P$_2$O$_5$)

	1900	1910	1920	1930	1940	1950	1960	1970	1980	1990	2000	2010
PK 22-22						19.9	10.3	8.9	7.2	7.7	8.3	8.9
AP						8.9	4.6	4.0	3.2	3.5	3.7	4.0
TSP						9.1	4.7	4.1	3.3	3.6	3.8	4.1
SSP						20.8	10.7	9.3	7.6	8.1	8.7	9.3
MAP						8.4	4.3	3.8	3.1	3.3	3.5	3.8
DAP						9.5	4.9	4.3	3.5	3.7	4.0	4.2
NPK						9.7	5.0	4.4	3.5	3.8	4.1	4.3
Slag						43.7	22.6	19.6	15.9	17.0	18.3	19.5
Ground rock						13.7	7.1	6.1	5.0	5.3	5.7	6.1
P fertilizers average	19.3		19.3	19.3	19.2	20.5	10.6	7.3	4.7	4.5	4.8	5.1

Table AII.6.5   Total Embodied Energy of P Fertilizers (MJ/kg P$_2$O$_5$)

	1900	1910	1920	1930	1940	1950	1960	1970	1980	1990	2000	2010
PK 22-22						107.1	77.5	61.9	49.2	41.3	36.2	32.1
AP						45.6	33.6	27.4	22.4	19.3	17.1	15.4
TSP						56.6	41.6	33.5	27.4	23.4	21.0	19.1
SSP						59.4	42.3	35.9	30.4	27.8	26.3	25.2
MAP						50.5	37.5	30.5	24.8	21.2	18.7	16.6
DAP						40.8	29.7	24.3	19.9	17.4	15.6	14.2
NPK						71.5	52.3	41.3	34.1	28.2	23.8	20.3
Slag						55.8	34.6	31.8	28.1	29.2	30.5	31.7

(*Continued*)

Section AII.6 (*Continued*) P Phosphorus (P) Fertilizer Production

Table AII.6.5 Total P Embodied Energy of P Fertilizers (MJ/kg P₂O₅)

	1900	1910	1920	1930	1940	1950	1960	1970	1980	1990	2000	2010
Ground rock						20.6	14.0	13.1	12.0	12.3	12.7	13.1
P fertilizers average	55.2	55.3	55.4	55.4	55.4	56.8	40.0	32.9	27.0	23.5	20.4	18.5

Table AII.6.6 Total Embodied Energy of Average P Fertilizers (MJ/kg P₂O₅)

	1900	1910	1920	1930	1940	1950	1960	1970	1980	1990	2000	2010
Mining and process	26.2	26.2	26.2	26.2	26.2	26.2	20.1	17.6	15.3	12.7	9.8	7.8
Fuel production	2.7	2.7	2.8	2.8	2.8	3.0	2.2	2.2	1.9	1.5	1.3	1.0
Buildings and packaging	7.1	7.1	7.1	7.1	7.1	7.1	7.1	5.8	5.2	4.6	4.6	4.6
Transport	19.3	19.3	19.3	19.3	19.2	20.5	10.6	7.3	4.7	4.5	4.8	5.1
Total	55.2	55.3	55.4	55.4	55.4	56.8	40.0	32.9	27.0	23.5	20.4	18.5

(Table AII.5.8) and transport (based on our own coefficients from Table AII.11.16, but taking into account the relative weight represented by N) (Table AII.5.9). In the case of complex fertilizers, we allocated transport energy to each nutrient based on their relative weight. We extended the trends exponentially up to 2010. We estimated ammonium sulphate (AS) energy intensity based on European data in Jenssen and Kongshaug (2003). For NPK complex fertilizers, we assumed that the proportion of NPK1 (based on ammonium nitrate, AN) and NPK2 (based on urea) depends on the proportion of AN and urea in world fertilizer production. We also estimated a weighted average of energy use in fertilizer production taking into account the relative shares of each fertilizer in world production, as reported by FAOSTAT (Food and Agriculture Organization of the United Nations [FAO], 2015). The sum of process direct and indirect energy requirements (Tables AII.5.6 and AII.5.7, respectively), embodied energy in buildings and packaging (Table AII.5.8) and transport energy (Table AII.5.9) results in the total embodied energy of fertilizers excluding N fixation energy (Table AII.5.10). If we sum this with the energy requirements of N fixation (Table AII.5.5) we obtain the total embodied energy of N fertilizers (Tables AII.5.11 through AII.5.13).

AII.6 PHOSPHORUS

We used the information in Ramirez and Worrell (2006) as the basis to estimate the evolution of the direct energy requirements of the most common phosphate fertilizers from 1950 to 2010 (Table AII.6.1). We extrapolated their 1990–2000 and 1960–1970 trends up to 2010 and 1950, respectively. We allocated the energy in compound fertilizers to N, P, and K based on their respective energy requirements. We added the energy required to produce the fuels (Table AII.6.2) using our own estimations of fuel energy intensity (Table AII.1.10). In order to calculate total energy requirements of phosphate fertilizers production we added the energy embodied in buildings and equipment (based on Nemecek et al., 2007), as well as the energy required to package the fertilizers (2.7 MJ/kg P_2O_5), taken from Helsel (1992) (Table AII.6.3). All these factors are assumed to be constant during the studied period. In the case of transport (Table AII.6.4), we used our own energy values (Table AII.11.6) taking into account the weight of P_2O_5 in relation to total fertilizer weight. The sum of direct mining and process energy (Table AII.6.1), fuel production energy (Table AII.6.2), buildings and packaging energy (Table AII.6.3) and transort energy (Table AII.6.4) results in the total embodied energy of P fertilizers (Table AII.6.5). We also calculated the energy embodied in average P fertilizers (summarized in Table AII.6.6), taking into account the relative share of each type according to consumption values in FAOSTAT (FAO, 2015).

AII.7 POTASSIUM

We estimated the evolution of direct energy use in potash production (Table AII.7.1) based on the data in Ramirez and Worrell (2006) and extrapolating the trends up to 1900 and 2010. We distinguished simple K fertilizers (primarily KCl) from complex fertilizers using the 1960–2012 data on world total K fertilizer consumption in FAOSTAT (FAO, 2015) and our own estimation of complex K fertilizer use, based on the average of complex N and P fertilizers. To the direct energy used in potash fertilizers production we added the energy consumed in fuels production (Table AII.7.2), buildings, equipment and packaging (Table AII.7.3), and transport (Table AII.7.4), to obtain the total embodied energy of K fertilizers (Table AII.7.5). Fuel production energy was estimated using our own coefficients (Table AII.1.10), buildings and equipment energy was obtained from ecoinvent database (Nemecek et al., 2007), packaging energy was obtained from Helsel (1992) and corrected for the relative mass represented by potash in complex fertilizers, and finally transport energy was estimated using our own assumptions. We also calculated the energy embodied in average K fertilizers (Table AII.7.6), taking into account the relative share of each type according to consumption values in FAOSTAT (FAO, 2015).

AII.8 PESTICIDES

We did not find information regarding the changes in the energy efficiency of the production of each type of pesticide. In fact, as noticed by Audsley et al. (2009), the work of Green (1987) has virtually been the only basis for assessing pesticide energy use and environmental impacts. Audsley et al. (2009) analyzed Green (1987) data, noticing that chemical families or use types did not explain the variability in pesticide energy requirements. On the contrary, they found a good correlation ($r^2 =$ 0.57) between the year of market release of the pesticide and its energy requirements. This approach was followed in this work to provide a 1940–2010 reconstruction of pesticide energy requirements (Table AII.8.1). In the same table, we also added a series with the estimation of average pesticides actually used in each period, based on the assumption that pesticides used in a given period are an even mixture of the pesticides released in all the previous decades. Formulation and packaging energy were taken from Green (1987). Transport energy was also added (Table AII.8.1) using our own transport energy coefficients (Table AII.11.6). We corrected the transport energy coefficients assuming 20% average content of active matter in pesticides.

The values of individual pesticides (Table AII.8.2) were taken from Green (1987) if the compound is included in that publication. Otherwise they were taken from other references which were also based on Green (1987), in this order of preference: Bhat (1994), Pimentel (1980), Audsley et al. (2009), and Alonso and Guzmán (2010).

AII.9 IRRIGATION

Tables AII.9.1 through AII.9.6 show basic data on the energy efficiency of water pumping for irrigation, as described in Chapter 4. **Irrigation infrastructure** embodied energy includes raw materials production and manufacture. This total embodied energy has to be divided by the number of years of useful life to estimate annual

Section AII.7   Potassium (K) Fertilizer Production

Table AII.7.1   Direct Energy Requirements K Fertilizer Mining and Process (MJ/kg K$_2$O)

	1900	1910	1920	1930	1940	1950	1960	1970	1980	1990	2000	2010
KCl	9.6	9.2	8.7	8.3	7.8	7.4	6.9	6.5	6.0	5.6	5.1	4.7
NPK						10.4	10.2	9.9	10.5	11.0	11.2	10.9
K fertilizers average	9.6	9.2	8.7	8.3	7.8	7.4	7.3	7.3	7.2	7.4	6.8	6.4

Table AII.7.2   Energy Requirements of the Production of Fuels Used in K Fertilizer Mining and Process (MJ/kg K$_2$O)

	1900	1910	1920	1930	1940	1950	1960	1970	1980	1990	2000	2010
KCl	1.0	0.9	0.9	0.9	0.8	0.8	0.7	0.8	0.7	0.7	0.7	0.6
NPK						1.2	1.1	1.2	1.3	1.3	1.4	1.4
K fertilizers average	1.0	0.9	0.9	0.9	0.8	0.8	0.8	0.9	0.9	0.9	0.9	0.8

Table AII.7.3   Embodied Energy of Buildings and Packaging Used in K Fertilizer Production (MJ/kg K$_2$O)

	1900	1910	1920	1930	1940	1950	1960	1970	1980	1990	2000	2010
KCl	3.9	3.9	3.9	3.9	3.9	3.9	3.9	3.9	3.9	3.9	3.9	3.9
NPK						2.7	2.7	2.7	2.7	2.7	2.7	2.7
K fertilizers average	3.9	3.9	3.9	3.9	3.9	3.9	3.8	3.6	3.6	3.5	3.6	3.6

Table AII.7.4   Embodied Energy in Transport of K Fertilizers (MJ/kg K$_2$O)

	1900	1910	1920	1930	1940	1950	1960	1970	1980	1990	2000	2010
KCl	6.9	6.9	6.9	6.9	6.8	7.3	3.8	3.3	2.7	2.8	3.0	3.3
NPK						9.7	5.0	4.4	3.5	3.8	4.1	4.3
K fertilizers average	6.9	6.9	6.9	6.9	6.8	7.3	3.9	3.5	2.9	3.2	3.3	3.6

(Continued)

Section AII.7 (*Continued*) Potassium (K) Fertilizer Production

Table AII.7.5 Total Embodied Energy of K Fertilizers (MJ/kg K₂O)

	1900	1910	1920	1930	1940	1950	1960	1970	1980	1990	2000	2010
KCl	21.3	20.8	20.4	19.9	19.4	19.4	15.3	14.4	13.3	13.0	12.7	12.4
NPK						24.0	19.0	18.2	18.0	18.8	19.4	19.4
K fertilizers average	21.3	20.8	20.4	19.9	19.4	19.4	15.7	15.3	14.6	14.9	14.6	14.4

Table AII.7.6 Total Embodied Energy of Average K Fertilizers (MJ/kg K₂O)

	1900	1910	1920	1930	1940	1950	1960	1970	1980	1990	2000	2010
Mining and process	9.6	9.2	8.7	8.3	7.8	7.4	7.3	7.3	7.2	7.4	6.8	6.4
Fuel production	1.0	0.9	0.9	0.9	0.8	0.8	0.8	0.9	0.9	0.9	0.9	0.8
Buildings	2.4	2.4	2.4	2.4	2.4	2.4	2.3	2.1	2.1	2.0	2.1	2.1
Packaging	1.5	1.5	1.5	1.5	1.5	1.5	1.5	1.5	1.5	1.5	1.5	1.5
Transport	6.9	6.9	6.9	6.9	6.8	7.3	3.9	3.5	2.9	3.2	3.3	3.6
Total	21.3	20.8	20.4	19.9	19.4	19.4	15.7	15.3	14.6	14.9	14.6	14.4

embodied energy. Table AII.9.9 shows the **material requirements** of typical irrigation systems studied by Batty and Keller (1980). We used these inventories as a reference to model the changes in the infrastructure energy of four typical irrigation systems categories (surface with or without IRRS, sprinkler, and drip irrigation) throughout history, taking into account the changes in the materials employed (Tables AII.9.10 through AII.9.13) and the changes in the embodied energy of the materials (Table AII.9.14). In our simplified classification, sprinkler systems are modeled as the average of all sprinkler systems in Batty ad Keller (1980). We modeled the changes in materials composition taking into account the main hits of irrigation technology history, considering the relative mass requirements of each material (Table AII.9.7) and the useful lives of the studied materials (Table AII.9.8), which were also from Batty and Keller (1980).

The embodied energy of **raw materials** was studied in Section 4.5, and the total embodied energy values of the ones used in irrigation infrastructure are shown again in Table AII.9.14. In the same table, we added the energy required for **manufacturing** of metallic components using our own estimations of manufacture energy requirements, assuming that these components can be classified as "Machinery type C," as defined in Section 4.6. We also added energy requirements of grading and ditching (Table AII.9.14), taken from Batty and Keller (1980). In the case of ditching, we assumed 535 kg nonreinforced concrete were used per linear meter ditch, corresponding to a ditch of 1-m bottom width and 1-m depth (Batty and Keller 1980).

The estimated material requirements were multiplied by the embodied energy of each material in each given year to obtain the annualized energy requirements of the infrastructure for each type of irrigation system (Tables AII.9.15 through AII.9.19). In the calculation of total embodied energy, we also added 20% maintenance energy and energy required for transport of irrigation materials to the farm, assuming our standard transport distances and modes. Last, Tables AII.9.20 through AII.9.26 show total embodied energy for 500 mm net irrigation with various irrigation systems and water lift heights.

AII.10 GREENHOUSES

We modeled the changes in the embodied energy of greenhouse infrastructure assuming constant material requirements, i.e., considering only the changes in the embodied energy of the materials. Heating systems and other equipment such as hydroponic systems or supplemental lighting are not reviewed here, but they should be included in energy balances of agricultural systems if they are present. Embodied energy in materials was studied in Section 4.5, while metallic components were assumed to belong to "Machinery type C" to take into account the energy required for manufacturing. As in the case of irrigation and machinery, useful life is a key parameter in the estimation of greenhouse infrastructure energy requirements, with common values ranging from 1 to 20 years depending on the type of material. Some common values of useful lives of the studied materials used in greenhouses, as well as our own choice, are shown in Table AII.10.2.

Section AII.8   Pesticides

Energy

Table AII.8.1   Total Embodied Energy of Synthetic Pesticides (MJ/kg Active Ingredient)

	1940	1950	1960	1970	1980	1990	2000	2010
New Pesticides								
Active matter	33	141	249	357	465	573	681	789
Formulation+Packaging	22	22	22	22	22	22	22	22
Transport	21	21	22	11	10	8	9	9
Total	76	184	293	390	497	603	712	820
Average Used Pesticides								
Active matter	33	87	141	195	249	303	357	411
Formulation+Packaging	22	22	22	22	22	22	22	22
Transport	21	21	21	19	17	15	14	14
Total	76	130	184	236	288	340	393	447

Table AII.8.2   Energy Requirements of Synthetic Pesticide Active Ingredients, per Type of Active Ingredient (MJ/kg Active Ingredient)

	Selected Value	Source
Herbicides		
2, 4, 5-T	160	Green (1987)
2,4-D	112	Green (1987)
Alachlor	303	Green (1987)
Atrazine	213	Green (1987)
Bentazon	459	Green (1987)
Bromoxynil	175	Bhat et al. (1994)
Butylate	166	Green (1987)
Carbetamide	307	Audsley et al. (2009)

(Continued)

Section AII.8 (*Continued*) Pesticides

Table AII.8.2 Energy Requirements of Synthetic Pesticide Active Ingredients, per Type of Active Ingredient (MJ/kg Active Ingredient)

	Selected Value	Source
Chloramben	195	Green (1987)
Chloridazon	296	Audsley et al. (2009)
Chlorotoluron	372	Audsley et al. (2009)
Chlorsulfuron	390	Green (1987)
Clopyralid	437	Audsley et al. (2009)
Cyanazine	226	Green (1987)
Dicamba	320	Green (1987)
Diflufenican	545	Audsley et al. (2009)
Dinoseb	105	Green (1987)
Diquat	425	Green (1987)
Diuron	300	Green (1987)
EPTC	185	Green (1987)
Ethofumesate	372	Audsley et al. (2009)
Flometuron	380	Green (1987)
Florasulam	696	Audsley et al. (2009)
Fluazifop-Methyl	543	Green (1987)
Flufenacet	653	Audsley et al. (2009)
Fluroxypyr	523	Audsley et al. (2009)
Glyphosate	479	Green (1987)
Iodosulfuron-methyl-sodium	696	Audsley et al. (2009)
Isopropalin	175	Bhat et al. (1994)
Isoproturon	383	Audsley et al. (2009)

(Continued)

Section AII.8 (Continued) Pesticides

Table AII.8.2 Energy Requirements of Synthetic Pesticide Active Ingredients, per Type of Active Ingredient (MJ/kg Active Ingredient)

	Selected Value	Source
Linuron	315	Green (1987)
MCPA	153	Green (1987)
Mecoprop-P	199	Audsley et al. (2009)
Mesosulfuron-methyl	664	Audsley et al. (2009)
Mesotrione	696	Audsley et al. (2009)
Metamitron	437	Audsley et al. (2009)
Metazachlor	393	Audsley et al. 2009
Methazole	175	Bhat et al. (1994)
Metolachlor	301	Green (1987)
Metribuzin	225	Bhat et al. (1994)
Metsulfuron-methyl	523	Audsley et al. (2009)
Molinate	175	Bhat et al. (1994)
Nicosulfuron	599	Audsley et al. (2009)
Norflurazon	175	Bhat et al. (1994)
Paraquat	484	Green (1987)
Pendimethalin	175	Bhat et al. (1994)
Phenmedipham	350	Audsley et al. (2009)
Prometryn	225	Bhat et al. (1994)
Propachlor	315	Green (1987)
Propanil	245	Green (1987)
Propaquizafop	566	Audsley et al. (2009)
Propyzamide	415	Audsley et al. (2009)
Prosulfuron	631	Audsley et al. (2009)

(Continued)

Section AII.8 (Continued)   Pesticides

Table AII.8.2   Energy Requirements of Synthetic Pesticide Active Ingredients, per Type of Active Ingredient (MJ/kg Active Ingredient)

	Selected Value	Source
Simazine	231	Audsley et al. (2009)
Thifensulfuronmethyl	545	Audsley et al. (2009)
Tribenuron-methyl	545	Audsley et al. (2009)
Triclopyr	437	Audsley et al. (2009)
Trifloxystrobin	685	Audsley et al. (2009)
Trifluralin	176	Green (1987)
Insecticides		
1,3-dichloropropene	231	Audsley et al. (2009)
Alpha-cypermethrin	523	Audsley et al. (2009)
Carbaryl	178	Green (1987)
Carbofuran	479	Green (1987)
Chlofpyrifos	275	Bhat et al. (1994)
Chlordimeform	275	Green (1987)
Chlorpyrifos	329	Audsley et al. (2009)
Cypermethrin	605	Green (1987)
Cypermethrin	605	Audsley et al. (2009)
DDT	126	Pimentel (1980)
Ethoprophos	339	Audsley et al. (2009)
Fensulfothion	225	Bhat et al. (1994)
Fonofos	225	Bhat et al. (1994)
Lambda-cyhalothrin	534	Audsley et al. (2009)
Lindane	83	Green (1987)

(Continued)

Section AII.8 (Continued) Pesticides

Table AII.8.2 Energy Requirements of Synthetic Pesticide Active Ingredients, per Type of Active Ingredient (MJ/kg Active Ingredient)

	Selected Value	Source
Malathion	254	Green (1987)
Metaldehyde	153	Audsley et al. (2009)
Methozychlor	95	Green (1987)
Methyl parathion	185	Green (1987)
Oxamyl	350	Audsley et al. (2009)
Parathion	163	Green (1987)
Tau-fluvalinate	491	Audsley et al. (2009)
Terbufos	225	Bhat et al. (1994)
Toxaphene	83	Green (1987)
Zeta-cypermethrin	620	Audsley et al. (2009)
Fungicides		
Azoxystrobin	620	Audsley et al. (2009)
Benomyl	422	Green (1987)
Boscalid	718	Audsley et al. (2009)
Captan	140	Pimentel (1980)
Carbendazim	415	Audsley et al. (2009)
Chlorothalonil	318	Audsley et al. (2009)
Cymoxanil	447	Audsley et al. (2009)
Cyproconazole	556	Audsley et al. (2009)
Cyprodinil	642	Audsley et al. (2009)
Epoxiconazole	631	Audsley et al. (2009)
Fenpropimorph	480	Audsley et al. (2009)

(Continued)

Section AII.8 (Continued) Pesticides

Table AII.8.2 Energy Requirements of Synthetic Pesticide Active Ingredients, per Type of Active Ingredient (MJ/kg Active Ingredient)

	Selected Value	Source
Ferbam	89	Pimentel (1980)
Fluazinam	599	Audsley et al. (2009)
Fluometuron	380	Green (1987)
Fluoxastrobin	642	Audsley et al. (2009)
Flusilazole	534	Audsley et al. (2009)
Kresoxim-methyl	523	Audsley et al. (2009)
Mancozeb	285	Audsley et al. (2009)
Maneb	124	Pimentel (1980)
Metconazole	620	Audsley et al. (2009)
Metrafenone	718	Audsley et al. (2009)
Prochloraz	458	Audsley et al. (2009)
Propamocarb hydrochloride	469	Audsley et al. (2009)
Prothioconazole	480	Audsley et al. (2009)
Pyraclostrobin	707	Audsley et al. (2009)
Spiroxamine	674	Audsley et al. (2009)
Sulfur	136	Pimentel (1980)
Tebuconazole	556	Audsley et al. (2009)
Growth Regulators		
Chlormequat (+/−chloride)	275	Audsley et al. (2009)
Imazaquin	523	Audsley et al. (2009)
Maleic hydrazine	156	Audsley et al. (2009)
Trinexapac-ethyl	588	Audsley et al. (2009)

(Continued)

Section AII.8 (*Continued*) Pesticides

Table AII.8.2 Energy Requirements of Synthetic Pesticide Active Ingredients, per Type of Active Ingredient (MJ/kg Active Ingredient)

	Selected Value	Source
Other Pesticides		
Ethephon	199	Audsley et al. (2009)
Metalaxyl-M	664	Audsley et al. (2009)
Tri-allate	275	Audsley et al. (2009)
Sulfur	3	Alonso and Guzmán (2010), based on fertilizers
Copper	13	Alonso and Guzmán (2010), based on fertilizers
Emulsifiable oil	20	Alonso and Guzmán (2010), based on Fluck (1992)
Liquid solutions	1	Alonso and Guzmán (2010), based on Fluck (1992)
Insect traps	2	Alonso and Guzmán (2010), based on Fluck (1992)
All Pesticides	*375*	

Section AII.9 Irrigation

Table AII.9.1 Pump Efficiency Coefficients

Pump efficiency	70%
Electric motor efficiency	88%

Table AII.9.2 Water Use Efficiency and Head Pressure of Irrigation Systems

	Efficiency (%)	Water Applied (mm)		Head Pressure (m)
		Net	Gross	
Surface no IRRS	50	500	1000	3
Surface IRRS	85	500	588	5
Solid set sprinkle	80	500	625	53
Permanent sprinkle	80	500	625	53
Hand-moved sprinkle	75	500	667	53
Side roll sprinkle	75	500	667	53
Center-pivot sprinkle	80	500	625	60
Traveler sprinkler	70	500	714	95
Trickle	90	500	556	35
Sprinkler, average	77	500	652	61

(Continued)

Section AII.9 (Continued) Irrigation

Table AII.9.3 Direct Energy Requirements of Electricity Consumption in Water Pumping, for 500 mm Net Irrigation (GJ/ha)

	0 m	50 m	100 m
Surface no IRRS	0.5	8.4	16.4
Surface IRRS	0.5	5.1	9.8
Solid set sprinkle	5.3	10.2	15.2
Permanent sprinkle	5.3	10.2	15.2
Hand-moved sprinkle	5.6	10.9	16.2
Side roll sprinkle	5.6	10.9	16.2
Center-pivot sprinkle	6.0	11.0	15.9
Traveler sprinkler	10.8	16.5	22.2
Trickle	3.1	7.5	11.9
Sprinkler, average	6	12	17

Table AII.9.4 Total Embodied Energy of Electricity Use in Pumping for 500 mm Net Irrigation, 0 m Lift (GJ/ha yr)

	1930	1940	1950	1960	1970	1980	1990	2000	2010
Surface no IRRS	2.5	2.0	1.7	1.2	1.2	1.3	1.4	1.4	1.4
Surface IRRS	2.5	2.0	1.6	1.2	1.2	1.3	1.3	1.4	1.3
Solid set sprinkle	27.9	22.2	18.4	13.6	13.5	14.7	15.1	15.5	15.0
Permanent sprinkle	27.9	22.2	18.4	13.6	13.5	14.7	15.1	15.5	15.0
Hand-moved sprinkle	29.8	23.7	19.6	14.5	14.4	15.7	16.1	16.5	16.0
Side roll sprinkle	29.8	23.7	19.6	14.5	14.4	15.7	16.1	16.5	16.0
Center-pivot sprinkle	31.6	25.1	20.8	15.4	15.2	16.7	17.1	17.5	17.0
Traveler sprinkler	57.1	45.4	37.6	27.8	27.6	30.2	30.9	31.7	30.7
Trickle				7.9	7.9	8.6	8.8	9.1	8.8
Sprinkler, average	34.0	27.1	22.4	16.5	16.4	18.0	18.4	18.8	18.3

(Continued)

Section AII.9 (Continued)   Irrigation

Table AII.9.5   Total Embodied Energy of Electricity Use in Pumping for 500 mm Net Irrigation, 50 m Lift (GJ/ha yr)

	1930	1940	1950	1960	1970	1980	1990	2000	2010
Surface no IRRS	44.7	35.5	29.4	21.7	21.6	23.6	24.1	24.8	24.0
Surface IRRS	27.2	21.6	17.9	13.2	13.1	14.4	14.7	15.1	14.6
Solid set sprinkle	54.2	43.1	35.7	26.3	26.1	28.6	29.3	30.0	29.1
Permanent sprinkle	54.2	43.1	35.7	26.3	26.1	28.6	29.3	30.0	29.1
Hand-moved sprinkle	57.8	46.0	38.1	28.1	27.9	30.5	31.3	32.1	31.1
Side roll sprinkle	57.8	46.0	38.1	28.1	27.9	30.5	31.3	32.1	31.1
Center-pivot sprinkle	57.9	46.1	38.1	28.2	27.9	30.6	31.3	32.1	31.1
Traveler sprinkler	87.2	69.4	57.4	42.4	42.1	46.0	47.1	48.3	46.9
Trickle				19.3	19.2	21.0	21.5	22.0	21.4
Sprinkler, average	61.5	48.9	40.5	29.9	29.7	32.5	33.3	34.1	33.1

Table AII.9.6   Total Embodied Energy of Electricity Use in Pumping for 500 mm Net Irrigation, 100 m Lift (GJ/ha yr)

	1930	1940	1950	1960	1970	1980	1990	2000	2010
Surface no IRRS	86.7	69.0	57.1	42.2	41.8	45.8	46.9	48.1	46.6
Surface IRRS	52.0	41.3	34.2	25.3	25.1	27.4	28.1	28.8	27.9
Solid set sprinkle	80.5	64.1	53.0	39.1	38.9	42.5	43.5	44.6	43.3
Permanent sprinkle	80.5	64.1	53.0	39.1	38.9	42.5	43.5	44.6	43.3
Hand-moved sprinkle	85.9	68.4	56.6	41.8	41.5	45.4	46.4	47.6	46.2
Side roll sprinkle	85.9	68.4	56.6	41.8	41.5	45.4	46.4	47.6	46.2
Center-pivot sprinkle	84.2	67.0	55.4	40.9	40.6	44.4	45.5	46.6	45.2
Traveler sprinkler	117.3	93.3	77.2	57.0	56.6	61.9	63.4	65.0	63.0
Trickle				30.7	30.5	33.3	34.1	35.0	33.9
Sprinkler, average	89.1	70.8	58.7	43.3	43.0	47.0	48.1	49.4	47.9

(Continued)

Section AII.9 (*Continued*) Irrigation

Table AII.9.7   Material Requirements of Irrigation Materials, Relative to PVC

PVC	1
Aluminium	1
Iron-based	4

Table AII.9.8   Useful Life of Irrigation Materials (Years)

	UL (yr)
Pumping unit, electric	12
Pumping unit, diesel	12
PE	10
PVC	40
Aluminium	20
Iron-based	20
Concrete	15
Grading	40
Ditching	40

Table AII.9.9   Material Requirements for Irrigation in 1970 (kg/ha)

	Surface No IRRS	Surface IRRS	Solid Set Sprinkle	Permanent Sprinkle	Hand-Moved Sprinkle	Side Roll Sprinkle	Center-Pivot Sprinkle	Traveler Sprinkler	Sprinkler Average	Trickle
Pumping unit	10	10	11.7	11.7	11.7	11.7	10.2	14.6	12	10
PE	0	0	0	0	0	0	0	0	0	191

(*Continued*)

Section AII.9 (Continued) Irrigation

Table AII.9.9 Material Requirements for Irrigation in 1970 (kg/ha)

	Surface No IRRS	Surface IRRS	Solid Set Sprinkle	Permanent Sprinkle	Hand-Moved Sprinkle	Side Roll Sprinkle	Center-Pivot Sprinkle	Traveler Sprinkler	Sprinkler Average	Trickle
PVC	20	35	95	404	95	95	56	129	145	247
Aluminium	0	66	506	0	37	63	0	0	101	0
Iron-based	0	0	126	140	9	37	232	110	109	12
Grading (1000 m3)	731	731	0	0	0	0	0	0	0	0
Ditching (m)	35	35	17	66	35	35	7	23	30	35

Table AII.9.10 Material Requirements of Surface without IRRS Irrigation System (kg/ha yr)

	1930	1940	1950	1960	1970	1980	1990	2000	2010
Pumping unit	0.8	0.8	0.8	0.8	0.8	0.8	0.8	0.8	0.8
PE	0.0	0.0	0.0	0.0	0.0	0.0	0.0	0.0	0.0
PVC	0.0	0.0	0.3	0.5	0.5	0.5	0.5	0.5	0.5
Aluminium	0.0	0.0	0.0	0.0	0.0	0.0	0.0	0.0	0.0
Iron-based	0.0	0.0	0.0	0.0	0.0	0.0	0.0	0.0	0.0
Grading (1000 m3)	36.5	36.5	31.1	23.8	18.3	18.3	18.3	18.3	18.3
Ditching (m)	0.9	0.9	0.9	0.9	0.9	0.9	0.9	0.9	0.9

Table AII.9.11 Material Requirements of Surface with IRRS Irrigation System (kg/ha yr)

	1930	1940	1950	1960	1970	1980	1990	2000	2010
Pumping unit	0.8	0.8	0.8	0.8	0.8	0.8	0.8	0.8	0.8
PE	0.0	0.0	0.0	0.0	0.0	0.0	0.0	0.0	0.0
PVC	0.0	0.0	0.6	0.8	0.9	0.9	0.9	0.9	0.9

(Continued)

Section AII.9 (Continued) Irrigation

Table AII.9.11 Material Requirements of Surface with IRRS Irrigation System (kg/ha yr)

	1930	1940	1950	1960	1970	1980	1990	2000	2010
Aluminium	0.0	0.0	2.1	3.0	3.3	3.3	3.3	3.3	3.3
Iron-based	20.2	20.2	7.5	2.0	0.0	0.0	0.0	0.0	0.0
Grading (1000 m3)	18.3	18.3	18.3	18.3	18.3	18.3	18.3	18.3	18.3
Ditching (m)	0.9	0.9	0.9	0.9	0.9	0.9	0.9	0.9	0.9

Table AII.9.12 Material Requirements of Average Sprinkler Irrigation System (kg/ha yr)

	1930	1940	1950	1960	1970	1980	1990	2000	2010
Pumping unit	1.0	1.0	1.0	1.0	1.0	1.0	1.0	1.0	1.0
PE	0.0	0.0	0.0	0.0	0.0	0.0	0.0	0.0	0.0
PVC	0.0	0.0	2.3	3.3	3.6	3.6	3.6	3.6	3.6
Aluminium	0.0	0.0	3.2	4.5	5.0	5.0	5.0	5.0	5.0
Iron-based	54.7	54.7	23.7	11.6	5.4	5.4	5.4	5.4	5.4
Grading (1000 m3)	0.0	0.0	0.0	0.0	0.0	0.0	0.0	0.0	0.0
Ditching (m)	0.8	0.8	0.8	0.8	0.8	0.8	0.8	0.8	0.8

Table AII.9.13 Material Requirements of Trickle Irrigation System (kg/ha yr)

	1930	1940	1950	1960	1970	1980	1990	2000	2010
Pumping unit				0.9	0.9	0.9	0.9	0.9	0.9
PE				19.1	19.1	19.1	19.1	19.1	19.1
PVC				5.6	6.2	6.2	6.2	6.2	6.2
Aluminium				0.0	0.0	0.0	0.0	0.0	0.0
Iron-based				6.8	0.6	0.6	0.6	0.6	0.6

(Continued)

Section AII.9 (*Continued*)　Irrigation

Table AII.9.13　Material Requirements of Trickle, Irrigation System (kg/ha yr)

	1930	1940	1950	1960	1970	1980	1990	2000	2010
Grading (1000 m3)				0.0	0.0	0.0	0.0	0.0	0.0
Ditching (m)				0.9	0.9	0.9	0.9	0.9	0.9

Energy

Table AII.9.14　Embodied Energy of Raw Materials Used in Irrigation (MJ/kg)

	1930	1940	1950	1960	1970	1980	1990	2000	2010
Polyethylene (HDPE)		265	206	160	124	97	75	58	
PVC		192	164	140	120	103	88	75	64
PVC-O							103	88	75
Aluminium		390	297	197	181	164	153	148	144
Iron-based	73	66	61	50	42	39	34	32	32
Manufacture	39	31	26	19	19	21	21	22	21
Concrete	1.8	1.6	1.4	1.2	1.1	1.0	0.9	0.8	0.7
Reinforced concrete	5.0	4.3	3.8	3.4	3.1	2.5	2.2	2.0	1.7
Grading (m3)	15	15	15	15	15	15	15	15	15
Ditching (m)	57	54	52	50	48	46	45	43	42

(Continued)

Section AII.9 (*Continued*) Irrigation

Table AII.9.15 Total Embodied Energy of Surface without IRRS Irrigation System Infrastructure (MJ/ha yr)

	1930	1940	1950	1960	1970	1980	1990	2000	2010
Pumping unit	64	64	64	64	64	64	64	64	64
PE	0	0	0	0	0	0	0	0	0
PVC	0	0	53	65	61	52	45	38	33
Aluminium	0	0	0	0	0	0	0	0	0
Iron-based	0	0	0	0	0	0	0	0	0
Manufacture	0	0	0	0	0	0	0	0	0
Transport	3	3	5	3	3	2	2	2	3
Maintenance	132	132	125	105	88	86	84	83	81
Grading	546	546	464	355	273	273	273	273	273
Ditching	50	47	45	43	42	40	39	38	37
Total	795	792	757	635	531	518	508	499	491

Table AII.9.16 Total Embodied Energy of Surface with IRRS Irrigation System Infrastructure (MJ/ha yr)

	1930	1940	1950	1960	1970	1980	1990	2000	2010
Pumping unit	64	64	64	64	64	64	64	64	64
PE	0	0	0	0	0	0	0	0	0
PVC	0	0	91	111	105	90	77	66	56
Aluminium	0	0	616	583	594	539	503	487	472
Iron-based	1480	1324	459	101	0	0	0	0	0
Manufacture	790	628	246	95	62	68	70	71	69
Transport	86	161	116	48	38	31	33	35	37
Maintenance	531	467	359	254	228	215	205	200	194
Grading	273	273	273	273	273	273	273	273	273

(*Continued*)

Section AII.9 (Continued) Irrigation

Table AII.9.16 Total Embodied Energy of Surface with IRRS Irrigation System Infrastructure (MJ/ha yr)

	1930	1940	1950	1960	1970	1980	1990	2000	2010
Ditching	50	47	45	43	42	40	39	38	37
Total	3274	2965	2269	1572	1406	1320	1264	1234	1204

Table AII.9.17 Total Embodied Energy of Average Sprinkler Irrigation System Infrastructure (MJ/ha yr)

	1930	1940	1950	1960	1970	1980	1990	2000	2010
Pumping unit	81	81	81	81	81	81	81	81	81
PE	0	0	0	0	0	0	0	0	0
PVC	0	0	376	459	436	373	319	272	233
Aluminium	0	0	944	894	911	827	772	747	724
Iron-based	4013	3589	1455	581	231	214	185	175	175
Manufacture	2142	1704	692	307	198	217	222	228	221
Transport	229	228	132	46	30	24	26	28	30
Maintenance	1256	1083	718	472	379	349	322	307	293
Grading	0	0	0	0	0	0	0	0	0
Ditching	43	41	39	38	36	35	34	33	32
Total	7764	6726	4437	2878	2301	2120	1960	1871	1788

Table AII.9.18 Total Embodied Energy of Trickle Irrigation System Infrastructure (MJ/ha yr)

	1930	1940	1950	1960	1970	1980	1990	2000	2010
Pumping unit	69	69	69	69	69	69	69	69	69
PE				3927	3053	2374	1845	1435	1115
PVC				780	741	633	541	463	395
Aluminium				0	0	0	0	0	0

(Continued)

Section AII.9 (*Continued*) Irrigation

Table AII.9.18 Total Embodied Energy of Trickle Irrigation System Infrastructure (MJ/ha yr)

	1930	1940	1950	1960	1970	1980	1990	2000	2010
Iron-based	339			339	25	23	20	19	19
Manufacture				129	11	12	12	13	12
Transport				73	52	43	46	49	52
Maintenance				1058	788	630	505	407	330
Grading				0	0	0	0	0	0
Ditching				43	42	40	39	38	37
Total				6418	4782	3825	3078	2492	2030

Table AII.9.19 Comparison of Total Embodied Energy of Irrigation System Infrastructure (GJ/ha yr)

	1930	1940	1950	1960	1970	1980	1990	2000	2010
Surface without IRRS	0.80	0.79	0.76	0.64	0.53	0.52	0.51	0.50	0.49
Surface with IRRS	3.27	2.96	2.27	1.57	1.41	1.32	1.26	1.23	1.20
Sprinkler average	7.76	6.73	4.44	2.88	2.30	2.12	1.96	1.87	1.79
Trickle				6.42	4.78	3.82	3.08	2.49	2.03

(Continued)

Section AII.9 (Continued) Irrigation

Table AII.9.20 Total Embodied Energy for 500 mm Net Irrigation with Surface without IRRS Irrigation System (GJ/ha yr)

	1930	1940	1950	1960	1970	1980	1990	2000	2010
Irrigation system	0.8	0.8	0.8	0.6	0.5	0.5	0.5	0.5	0.5
Pumping 0 m	2.5	2.0	1.7	1.2	1.2	1.3	1.4	1.4	1.4
Pumping 50 m	44.7	35.5	29.4	21.7	21.6	23.6	24.1	24.8	24.0
Pumping 100 m	86.7	69.0	57.1	42.2	41.8	45.8	46.9	48.1	46.6

Table AII.9.21 Total Embodied Energy for 500 mm Net Irrigation with Surface with IRRS Irrigation System (GJ/ha yr)

	1930	1940	1950	1960	1970	1980	1990	2000	2010
Irrigation system	3.3	3.0	2.3	1.6	1.4	1.3	1.3	1.2	1.2
Pumping 0 m	2.5	2.0	1.6	1.2	1.2	1.3	1.3	1.4	1.3
Pumping 50 m	27.2	21.6	17.9	13.2	13.1	14.4	14.7	15.1	14.6
Pumping 100 m	52.0	41.3	34.2	25.3	25.1	27.4	28.1	28.8	27.9

Table AII.9.22 Total Embodied Energy for 500 mm Net Irrigation with Average Sprinkler Irrigation System (GJ/ha yr)

	1930	1940	1950	1960	1970	1980	1990	2000	2010
Irrigation system	7.8	6.7	4.4	2.9	2.3	2.1	2.0	1.9	1.8
Pumping 0 m	34.0	27.1	22.4	16.5	16.4	18.0	18.4	18.8	18.3
Pumping 50 m	61.5	48.9	40.5	29.9	29.7	32.5	33.3	34.1	33.1
Pumping 100 m	89.1	70.8	58.7	43.3	43.0	47.0	48.1	49.4	47.9

(Continued)

Section AII.9 (*Continued*)   Irrigation

Table AII.9.23   Total Embodied Energy for 500 mm Net Irrigation with Trickle Irrigation System (GJ/ha yr)

	1930	1940	1950	1960	1970	1980	1990	2000	2010
Irrigation system				6.4	4.8	3.8	3.1	2.5	2.0
Pumping 0 m				7.9	7.9	8.6	8.8	9.1	8.8
Pumping 50 m				19.3	19.2	21.0	21.5	22.0	21.4
Pumping 100 m				30.7	30.5	33.3	34.1	35.0	33.9

Table AII.9.24   Comparison of Total Embodied Energy for 500 mm Net Irrigation with Various Irrigation Systems, 0 m Lift (GJ/ha yr)

	1930	1940	1950	1960	1970	1980	1990	2000	2010
Surface without IRRS	3.3	2.8	2.4	1.9	1.8	1.9	1.9	1.9	1.9
Surface with IRRS	5.7	4.9	3.9	2.8	2.6	2.6	2.6	2.6	2.5
Sprinkler average	41.8	33.8	26.8	19.4	18.7	20.1	20.3	20.7	20.1
Trickle				14.4	12.7	12.5	11.9	11.5	10.8

Table AII.9.25   Comparison of Total Embodied Energy for 500 mm Net Irrigation with Various Irrigation Systems, 50 m Lift (GJ/ha yr)

	1930	1940	1950	1960	1970	1980	1990	2000	2010
Surface without IRRS	45.5	36.3	30.2	22.3	22.1	24.1	24.7	25.3	24.5
Surface with IRRS	30.5	24.6	20.2	14.8	14.5	15.7	16.0	16.3	15.8
Sprinkler average	69.3	55.7	45.0	32.8	32.0	34.6	35.2	36.0	34.8
Trickle				25.7	24.0	24.8	24.6	24.5	23.4

(*Continued*)

Section AII.9 (*Continued*) Irrigation

Table AII.9.26 Comparison of Total Embodied Energy for 500 mm Net Irrigation with Various Irrigation Systems, 100 m Lift (GJ/ha yr)

	1930	1940	1950	1960	1970	1980	1990	2000	2010
Surface without IRRS	87.5	69.8	57.9	42.8	42.4	46.3	47.4	48.6	47.1
Surface with IRRS	55.3	44.3	36.5	26.8	26.5	28.8	29.4	30.0	29.1
Sprinkler average	96.8	77.6	63.1	46.2	45.3	49.1	50.1	51.2	49.6
Trickle				37.1	35.3	37.2	37.2	37.5	36.0

We compiled four examples of greenhouses from the literature: Almeria "Parral" type ("Almeria vineyard type" in Alonso ad Guzman, 2010), Glass greenhouse in Austria, Tunnel greenhouse in Austria and Multitunnel in Spain (Theurl et al., 2013). The annual materials and process requirements are calculated dividing total requirements by useful life, and are shown in Table AII.10.1. The useful lives of the materials are those of the original papers. We could only show simplified inventories of example greenhouses, but more detailed information on greenhouse material requirements can be found in specific studies, such as Torrellas et al. (2012). A recent comprehensive study (Anton et al., 2014) offers equations for calculating detailed material requirements of the four types of greenhouses studied here, as a function the main greenhouse dimensions.

With the information of systems characteristics in Table AII.10.1, and the embodied energy coefficient of each material and process given in Table AII.10.3, we calculated total energy requirements of each type of greenhouse during the period 1950–2010 (Table AII.10.4 through AII.10.8). We included a 20% repair and maintenance rate for greenhouse infrastructure (excluding plastic). We also included transport energy, assuming our standard embodied energy coefficients for farm inputs (Table AII.11.16) for all materials except concrete, for which a 200 km road transport distance was assumed.

AII.11 TRANSPORT

Our estimations of **direct energy** consumption of transport modes are shown in Table AII.11.1. We constructed a series of **rail freight** transport energy consumption taking Hirst (1973) data for 1950 and 1960. For 1980 onward, we used Kamakaté and Schipper (2009) values for selected OECD countries in 1973 and 2005. We calculated a weighted average (weighting by total primary energy consumption in each country) of energy efficiencies in those two time points, and assumed a constant rate of efficiency gain in the period, extrapolating up to 1970 and 2010. For 1970 we used the average of Hirst (1973) and our own elaboration of Kamakaté and Schipper (2009) data. Given the lack of information, we assumed that the energy efficiency of rail freight transport remained constant in the decades previous to 1950. For **road freight** transport energy, we combined the 1950–1970 U.S. data of Hirst (1973) with the OECD 1973–2005 data of Kamakaté and Schipper (2009) to construct a 1950–2010 series of direct energy use (Table AII.11.1). There is a wide disparity between different types of road freight transport, ranging from 1.5 MJ/Mg-km for highest capacity lorries to about 16 MJ/Mg-km for delivery vans (Spielmann and Scholz, 2005; Spielmann et al. 2007). We constructed the series of **water transport** using the data in Stopford (2009) for water container and bulk freight transport (Table AII.11.1). In the case of international tanker water transport, we took the value of 0.1 MJ/Mg-km for around year 2000 and assumed that its efficiency had followed the trend that can be derived from Kamakaté and Schipper (2009) data, of –1.2% yearly change.

Section AII.10   Greenhouses

Table AII.10.1   Materials Used in Typical Greenhouse Types (unit/ha yr)

	Unit	Almeria Vineyard Type	Glass Greenhouse, Austria	Tunnel, Austria	Multitunnel, Spain
Plastic	kg	1,208			406	2,624
Glass	kg	0	6,700		
Pexiglass	kg	0	583		
Iron-based	kg	411	5,500	781	4,563
Aluminium	kg	0	1,250		
Manufacture	kg	411	6,750	781	4,563
Concrete	kg	6,075	25,203		6,377
Bulldozer	h	1			
Rockwool	kg		4,390		
Heating	kg		7,906		

Table AII.10.2   Useful Lives of Common Components of Greenhouses (Years)

	Alonso and Guzman (2010)	Theurl (2008)	Own Selection
Plastic	2	1.5	2
Glass		15	15
Iron-based	20	15–20	20
Aluminium		20	20
Concrete	20	15	20

(Continued)

Section AII.10 (*Continued*) Greenhouses

Table AII.10.3   Embodied Energy of Raw Materials Used in Greenhouses (MJ/unit)

	1950	1960	1970	1980	1990	2000	2010
Plastic	265	206	160	124	97	75	58
Glass	26	21	18	16	14	12	10
Plexiglass	315	261	216	179	148	123	102
Iron-based	61	50	42	39	34	32	32
Aluminium	297	197	181	164	153	148	144
Manufacture	26	19	19	21	21	22	21
Concrete	1	1	1	1	1	1	1
Bulldozer	652	652	652	652	652	652	652

Table AII.10.4   Total Embodied Energy of Almeria Vineyard Greenhouse (GJ/ha yr)

	1950	1960	1970	1980	1990	2000	2010
Plastic	320	249	193	150	117	91	71
Glass	0	0	0	0	0	0	0
Plexiglass	0	0	0	0	0	0	0
Iron-based	25	21	17	16	14	13	13
Aluminium	0	0	0	0	0	0	0
Manufacture	11	8	8	8	9	9	9
Concrete	8	8	7	6	5	5	4
Bulldozer	1	1	1	1	1	1	1
Transport	14	11	10	8	8	8	8
Maintenance	9	7	7	6	6	6	5
Total	388	304	243	196	160	132	111

(*Continued*)

Section AII.10 (Continued) Greenhouses

Table AII.10.5 Total Embodied Energy of Glass Austria Greenhouse (GJ/ha yr)

	1950	1960	1970	1980	1990	2000	2010
Plastic	0	0	0	0	0	0	0
Glass	174	141	120	105	91	79	69
Plexiglass	183	152	126	104	86	71	59
Iron-based	338	276	233	216	186	177	177
Aluminium	371	246	226	205	191	185	179
Manufacture	174	128	127	140	143	146	142
Concrete	35	31	28	25	22	19	17
Bulldozer	0	0	0	0	0	0	0
Transport	91	62	57	46	46	47	49
Maintenance	255	195	172	159	144	136	129
Total	1623	1231	1089	999	910	862	821
Rockwool	268	225	189	159	134	113	95
Heating	7906	7906	7906	7906	7906	7906	7906

Table AII.10.6 Total Embodied Energy of Tunnel Austria Greenhouse (GJ/ha yr)

	1950	1960	1970	1980	1990	2000	2010
Plastic	107	84	65	51	39	31	24
Glass	0	0	0	0	0	0	0
Plexiglass	0	0	0	0	0	0	0
Iron-based	48	39	33	31	26	25	25
Aluminium	0	0	0	0	0	0	0
Manufacture	20	15	15	16	17	17	16
Concrete	0	0	0	0	0	0	0

(Continued)

Section AII.10 (*Continued*) Greenhouses

Table AII.10.6 Total Embodied Energy of Tunnel Austria Greenhouse (GJ/ha yr)

	1950	1960	1970	1980	1990	2000	2010
Bulldozer	0	0	0	0	0	0	0
Transport	5	3	2	2	2	2	2
Maintenance	14	11	10	9	9	8	8
Total	194	151	125	109	93	83	76

Table AII.10.7 Total Embodied Energy of Multitunnel Spain Greenhouse (GJ/ha yr)

	1950	1960	1970	1980	1990	2000	2010
Plastic	695	540	420	326	254	197	153
Glass	0	0	0	0	0	0	0
Plexiglass	0	0	0	0	0	0	0
Iron-based	280	229	194	180	155	147	147
Aluminium	0	0	0	0	0	0	0
Manufacture	118	87	86	94	97	99	96
Concrete	9	8	7	6	6	5	4
Bulldozer	0	0	0	0	0	0	0
Transport	39	24	21	17	18	19	19
Maintenance	81	65	57	56	51	50	49
Total	1222	952	786	680	580	517	469

(*Continued*)

Section AII.10 (*Continued*) Greenhouses

Table AII.10.8 Comparison of Total Embodied Energy of Various Greenhouse Types (GJ/ha yr)

	1950	1960	1970	1980	1990	2000	2010
Almeria vineyard	388	304	243	196	160	132	111
Glass, Austria	1623	1231	1089	999	910	862	821
Tunnel, Austria	194	151	125	109	93	83	76
Multitunnel, Spain	1222	952	786	680	580	517	469

Section AII.11 Transport

Energy

Table AII.11.1 Direct Energy Requirements of Transport Modes (MJ/t-km)

	1930	1940	1950	1960	1970	1980	1990	2000	2010
Truck	4.33	4.33	4.33	4.33	4.14	3.20	3.06	2.93	2.80
Rail	5.26	5.26	5.26	1.39	0.82	0.41	0.34	0.28	0.23
Water (container and bulk)	0.46	0.41	0.36	0.32	0.28	0.25	0.22	0.20	0.18
Water (tanker)	0.23	0.20	0.18	0.16	0.14	0.13	0.11	0.10	0.09

Table AII.11.2 Fuel Production Energy of Transport Modes (MJ/t-km)

	1930	1940	1950	1960	1970	1980	1990	2000	2010
Truck	0.87	0.86	0.87	0.91	0.90	0.68	0.65	0.65	0.68
Rail	0.52	0.50	1.04	0.66	0.72	0.40	0.34	0.29	0.24
Water (container and bulk)	0.07	0.06	0.05	0.05	0.05	0.04	0.04	0.03	0.03
Water (tanker)	0.03	0.03	0.03	0.03	0.02	0.02	0.02	0.02	0.02

Table AII.11.3 Vehicle and Infrastructure Production and Maintenance Energy of Transport Modes (MJ/t-km)

	2000
Truck	0.73
Rail	0.07
Water (container and bulk)	0.02
Water (tanker)	0.01

(Continued)

Section AII.11 (Continued) Transport

Table AII.11.4 Total Embodied Energy of Transport Modes (MJ/t-km)

	1930	1940	1950	1960	1970	1980	1990	2000	2010
Truck	5.93	5.92	5.93	5.98	5.77	4.61	4.44	4.31	4.21
Rail	5.86	5.84	6.37	2.12	1.61	0.88	0.75	0.64	0.53
Water (container and bulk)	0.55	0.49	0.44	0.40	0.36	0.32	0.28	0.26	0.24
Water (tanker)	0.27	0.24	0.22	0.20	0.18	0.16	0.14	0.13	0.12

Table AII.11.5 Distances Traveled by Farm Inputs per Transport Modes (km)

	1900	1910	1920	1930	1940	1950	1960	1970	1980	1990	2000	2010
Farm Inputs												
Truck				200	200	200	200	200	250	300	350	400
Rail				500	500	500	500	500	500	500	500	500
Refined Oil Products												
Truck				200	200	200	200	200	200	200	200	200
Guano and Saltpeter												
Water	16,000	16,000	12,000	10,000	10,000	10,000	10,000	10,000	10,000	10,000	10,000	10,000

Table AII.11.6 Total Embodied Energy of Farm Inputs Transport (MJ/kg Farm Input)

	1900	1910	1920	1930	1940	1950	1960	1970	1980	1990	2000	2010
Farm Inputs												
Truck				1.19	1.18	1.19	1.20	1.15	1.15	1.33	1.51	1.68
Rail				2.93	2.92	3.18	1.06	0.81	0.44	0.38	0.32	0.27
Total				4.11	4.10	4.37	2.26	1.96	1.59	1.71	1.83	1.95
Refined Oil Products												
Truck				1.19	1.18	1.19	1.20	1.16	0.93	0.90	0.87	0.85
Guano and Saltpeter												
Sea	8.89	7.95	5.52	4.54	4.46	4.31	4.02	3.54	3.16	2.92	2.70	2.49

Indirect energy in transport is consumed in the production of fuels and electricity, the production and maintenance of vehicle and the construction and maintenance of infrastructure such as ports, roads, and railways. As a reference case around year 2000, we estimated the distribution of total energy requirements of different transport modes averaging the data provided by Spielmann and Scholz (2005) and Khan Ribeiro et al. (2012) (Figure AII.3). For the calculation of this figure, electricity in rail transport was converted to primary fuel equivalents.

We estimated the historical evolution of transport fuel production energy (Table AII.11.2) using our own estimations of fuel production energy efficiencies (Section 4.4). We assumed that trucks are fueled by diesel fuel and ships by fuel oil. In the case of trains, we assumed that coal was the main fuel in 1930 and 1940—that it represented 70% and 50% in 1950 and 1960 and it had disappeared in 1970. It was substituted by 50% diesel fuel 50% electricity. In the case of vehicle and infrastructure production and maintenance, we assumed fixed values of 0.73, 0.07, and 0.02 MJ/Mg-km for truck, rail and ship freight transport, respectively (Table AII.11.3), resulting from the application of the above calculated percentages to direct fuel energy consumption in year 2000. The sum of direct energy requirements (Table AII.11.1), fuel production energy (Table AII.11.2) and infrastructure energy (Table AII.11.3) results in the total embodied energy values of transport modes shown in Table AII.11.4.

With regard to the **distances traveled by agricultural inputs** up to the farm, Pimentel (1980) reports that farm supplies are transported an average of 640 km, 60% by rail and 40% by truck. According to Audsley et al. (2003), farm supplies are transported 1200 km, 83% by rail and 17% by truck. In ecoinvent database (ecoinvent Centre, 2007) there is a wide variability of transport distances of agricultural inputs. For example, phosphate fertilizers are assumed to travel many thousand kilometers by sea, while the values for nitrogen fertilizers are in the range of those of

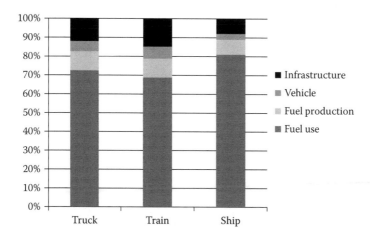

Figure AII.3   Partitioning of total energy inputs of selected transport modes, around year 2000 (%). (Average of the data in Spielmann, M. and Scholz, R.W., *Int. J. Life Cycle Assess.*, 2005, 10(1), 85–94, doi:10.1065/lca.10.181.10; Kahn Ribeiro et al., 2012.)

Audsley et al. (2003). These differences between sources partially represent different situations in United States (Pimentel, 1980), United Kingdom (Audsley et al. 2003), and the European Union (ecoinvent Centre, 2007) in the different periods and for the different products considered. The distances traveled and the transport modes differ between United States and EU and along history. In 1970, 30% of the transport was by road and 20% by rail in the EU. By 1998, these shares were 44% and 8% in the EU and 28% and 37% in the United States, respectively (Caldwell et al., 2002).

We assumed a constant distance of 500 km by rail and 0 km by water. In the case of road transport, we assumed 200 km up to 1970, and a linear growth since that date up to 400 km in 2000 (Table AII.11.5). In the case of refined oil products, we assumed that they were transported only by truck at a distance of 200 km during the whole period, whereas guano and saltpeter were assumed to travel only by sea. The results of the multiplication of total energy inputs by total distance traveled are given in Table AII.11.6.

References

Adams, R. N. 1975. *Energy and Structure. A Theory of Social Power.* Austin, TX: University of Texas Press.

Adams, R. N. 1988. *The Eight Day: Social Evolution as the Self-Organization of Energy.* Austin, TX: University of Texas Press.

Adriaensen, F., J. P. Chardon, G. De Blust, et al. 2003. The application of "least-cost" modelling as a functional landscape model. *Landscape and Urban Planning* 64:233–247.

Agencia de Medio Ambiente (AMA). 1991. *Informe general sobre el medio ambiente, 1990.* Sevilla, Spain: Agencia de Medio Ambiente.

Agergaard, J., N. Fold, and K. V. Gough. 2009. Global-local interactions: Socioeconomic and spatial dynamics in Vietnam's coffee frontier. *The Geographical Journal* 175(2):133–145.

Agnoletti, M. ed. 2006. *The Conservation of Cultural Landscapes.* Wallingford/Cambridge, MA: CAB International.

Agrobit. 2013. Topinambur (*Helianthus tuberosus*). Una forrajera extraordinaria y alternativa económica al alcance de productores agrícolas y ganaderos. Available at http://www.agrobit.com/Info_tecnica/agricultura/forraje_past/AG_000022fp.htm (accessed on April 20, 2014).

Aguilera, E., G. I. Guzmán, and A. M. Alonso. 2015a. Greenhouse gas emissions from conventional and organic cropping systems in Spain. I. Herbaceous crops. *Agronomy for Sustainable Development* 35:713–724.

Aguilera, E., G. I. Guzmán, and A. M. Alonso. 2015b. Greenhouse gas emissions from conventional and organic cropping systems in Spain. II. Fruit tree orchards. *Agronomy for Sustainable Development* 35:725–737.

Aguilera, E., G. I. Guzmán, J. Infante-Amate, et al. 2015c. *Embodied Energy in Agricultural Inputs. Incorporating a Historical Perspective. Sociedad Española de Historia Agraria. DT-SEHA* 1507. Available at www.seha.info (accessed on October 28, 2015).

Aguilera, E., J. Infante-Amate, D. Soto, G. Guzmán, and M. González de Molina. (forthcoming). Industrial inputs in Spanish agriculture, 1900–2008. Agro-ecosystems History Laboratory Working Paper. Seville, Spain: Pablo de Olavide University.

Aguilera, E., L. Lassaletta, A. Gattinger, and B. S. Gimeno. 2013b. Managing soil carbon for climate change mitigation and adaptation in Mediterranean cropping systems. A meta-analysis. *Agriculture, Ecosystems and Environment* 168:25–36.

Aguilera, E., L. Lassaletta, A. Sanz-Cobena, J. Garnier, and A. Vallejo. 2013a. The potential of organic fertilizers and water management to reduce N_2O emissions in Mediterranean climate cropping systems. A review. *Agriculture, Ecosystems and Environment* 164:32–52.

Alcántara, C., S. Sánchez, A. Pujadas, and M. Saavedra. 2009. *Brassica* species as winter cover crops in sustainable agricultural systems in southern Spain. *Journal of Sustainable Agriculture* 33:619–635.

Alexandratos, N., and J. Bruinsma. 2012. *World Agriculture Towards 2030/2050: The 2012 Revision*. FAO, Food and Agriculture Organization of the United Nations. Available at http://www.fao.org/economic/esa (accessed on May 30, 2014).

Almagro, M., J. López, C. Boix-Fayos, J. Albaladejo, and M. Martínez-Mena. 2010. Belowground carbon allocation patterns in a dry Mediterranean ecosystem: A comparison of two models. *Soil Biology and Biochemistry* 42:1549–1557.

Almoguera Millán, J. 2007. *Modelo dehesa sobre las relaciones pastizal-encinar-ganado. Trabajo Fin de Carrera*. Madrid, Spain: Universidad Politécnica de Madrid.

Alonso, A. M., and G. I. Guzmán. 2010. Comparison of the efficiency and use of energy in organic and conventional farming in Spanish agricultural systems. *Journal of Sustainable Agriculture* 34 (3):312–338. doi:10.1080/10440041003613362.

Alonso, A. M., G. I. Guzmán, L. Foraster Pulido, and R. González Lera. 2008. Impacto socioeconómico y ambiental de la agricultura ecológica en el desarrollo rural. In *Producción ecológica. Influencia en el desarrollo rural,* eds. G. I. Guzmán, A. R. García, A. M. Alonso, and J. M. Perea, pp. 71–266. Madrid, Spain: MARM.

Alonso, J. M., J. L. Espada, and R. Socias i Company. 2012. Short communication. Major macroelement exports in fruits of diverse almond cultivars. *Spanish Journal of Agricultural Research* 10(1):175–178.

Altieri, M. A. 1989. *Agroecology: The Science of Sustainable Agriculture*. Boulder, CO: Westview Press.

Altieri, M. A., and C. I. Nicholls. 2007. *Biodiversidad y manejo de plagas en agroecosistemas*. Barcelona, Spain: Icaria Editorial.

Amaducci, S., A. Zatta, M. Raffanini, and G. Venturi. 2008. Characterisation of hemp (*Cannabis sativa* L.) roots under different growing conditions. *Plant and Soil* 313:227–235.

Ambrose, M. D., G. D. Salomonsson, and S. Burn. 2002. Piping systems embodied energy analysis. CMIT Doc. No. 02/302. Highett, Australia: CSIRO Manufacturing and Infrastructure Technology (CMIT).

Anderson, E. L. 1988. Tillage and N-fertilization effects on maize root growth and shoot-root ratio. *Plant and Soil* 108:245–251.

Andrews, M., J. A. Raven, and J. I. Sprent. 2001. Environmental effects on dry matter partitioning between shoot and root of crop plants: Relations with growth and shoot protein concentration. *Annals of Applied Biology* 138:57–68.

Anton, A., M. Torrellas, V. Raya, and J. I. Montero. 2014. Modelling the amount of materials to improve inventory datasets of greenhouse infrastructures. *International Journal of Life Cycle Assessment* 19(1):29–41. doi:10.1007/s11367-013-0607-z.

Andrieu, J., Demarquilly, C., 1987. Valeur nutritive des fourrages: Tables et prévision. *Bull. Tech. C.R.Z.V. Theix*, INRA. 70:61–73.

Apel, P. 1984. Photosyntesis and assimilate partitioning in relation to plant breeding. In *Crop Breeding. A Contemporary Basis*, eds. P. B. Vose and S. G. Blix, pp. 163–184. Oxford, England: Pergamon Press.

Aranguiz, F. 2006. *Las rastrojeras: su reinserción en el ecosistema suelo y su uso en la alimentación del ganado*. Valparaíso, Chile: Pontificia Universidad Católica de Valparaíso.

ASAE. 2000. Agricultural machinery and management data. In *ASAE Standards 2000*. St. Joseph's, MI: American Society of Agricultural Engineers.

Asdrubali, F., G. Baldinelli, F. D'Alessandro, and F. Scrucca. 2015. Life cycle assessment of electricity production from renewable energies: Review and results harmonization. *Renewable and Sustainable Energy Reviews* 42:1113–1122.

Asner, G. P., A. J. Elmore, L. P. Olander, R. E. Martin, and A. T. Harris. 2004. Grazing systems, ecosystem responses, and global change. *Annual Review of Environment and Resources*, 29:261–299.

Asociación Española para la Valorización Energética (AVEBIOM). 2009. Sarmientos de viña. Aprovechamiento energético. *Revista Bioenergy International España*. Available at http://www.bioenergyinternational.es/noticias/News/show/sarmientos-de-vina-aprovechamiento-energetico-194 (accessed on April 25, 2012).

Astier, M. 2002. *El efecto de las leguminosas en la calidad de suelos de Ando en sistemas de maíz de la Cuenca del Lago de Zirahuén, México. Tesis de Doctorado, Facultad de Ciencias.* Mexico: Universidad Nacional Autónoma de México.

Astier, M., Y. Merlin-Uribe, L. Villamil-Echeverri, A. Garciarreal, M. E. Gavito, and O. R. Masera. 2014. Energy balance and greenhouse gas emissions in organic and conventional avocado orchards in Mexico. *Ecological Indicators* 43:281–287.

Audsley, E., K. Stacey, D. J. Parsons, and A. G. Williams. 2009. *Estimation of the Greenhouse Gas Emissions from Agricultural Pesticide Manufacture and Use.* Cranfield, UK: Cranfield University.

Audsley, E., S. Alber, R. Clift, et al. 2003. Harmonisation of environmental life cycle assessment for agriculture. In *Final Report Concerted Action AIR3-CT94-2028*, ed. European Commission DG VI Agriculture.

Ayres, R. U. 2007. On the practical limits to substitution. *Ecological Economics* 61:115–128.

Ayres, R. U., and A. V. Kneese. 1969. Production, consumption and externalities. *American Economy Review* 59:282–297.

Ayres, R. U., and U. E. Simonis (eds). 1994. *Industrial Metabolism: Restructuring for Sustainable Development.* New York, NY: UNU Press.

Babbar, L. I., and D. R. Zak. 1995. Nitrogen loss from coffee agroecosystems in Costa Rica: Leaching and denitrification int the presence and absence of shade trees. *Journal of Environmental Quality* 24:227–233.

Bairoch, P. 1973. Agriculture and the industrial revolution, 1700–1914. In *The Industrial Revolution. Fontana Economic History of Europe*, Vol. III, ed. C. Cipolla, pp. 452–506. London, UK: Collins/Fontana.

Bairoch, P. 1999. *L'Agriculture des pays développés, 1800 à nos jours: Productivité, rendements.* Paris, France: Economica.

Baker, J. F., T. S. Stewart, C. R. Long, and T. C. Cartwright. 1988. Multiple regression and principal components analysis of puberty and growth in cattle. *Journal of Animal Science* 66:2147–2158.

Baraja Rodríguez, E. 1994. *La expansión de la industria azucarera y el cultivo remolachero del Duero en el contexto nacional.* Serie Estudios. Madrid, Spain: Ministerio de Agricultura, Pesca y Alimentación.

Barkin, D., R. Batt, and B. R. DeWalt. 1991. *Alimentos versus Forrajes: La sustitución entre granos a escala mundial.* Mexico D.F. (Mexico): Siglo XXI Editores.

Barral, M. P., J. M. Rey Benayas, P. Meli, and N. O. Maceira. 2015. Quantifying the impacts of ecological restoration on biodiversity and ecosystem services in agro-ecosystems: A global meta-analysis. *Agriculture, Ecosystems and Environment* 202:223–231.

Basalla, G. 1988. *The Evolution of Technology.* New York, NY: Cambridge University Press.

Batty, J. C. and J. Keller. 1980. Energy requirements for irrigation. In *Handbook of Energy Utilization in Agriculture*, ed. D. Pimentel. Boca Raton, FL: CRC Press.

Bayliss-Smith, T. 1982. *The Ecology of Agricultural Systems, Cambridge Topics in Geography. Second Series.* Cambridge, MA: Cambridge University Press.

Beccaro, G. L., A. K. Cerutti, I. Vandecasteele, L. Bonvegna, D. Donno, and G. Bounous. 2014. Assessing environmental impacts of nursery production: Methodological issues and results from a case study in Italy. *Journal of Cleaner Production* 80:159–169.

Beer, J. W. 1988. Litter production and nutrient cycling in coffee (*Coffea arabica*) or cacao (*Theobroma cacao*) plantations with shade trees. *Agroforestry System* 7:103–114.

Beer, J., R. Muschler, D. Kass, and E. Somarriba. 1998. Shade management in coffee and cacao plantations. In *Directions in Tropical Agroforestry Research*, eds. P. K. R. Nair and C. R. Latt, pp. 139–164. New York, NY: Springer.

Bengtsson, J., J. Ahnström, and A. Weibull. 2005. The effects of organic agriculture on bio-diversity and abundance: A meta-analysis. *Journal of Applied Ecology* 42:261–269.

Benítez, E., R. Nogales, M. Campos, and F. Ruano. 2006. Biochemical variability of olive orchard soils under different management systems. *Applied Soil Ecology* 32:221–231.

Benton, T. G., J. A. Vickery, and J. D. Wilson. 2003. Farmland biodiversity: Is habitat hetero-geneity the key? *Trends in Ecology and Evolution* 18(4):182–188.

Benyus, J. M. 1997. *Biomimicry: Innovation Inspired by Nature.* New York, NY: William Morrow & Company.

Berkes, F., J. Colding, and C. Folke. 2000. Rediscovery of traditional ecological knowledge as adaptive management. *Ecological Applications* 10(5):1251–1262.

Berlin, D. and H.-E. Uhlin. 2004. Opportunity cost principles for life cycle assessment: Toward strategic decision making in agriculture. *Progress in Industrial Ecology* 1:187–202.

Bernués, A., I. Casasús, N. Flores, A. Olaizola, A. Ammar, and E. Manrique. 2004. Livestock farming systems and conservation of Spanish Mediterranean mountain areas: The case of the 'Sierra de Guara' Natural Park. 1. Characterisation of farming systems. In *Réhabilitation des pâturages et des parcours en milieux méditerranéens*, eds. A. Ferchichi, pp. 195–198. Zaragoza, Spain: CIHEAM.

Bettwy, M. 2006. Growth in Amazon Cropland May Impact Climate and Deforestation Patterns. Greenbelt, MD: NASA. Available at https://web.archive.org/web/20081024151206/http://www.nasa.gov/centers/goddard/news/topstory/2006/amazon_crops.html (accessed on April 29, 2016).

Bevilacqua, P. (ed.) 1989–91. *Storia dell'agricoltura italiana in età contemporanea*, 3 vols. Venezia, Italy: Marsilio.

Bevilacqua, P. and M. Rossi-Doria. 1984. *Le bonifiche in Italia dal '700 a oggi.* Bari-Roma, Italy: Laterza.

Bhat, M. G., B. C. English, A. F. Turhollow, and H. O. Nyangito. 1994. Energy in synthetic fertilizers and pesticides: Revisited. Final project report. Knoxville, TN: Department of Agricultural Economics and Rural Sociology, Tennessee University.

Bilandzija, N., N. Voca, T. Kricka, A. Matin, and V. Jurisic. 2012. Energy potential of fruit tree pruned biomass in Croatia. *Spanish Journal of Agricultural Research* 10(2):292–298.

Bindraban, P. S. and R. Rabbinge. 2012. Megatrends in agriculture—Views for discontinui-ties in past and future developments. *Global Food Security* 1(2):99–105.

Bolinder, M. A., D. A. Angers, G. Belanger, R. Michaud, and M. R. Laverdiere. 2002. Root biomass and shoot to root ratios of perennial forage crops in eastern Canada. *Canadian Journal of Plant Science* 82:731–737.

Bolinder, M. A., H. H. Janzen, E. G. Gregorich, D. A. Angers, and A. J. VandenBygaart. 2007. An approach for estimating net primary productivity and annual carbon inputs to soil for common agricultural crops in Canada. *Agriculture, Ecosystems and Environment* 118:29–42.

Bolinder, M. A., D. A. Angers, and J. P. Dubuc. 1997. Estimating shoot to root ratios and annual carbon inputs in soils for cereal crops. *Agriculture, Ecosystems and Environment* 63:61–66.

Boserup, E. 1981. *Population and Technological Change. A Study of Long-term Trends.* Chicago, IL: University of Chicago Press.

Boulding, K. E. 1966. The economics of the coming spaceship Earth. In *Environmental Quality in a Growing Economy*, ed. H. Jarret, pp. 3–14. Baltimore, MD: John Hopkins University Press.

Bowren, K. E., D. A. Cooke, D. A., and R. K. Downey. 1969. Yield of dry matter and nitrogen from tops and roots of sweetclover alfalfa and red clover at five stages of growth. *Canadian Journal of Plant Science* 49:61–68.

Boza, J., A. B. Robles, M. P. Fernández-García, and J. L. González-Rebollar. 2000. Impacto ambiental en las explotaciones ganaderas del extensivo mediterráneo. In *Globalización medioambiental. Perspectivas agrosanitarias y urbanas*, eds. F. Férnandez-Buendía, M. V. Pablos, and J. V. Tarazona, pp. 257–268. Madrid, Spain: MAPA.

BP. 2011. *BP Statistical Review of World Energy 2011.* ed. BP. http://www.bp.com /en/global/corporate/energy-economics/statistical-review-of-world-energy.html (accessed September 12, 2016).

BP. 2014. *BP Statistical Review of World Energy 2014.* ed. BP. http://www.bp.com /en/global/corporate/energy-economics/statistical-review-of-world-energy.html (accessed September 12, 2016).

Bredeson, L., R. Quiceno-Gonzalez, X. Riera-Palou, and A. Harrison. 2010. Factors driving refinery CO_2 intensity, with allocation into products. *International Journal of Life Cycle Assessment* 15(8):817–826. doi:10.1007/s11367-010-0204-3.

Bruton, H. J. 1998. A reconsideration of import substitution. *Journal of Economic Literature* 36:903–936.

Bulatkin, G. A. 2012. Analysis of energy flows in agro-ecosystems. *Herald of the Russian Academy of Sciences* 82(4):326–334.

Burkhard, B., B. D. Fath, and F. Müller. 2011. Adapting the adaptive cycle: Hypotheses on the development of ecosystem properties and services. *Ecological Modelling* 222:2878–2890.

Burkhard, B., F. Kroll, S. Nedkov, and F. Müller. 2012. Mapping ecosystem service supply, demand and budgets. *Ecological Indicators* 21:17–29.

Butzer, K. W. 1976. *Early Hydraulic Civilization in Egypt.* Chicago, IL: University of Chicago Press.

Buyanovsky, G. A. and G. H. Wagner. 1986. Postharvest residue input to cropland. *Plant and Soil* 93:57–65.

Cáceres Díaz, R. O. 2012. Respuesta a la fertilización orgánica e inorgánica del Algodón en el Suroeste de Chaco, Argentina. *Actas V Congreso Iberoamericano sobre Desarrollo y Ambiente de REDIBEC (CISDA),* Septiembre 2011, Santa Fe (Argentina). Available at http://fich.unl.edu.ar/CISDAV/upload/Ponencias_y_Posters/Eje02/Caceres_Diaz _Raul_Omar/FERTILIZACION%20ORGANICA%20E%20INORGANICA%20 EN%20ALGODON.pdf (accessed on June 12, 2013).

CAFETICO 1992. Aplicación de fertilizante (Caída histórica). *CAFETICO* 9:6.

Cairns, M. A., S. Brown, E. H. Helmer, and G. A. Baumgardner. 1997. Root biomass allocation in the world's upland forests. *Oecologia* 111(1):1–11.

Calatayud, S. and J. M. Martínez Carrión. 1999. El cambio técnico en los sistemas de captación e impulsión de aguas subterráneas para riego en la España Mediterránea. In *El agua en los sistemas agrarios. Una perspectiva histórica*, eds. R. Garrabou and J. M. Naredo, pp. 15–39. Madrid, Spain: Argentaria/Visor.

Calderón Espinosa, E. 2002. Manejos tradicionales del olivar en la Comarca de los Montes Orientales (Granada). Tesis de Maestría en Agroecología y Desarrollo Rural Sostenible, Universidad Internacional de Andalucía.

Caldwell, H., O. de Buen, M. D. Meyer, et al. 2002. Freight transportation in the European market. *FHWA International Technology Exchange Reports*.

Calle, Z., E. Murgueitio, J. Chará, C. H. Molina, A. F. Zuluaga, and A. Calle. 2013. A strategy for scaling-up intensive silvopastoral systems in Colombia. *Journal of Sustainable Forestry* 32:677–693.

Camacho, J. L., J. M. Urbano, I. M. Pedraza, and G. Pardo. 2011. Malas hierbas en arroz ecológico: ¿son realmente un factor limitante del rendimiento en Andalucía? Plantas invasoras, resistencia a herbicidas y detección de malas hierbas. *Actas del XIII Congreso de la Sociedad Española de Malherbología*, pp. 195–198. San Cristóbal de La Laguna, 22–24 de Noviembre de 2011xs.

Campbell, C. A. and R. de Jong. 2001. Root-to-straw ratios—Influence of moisture and rate of N fertilizer. *Canadian Journal of Plant Science* 81:39–43.

Campos, P. and J. M. Naredo. 1980. La energía en los sistemas agrarios. *Agricultura y Sociedad* 15:17–113.

Cannavo, P., J. –M. Harmand, B. Zeller, P. Vaast, J. E. Ramírez, and E. Dambrine. 2013. Low nitrogen use efficiency and high nitrate leaching in a highly fertilized *Coffea arabica–Inga densiflora* agroforestry system: A 15N labeled fertilizer study. *Nutrient Cycling in Agroecosystems* 95:377–394.

Cannell, M. G. R. and S. C. Willett. 1976. Shoot growth phenology, dry matter distribution and root: Shoot ratios of provenances of *Populus trichocarpa, Picea sitchesis* and *Pinus contorta* growing in Scotland. *Silvae Genetica* 25(2):49–59.

Cao, S., G. Xie, and L. Zhen. 2010. Total embodied energy requirements and its decomposition in China's agricultural sector. *Ecological Economics* 69(7):1396–1404. doi:10.1016/j.ecolecon.2008.06.006.

Cardoso, A. S., A. Berndt, A. Leytem, et al. 2016. Impact of the intensification of beef production in Brazil on greenhouse gas emissions and land use. *Agricultural Systems* 143:86–96.

Carpintero, O. and J. M. Naredo. 2006. Sobre la evolución de los balances energéticos de la agricultura española, 1950–2000. *Historia Agraria* 40:531–556.

Carter, M. R., H. T. Kunelius, J. B. Sanderson, J. Kimpinski, H. W. Platt, and M. A. Bolinder. 2003. Productivity parameters and soil health dynamics under long-term 2-year potato rotations in Atlantic Canada. *Soil and Tillage Research* 72:153–168.

Castro, F., E. Montes, and M. Raine. 2004. *Centroamérica. La crisis cafetalera: Efectos y estrategias para hacerle frente*. Roma: FAO.

Castro, J., E. Fernandez-Ondoño, C. Rodríguez, A. M. Lallena, M. Sierra, and J. Aguilar. 2008. Effects of different olive-grove management systems on the organic carbon and nitrogen content of the soil in Jaen (Spain). *Soil and Tillage Research* 98:56–67.

Castro-Tanzia, S., T. Dietsch, N. Urena, L. Vindasa, and M. Chandler. 2012. Analysis of management and site factors to improve the sustainability of smallholder coffee production in Tarrazú, Costa Rica. *Agriculture, Ecosystems and Environment* 155:172–181.

Centeno, J. 2009. Tema 2. Enología. Universidad de Vigo. Available at http://webs.uvigo.es /jcenteno/Documentacion_Tema_2_2008-09.pdf (accessed on March 9, 2013).

Centre International de Hautes Etudes Agronomiques Méditerranéennes (CIHEAM) 1990. *Tableaux de la valeur alimentaire pour les ruminants des fourrages et sous-produits d'origine méditerranéenne.* Zaragoza: CIHEAM; Editem scientifiques: X. Alibes and J. L. Tisserand.

Centro de Comercio Internacional. 2007. Almacenamiento y manipulación del algodón en rama. Available at http://www.guiadealgodon.org/guia-de-algodon /almacenamiento-y-manipulacion-del-algodon-en-rama/ (accessed on May 10, 2013).

Centro de Investigación y Formación Agrarias de Cantabria (CIFA) 2007. *Los pastos de Cantabria y su aprovechamiento, Producción y calidad: Anexo II,* Santander: Consejería de Desarrollo Rural, Ganadería, Pesca y Biodiversidad.

Cerón-Salazar, I., and C. Cardona-Alzate. 2011. Evaluación del proceso integral para la obtención de aceite esencial y pectina a partir de cáscara de naranja. *Ingeniería y Ciencia* 7(13):65–86.

Chao, J., C. Lacasta, R. Estalrich, R. Meco, and R. González Ponce. 2002. Estudio de la fora arvense asociada a los cereales de ambientes semiáridos en rotación de cultivos de secano. *Actas V Congreso de la Sociedad Española de Agricultura Ecológica.* Gijón, Septiembre de 2002. 733–740.

Chirinda, N., J. E. Olesen, and J. R. Porter. 2012. Root carbon input in organic and inorganic fertilizer-based systems. *Plant Soil* 359:321–333.

Chirinda, N., M. S. Carter, K. R. Albert, et al. 2010. Emissions of nitrous oxide from arable organic and conventional cropping systems on two soil types. *Agriculture, Ecosystems and Environment* 136:199–208.

Choi, H.L., Sudiarto, S.I.A., Renggaman, A., 2014. Prediction of livestock manure and mixture higher heating value based on fundamental analysis. *Fuel* 116, 772–780.

Civantos, L., and M. Olid. 1982. Los ramones de los olivos. *Agricultura* 605:978–980.

Clar, E., V. Pinilla, and R. Serrano. 2014. El comercio agroalimentario español en la segunda globalización, 1951–2011. *DT-AEHE,* 1414.

Cleveland, C. J., R. Costanza, C. A. S. Hall, and R. Kaufmann. 1984. Energy and the U.S. Economy: A biophysical perspective. *Science* 225(4665):890–897.

Cleveland, C. J. 1992. Energy quality and energy surplus in the extraction of fossil fuels in the U.S. *Ecological Economics* 6:139–162.

Cleveland, C. J. 1995. The direct and indirect energy use of fossil-fuels and electricity in USA agriculture, 1910–1990. *Agriculture Ecosystems and Environment* 55(2):111–121. doi:10.1016/0167-8809(95)00615-y.

Comisión Nacional de Emergencias (CNA). 1976. *Balance energético nacional: Resultados parciales preliminares.* San José, Costa Rica.

Connor, D. J., and E. Fereres. 2005. The physiology of adaptation and yield expression in olive. In *Horticultural Reviews,* ed. J. Janick, 31:155–229. New Jersey: John Wiley & Sons.

Consejería de Agricultura y Pesca (CAP) 2008. *Potencial energético de la biomasa residual agrícola y ganadera en Andalucía.* Sevilla, Spain: Junta de Andalucía.

Consejería de Medio Ambiente y Ordenación del Territorio de Andalucía (CMAOT) 2014. Biomasa Forestal de Andalucía. Available at http://www.juntadeandalucia.es/medio-ambiente/bioforan/plantillas/biomasa/index.html?especie=halepensis (accessed on October 22, 2014).

Cook, S. 1973. *Zapotec Stoneworkers: The Dynamics of Rural Simple Commodity Production in Modern Mexican Capitalism.* Lanham, MD: University Press of America.

Cordell, D., J. O. Drangert, and S. White. 2009. The story of phosphorus: Global food security and food for thought. *Global Environmental Change* 19:292–305. doi:10.1016/j.gloenvcha.2008.10.009.

Cornell, S. 2010. Valuing ecosystem benefits in a dynamic world. *Climate Research* 45:261–272.

Correal, E., A. Robledo, and M. Erena (coords.) 2007. *Tipificación, cartografía y evaluación de los recursos de la Región de Murcia. Informe 18*. Murcia: Consejería de Agricultura.

Costa Batllori, P. 1978. Energía y fibra bruta en alimentación del conejo. p. 20. (dialnet.unirioja.es/descarga/articulo/2915578.pdf).

Costa Pérez, J. C., Ángel Martín Vicente, Rocío Fernández Alés, and María Estirado Oliet. 2006. *Dehesas de Andalucía. Caracterización ambiental*. Sevilla, Spain: Consejería de Medio Ambiente. Junta de Andalucía.

Costanza, R. 2012. Ecosystem health and ecological engineering. *Ecological Engineering* 45:24–29.

Crawford, R. H. 2009. Life cycle energy and greenhouse emissions analysis of wind turbines and the effect of size on energy yield. *Renewable and Sustainable Energy Reviews* 13(9):2653–2660. doi:10.1016/j.rser.2009.07.008.

Cuadra, M., and T. Rydberg. 2006. Energy evaluation on the production, processing and export of coffee in Nicaragua. *Ecological Modelling* 196(3):421–433.

Cusso, X., R. Garrabou, and E. Tello. 2006. Social metabolism in an agrarian region of Catalonia (Spain) in 1860–1870: Flows, energy balance and land use. *Ecological Economics* 58(1):49–65. doi:10.1016/j.ecolecon.2005.05.026.

Cussó, X., R. Garrabou, J. R. Olarieta, and E. Tello. 2006a. Balances energéticos y usos del suelo en la agricultura catalana: Una comparación entre mediados del siglo XIX y finales del siglo XX. *Historia Agraria* 40:471–500.

D'Attorre, P. P., and D. B. Alberto (eds.). 1994. *Studi sull'agricoltura italiana. Società rurale e modernizzazione*. Milano: Feltrinelli.

Daccache, A., J. S. Ciurana, J. A. Rodriguez Diaz, and J. W. Knox. 2014. Water and energy footprint of irrigated agriculture in the Mediterranean region. *Environmental Research Letters* 9(12):124014.

Dahmus, J. B. 2014. Can efficiency improvements reduce resource consumption? *Journal of Industrial Ecology* 18(6):883–897. doi:10.1111/jiec.12110.

De Boer, I. J. M. 2003. Environmental impact assessment of conventional and organic milk production. *Livestock Production Science* 80:69–77.

De Groot, R. S., M. A. Wilson, and R. M. J. Boumans. 2002. A typology for the classification, description and valuation of ecosystem functions, goods and services. *Ecological Economics* 41:393–408.

De Groot, R., J. Van der Perk, A. Chiesura, and A. Van Vliet. 2003. Importance and threats determining factors for criticality of natural capital. *Ecological Economics* 44:187–204.

De Groot, R., R. Alkemade, L. Braat, L. Hein, and L. Willemen. 2010. Challenges in integrating the concept of ecosystem services and values in landscape planning, management and decision making. *Ecological Complexity* 7:260–272.

De Masson, L. 1997. Métodos analíticos para la determinación de humedad, alcohol, energía, materia grasa y colesterol en alimentos. In *Producción y manejo de datos de composición química de alimentos en nutrición*, eds. C. Morón, I. Zacarías, and S. de Pablo, pp. 147–163. Roma: FAO.

Debier, J. C., J. P. Déleage, and D. Hémery. 1986. *Les Servitudes de la Puissance: Une Histoire de L'Energie*. Paris: Flammarion.

Della Porta, G. 2015. *Todo sobre el aceite de oliva extra virgen*. Spain: Babelcube.

Denevan, W. M. 1982. Hydraulic agriculture in the American tropics: Forms, measures and recent research. In *Maya Subsistence*, Studies in Memory of Dennis E. Puleston, ed. K. V. Flannery, pp. 181–204. New York, NY: Academic Press.

Detlefsen, G, and E. Somarriba. 2012. *Producción de madera en sistemas agroforestales de Centroamérica*. Turrialba (Costa Rica): Ministry for Foreing Affairs of Finland and CATIE.

Deugd, M. 2003. *Crisis del café: Nuevas estrategias y oportunidades*. San José: Ruta.

DGEC. 1953. *Censo Agropecuario de 1950*. San José, Costa Rica: Dirección General de Estadística y Censos.

DGEC. 1965. *Censo Agropecuario de 1963*. San José, Costa Rica: Dirección General de Estadística y Censos.

DGEC. 1974. *Censo Agropecuario de 1973*. San José, Costa Rica: Dirección General de Estadística y Censos.

DGEC. 1985. *Censo Agropecuario de 1984*. San José, Costa Rica: Dirección General de Estadística y Censos.

Di Blasi, C., V. Tanzi, and M. Lenzetta. 1997. A study on the production of agricultural residues in Italy. *Biomass and Bioenergy* 12(5):321–331.

Díaz del Cañizo, M. A., G. I. Guzmán, and A. Lora González. 1998. Control de la flora arvense en dos cultivos hortícolas en función del período crítico de competencia. *Actas II Congreso de la Sociedad Española de Agricultura Ecológica*. Pamplona, Septiembre de 1996. pp. 65–76.

Díaz Gaona, C., V. Rodríguez Estévez, M. Sánchez Rodríguez, J. M. Ruz Luque, C. Hervás Castillo, and C. Mata Moreno. 2014. Estudio del aprovechamiento de los pastos en Andalucía y Castilla-La Mancha. *Actas XI Congreso de Sociedad Española de Agricultura Ecológica*, Vitoria-Gasteiz (Álava), 1–4 Octubre 2014.

Dicum, G, and N. Luttinger. 1999. *The Coffee Book: Anatomy of an Industry from the Crop to the Last Drop*. New York, NY: The New Press.

Dirección General de Aduanas. 1899, 1900, 1903a, 1903b, 1903c. *Estadística general del comercio exterior de España en 1898, 1899, 1900, 1901, 1902*. Madrid, Spain: Dirección General de Aduanas.

Dittirch, M., and S. Bringezu. 2010. The physical dimension of international trade. Part 1: Direct global flows between 1962 and 2005. *Ecological Economics* 69:1838–1847.

Dittrich, M., S. Bringezu, and H. Schütz. 2012. The physical dimension of international trade. Part 2: Indirect global resource flows between 1962 and 2005. *Ecological Economics* 79:32–43.

Doering, O. C. 1980. Accounting for energy in farm machinery and buildings. In *Handbook of Energy Utilization in Agriculture*, ed. D. Pimentel. Boca Raton, FL: CRC Press.

Dordas, C. A., and C. Sioulas. 2009. Dry matter and nitrogen accumulation, partitioning, and retranslocation in safflower (*Carthamus tinctorius* L.) as affected by nitrogen fertilization. *Field Crops Research* 110:35–43.

Dovring, F. 1985. Energy use in United States agriculture: A critique of recent research. *Energy in Agriculture* 4:79–86.

Du, F., G. J. Woods, D. Kang, K. E. Lansey, and R. G. Arnold. 2013. Life cycle analysis for water and wastewater pipe materials. *Journal of Environmental Engineering-Asce* 139(5):703–711. doi:10.1061/(asce)ee.1943-7870.0000638.

Duque-Orrego, H., and L. C. Dussán. 2004. Productividad de la mano de obra en la cosecha de café en cuatro municipios de la región cafetera central de Caldas. *Cenicafé* 55(3):246–258.

Durán Zuazo, V. H., C. R. Rodríguez Pleguezuelo, L. Arroyo Panadero, A. Martínez Raya, J. R. Francia Martínez, and B. Cárceles Rodríguez. 2009. Soil conservation measures in rainfed olive orchard in south-eastern Spain: Impacts of plant strips on soil water dynamics. *Pedosphere* 19:453–464.

ecoinvent-Centre. 2007. ecoinvent data v2.0. ecoinvent reports No. 1–25. Dübendorf, Switzerland: Swiss Centre for Life Cycle Inventories.

EIA. 2014. Average Operating Heat Rate for Selected Energy Sources. U.S. Energy Information Administration. Available at http://www.eia.gov/electricity/annual/html /epa_08_01.html (accessed on September 20, 2015).

Ekins, P., S. Simon, L. Deutsch, C. Folke, and R. de Groot. 2003. A framework for the practical application of the concepts of critical natural capital and strong sustainability. *Ecological Economics* 44:165–185.

Erb, K.-H., C. Lauk, T. Kastner, A. Mayer, M. C. Theurl, and H. Haberl. 2016. Exploring the biophysical option space for feeding the world without deforestation. *Nature Communications* 7:11382.

Erb, K.-H., H. Haberl, F. Krausmann, et al. 2009. *Eating the Planet: Feeding and Fuelling the World Sustainably, Fairly and Humanely—A Scoping Study.* Institute of Social Ecology and PIK Potsdam. Vienna: Social Ecology Working Paper No. 116.

Erley, G. S. E. R. auf'm., Dewi, O. Nikus, and W. J. Horst. 2010. Genotypic differences in nitrogen efficiency of white cabbage (*Brassica oleracea* L.). *Plant and Soil* 328(1–2):313–325.

Esguerra, M. P. 1991. Colombia, Guatemala y Costa Rica: Países cafeteros de la Cuenca del Caribe. *Coyuntura Económica* 21(1):111–137.

Espada Carbó, J. L. 2011. Nuevas técnicas de producción en el cultivo del almendro. *Jornadas Técnicas del almendro*, Logroño 27 de Octubre de 2011.

European Bioenergy Networks (Eubionet) 2003. Biomass survey in Europe. Country report of Greece. Available at http://www.afbnet.vtt.fi/greece_biosurvey.pdf (accessed on September 22, 2012).

European Commission 2011. Soil organic matter management across the EU—Best practices, constraints and trade-offs. Available at http://ec.europa.eu/environment/soil /som_en.htm (accessed on April 21, 2014).

European Commission 2013. Report from the commission to the council and the European parliament on the implementation of Council Directive 91/676/EEC concerning the protection of waters against pollution caused by nitrates from agricultural sources based on Member State reports for the period 2008–2011. Brussels, 4-10-2013.

Evans, J. 2004. Nebraska tractor test data. Accessed on February 15, 2015.

Evenson, R. E. and D. Gollin. 2003. Assessing the impact of the Green Revolution, 1960 to 2000. *Science* 300(5620):758–762.

FAO. 1971. The adoption of Joules as units of energy. FAO/Who ad hoc committee of experts on energy and protein: Requirements and recommended intakes. Available at http:// www.fao.org/docrep/meeting/009/ae906e/ae906e17.htm (accessed on December 11, 2013).

FAO. 1991. Conservación de energía en las industrias mecánicas forestales. Estudio FAO Montes 93. Roma, Italy: FAO. Available at http://books.google.es/books?id=dW1jtv vVB0UC=PA86=valor+calorifico=es=zjPYTdScDYyctwfFro3pDg=X=book_result =book-preview-link=1=0CC4QuwUwAA#v=onepage=valor%20calorifico=false (accessed on December 6, 2013).

FAO. 2009. *Livestock in the Balance. The State of Food and Agriculture.* Roma, Italy: FAO.

FAO. 2015. FAOSTAT—FAO database for food and agriculture, Food and agriculture Organisation of United Nations (FAO). Rome, Italy: FAO. Available: http://faostat3.fao.org/ (accessed on January 15, 2015).

FAO. 2016. FAOSTAT. Available at http://faostat.fao.org/ (accessed on July 20, 2016).

Farfán-Valencia, F. 2005. Producción de café en un sistema intercalado con plátano dominico hartón con y sin fertilización química. *Cenicafé* 56(3):269–280.

Fath, B. D., S. E. Jørgensen, B. C. Patten, and M. Straskraba. 2004. Ecosystem growth and development. *Biosystems* 77:213–228.

Federico, G. 2008. *Feeding the World: An Economic History of Agriculture, 1800–2000.* Princeton, IL: Princeton University Press.

Fernandes, S. D., N. M. Trautmann, D. G. Streets, C. A. Roden, and T. C. Bond. 2007. Global biofuel use, 1850–2000. *Global Biogeochemical Cycles* 21(2):GB2019. doi:10.1029/2006gb002836.

Fernández, M. D., M. Gallardo, S. Bonachela, F. Orgaz, R. B. Thompson, and E. Fereres. 2005. Water use and production of a greenhouse pepper under optimum and limited water supply. *Journal of Horticultural Science and Biotechnology* 80:87–96.

Ferreira, J., A. García, L. Frias, and A. Fernández. 1986. Los nutrientes N-P-K en la fertilización del olivar. *Olea* 17:141–152.

Firbank, L. G., S. Petit, S. Smart, A. Blain, and R. J. Fuller. 2008. Assessing the impacts of agricultural intensification on biodiversity: A British perspective. *Philosophical Transactions of Royal Society B* 363:777–787.

Fischer-Kowalski, M. 1997. Society's metabolism: On the childhood and adolescence of a rising conceptual star. In *The International Handbook of Environ-mental Sociology*, eds. M. Redclift and G. Woodgate (eds), 119–137. Chelten-ham, UK: Edward Elgar.

Fischer-Kowalski, M. 1998. Society's metabolism: The intellectual history of materials flow analysis, part I, 1860–1970. *Journal of Industrial Ecology* 2:61–77.

Fischer-Kowalski, N. 2003. On the history of industrial metabolism. *Perspectives on Industrial Ecology* 2:35–45.

Fischer-Kowalski, M, and H. Haberl. 2007. *Socioecological Transitions and Global Change: Trajectories of Social Metabolism and Land Use.* Cheltenham: Edward Elgar Publishing.

Fisher-Kowalski, M., and H. Haberl. 1997. Tons, joules, and money: Modes of production and their sustainability problems. *Society and Natural Resources* 10:61–85.

Fisher-Kowalski, M., and W. Hüttler 1999. Society's metabolism: The intellectual history of materials flow analysis, part II, 1970–1998. *Journal of Industrial Ecology* 2:107–129.

Fisher-Kowalski, M., F. Krausmann, and I. Pallua. 2014. A sociometabolic reading of the Anthropocene. *The Anthropocene Review* 1(1):8–33. doi:10.1177/2053019613518033.

Fishman, T., H. Schandl, H. Tanikawa, P. Walker, and F. Krausmann. 2014. Accounting for the Material Stock of Nations. *Journal of Industrial Ecology* 18(3):407–420.

Flannery, K. V. 1972. The cultural evolution of civilizations. *Annual Review of Ecology and Systematics* 3:399–426.

Flauzino, A., J. J. Fonseca, R. Jiménez, A. Ribeiro, and E. de Castro Melo. 2014. Energy balance in the production of mountain coffee. *Renewable and Sustainable Energy Reviews* 39:1208–1213.

Flessa, H., R. Ruser, P. Dörsch, et al. 2002. Integrated evaluation of greenhouse gas emissions (CO_2, CH_4, N_2O) from two farming systems in southern Germany. *Agriculture, Ecosystems and Environment* 91:175–189.

Flores Mengual, M. P., and M. Rodríguez Ventura. 2013. Nutrición animal. Ed: Universidad de las Palmas de Gran Canaria, Las Palmas, Gran Canaria. Available at http://www.webs.ulpgc.es/nutranim (accessed on January 28, 2013).

Fluck, R.C., 1981. Net energy sequestered in agricultural labour. *Transactions of the ASABE* 24, 1449–1455.

Fluck, R. C. (ed.) 1992. *Energy in Farm Production.* Amsterdam, the Netherlands: Elsevier.

Fluck, R. C., and C. D. Baird. 1980. *Agricultural Energetics.* Westport, CT: The AVI Publishing Company.

Food and Agriculture Organization. 2007. *The State of Food and Agriculture 2007.* FAO: Rome.

Folke, C., Å. Jansson, J. Rockström, et al. 2011. Reconnecting to the biosphere. *AMBIO* 40:719–738.

Foraster, L., M. J. Lorite, I. Mudarra, A. M. Alonso, A. Pujadas-Salvá, and G. I. Guzmán. 2006a. Evaluación de distintos manejos de las cubiertas vegetales en olivar ecológico. In *VII Congreso de la Sociedad Española de Agricultura Ecológica,* N 14 (CD edition), pp. 18–23 de Septiembre de 2006. Zaragoza: Sociedad Española de Agricultura Ecológica.

Foraster, L., P. Rodríguez, G. I. Guzmán, and A. Pujadas-Salvá. 2006b. Ensayo de diferentes cubiertas vegetales en olivar ecológico en Castril (Granada). In *VII Congreso de la Sociedad Española de Agricultura Ecológica,* 18–23 de Septiembre de 2006. Sociedad Española de Agricultura Ecológica, Zaragoza. N 16 (CD edition).

Fortunel, F. 2000. *Le Cafeé au Vieêtnam: de la Colonisation a 'l'Essor d'un Grand Producteur Mondial.* Paris: L'Harmattan.

Francescato, V., E. Antonini, and L. Zuccoli. 2008. Manual de combustibles de madera producción, requisitos de calidad, comercialización. Valladolid: Asociación Española de Valorización Energética de la Biomasa (AVEBIOM).

Frischknecht, R., N. Jungbluth, H. –J. Althaus, et al. 2007a. Implementation of Life Cycle Impact Assessment Methods. In *ecoinvent report No. 3.* Dübendorf: Swiss Centre for Life Cycle Inventories.

Frischknecht, R., N. Jungbluth, H.-J Althaus, et al. 2007b. Overview and methodology. In *ecoinvent report No. 1.* Dübendorf: Swiss Centre for Life Cycle Inventories.

Fuertes, A. 2009. Posibilidades técnicas del uso de la biomasa no alimentaria para la obtención de energía en España. Bachelor, Universidad Politécnica de Madrid, Spain: Escuela Técnica Superior de Ingenieros Agrónomos.

Fujiyoshi, P. T., S. R. Gliessman, and J. H. Langenheim. 2007. Factors in the suppression of weeds by squash interplanted in corn. *Weed Biology and Management* 7:105–114.

Fundación Abertis. 2005. La biomassa com a font de matèries primeres i d'energia: Estudi de viabilitat al Montseny i Montnegre-corredor (Memòria final) Available at http://www.fundacioabertis.org/es/actividades/estudio.php?id=28 (accessed on January 28, 2011).

Fundación Española para el Desarrollo de la Nutrición Animal (FEDNA) 2010. Tablas FEDNA de composición y valor nutritivo de alimentos. Available at http://www.fundacionfedna.org/tablas-fedna-composicion-alimentos-valor-nutritivo (accessed on March 20, 2015).

Funt, R. C. 1980. Energy use in low, medium and high denisty apple orchards—Eastern U.S. In *Handbook of energy utilization in agriculture,* ed. D. Pimentel. Boca Raton, FL: CRC Press.

Gagnon, N., C. A. S. Hall, and L. Brinker. 2009. A preliminary investigation of energy return on energy investment for global oil and gas production. *Energies* 2(3):490–503. doi:10.3390/en20300490.

Gama, T. C. M., E. Volpe, and B. Lempp. 2014. Biomass accumulation and chemical composition of Massai grass intercropped with forage legumes on an integrated crop-livestock-forest system. *Revista Brasileira de Zootecnia* 43(6):279–288.

García-Gómez, K. I. 2011. *Estimación de la acumulación de biomasa y extracción estacional de nitrógeno, fósforo, potasio, calcio y magnesio en plantas de granado (Punica granatum L.). Tesis de máster.* Universidad de Chile: Facultad de Ciencias Agronómicas.

García-González, R., and A. Marinas. 2008. Bases ecológicas para la ordenación de superficies pastorales. In *Pastos del Pirineo*, eds. F. Fillat, R. García-González, D. Gómez, R. Reiné, pp. 229–253. Madrid, Spain: CSIC.

García-Martín, A., R. J. López-Bellido, and J. M. Coleto. 2007. Fertilisation and weed control effects on yield and weeds in durum wheat grown under rain-fed conditions in a Mediterranean climate. *Weed Research* 47:140–148.

García-Ruiz, J. M., J. I. Lopez-Moreno, S. M. Vicente-Serrano, T. Lasanta-Martínez, and S. Beguería. 2011. Mediterranean water resources in a global change scenario. *Earth-Science Reviews* 105:121–139.

García-Ruiz, R., V. Ochoa, B. Vinegla, et al. 2009. Soil enzymes, nematode community and selected physico-chemical properties as soil quality indicators in organic and conventional olive oil farming: Influence of seasonality and site features. *Applied Soil Ecology* 41:305–314.

Garrido, A. (coord.) 2012. *Indicadores de sostenibilidad de la agricultura y ganadería españolas.* Almería: Fundación Cajamar.

Gaspar García, P., F. J. Mesías Díaz, M. Escribano Sánchez, and F. Pulido García. 2009. Evaluación de la sostenibilidad en explotaciones de dehesa en función de su tamaño y orientación ganadera. *ITEA* 105:117–141.

Gaston, K. J. 2000. Global patterns in biodiversity. *Nature* 405:220–227.

Gattinger, A., A. Muller, M. Haeni, et al. 2012. Enhanced top soil carbon stocks under organic farming. *Proceedings of the National Academy of Sciences of the USA* 109:18226–18231.

Geels, F. W. 2005. *Technological Transitions and Systems Innovations. A Co-Evolutionary and Socio-Technical Analysis.* Cheltenham: Edward Elgar.

GEHR 1991. *Estadísticas Históricas de la producción agraria española, 1859–1935.* Madrid, Spain: MAPA.

Georgescu-Roegen, N. 1975. Energy and the economic myths. *The Southern Economic Journal* 41(3):347–381.

Georgescu-Roegen, N. 1971. *The Entropy Law and the Economic Process.* Cambridge, MA: Harvard University Press.

Gewald, N., and L. Ugalde. 1981. *Informe del Seminario Móvil del Proyecto Leña realizado en Costa Rica y Nicaragua.* Turrialba: CATIE.

Giambalvo, D., G. Alfieri, G. Amato, A. S. Frenda, P., Iudicello, and L. Stringi. 2009. Energy use efficiency of livestock farms in a mountain area of Sicily. *Italian Journal of Animal Science* 8(2):3007–3309.

Giampietro, M., and D. Pimentel. 1990. Assessment of the energetics of human labour. *Agriculture Ecosystems and Environment* 32(3–4):257–272. doi:10.1016/0167-8809(90)90164-9.

Giampietro, M., and K. Mayumi. 2000. Multiple-scale integrated assessment of societal metabolism: Introducing the approach. *Population and Environment* 22:109–154.

Giampietro, M., K. Mayumi, and A. H. Sorman. 2010. Assessing the quality of alternative energy sources: Energy return on the investment (EROI). The Metabolic Pattern of Societies and Energy Statistics. Working Papers on Environmental Sciences, ICTA, Barcelona, Spain.

Giampietro, M., K. Mayumi, and J. Ramos-Martin. 2008a. Multi-scale integrated analysis of societal and ecosystem metabolism (MUSIASEM). An Outline of Rationale and Theory. Document de Treball. Barcelona, Spain: Departament d'Economia Aplicada.

Giampietro, M., K. Mayumi, and J. Ramos-Martin. 2008b. Multi-scale integrated analysis of societal and ecosystem metabolism (MuSIASEM): Theoretical concepts and basic rationale. *Energy* 34:313–322.

Giampietro, M., S. G. F. Bukkens, and D. Pimentel. 1999. General trends of technological changes in agriculture. *Critical Reviews in Plant Sciences* 18(3):261–282. doi:10.1080/07352689991309225.

Giampietro, M., T. F. H. Allen, and K. Mayumi. 2006. The epistemological predicament associated with purposive quantitative analysis. *Ecological Complexity* 3(4):307–327.

Giampietro, M., S. Bukkens, and D. Pimentel. 1997. Models of energy analysis to assess the performance of food systems. *Agricultural Systems* 45(1):19–41.

Giampietro, M., K. Mayumi, and A. H. Sorman. 2012. *The Metabolic Pattern of Society*. New York, NY: Routledge.

Giampietro, M., R. J. Aspinall, J. Ramos-Martin, and S. G. F. Bukkens (eds). 2014. *Resource Accounting for Sustainability Assessment: The Nexus between Energy, Food, Water and Land Use*. London: Routledge.

Giampietro, M. 2004. *Multi-Scale Integrated Analysis of Agroecosystems*. Boca Raton, FL: CRC Press.

Giannetti, B. F., Y. Ogura, S. H. Bonilla, and C. M. V. B. Almeida. 2011a. Accounting energy flows to determine the best production model of a coffee plantation. *Energy Policy* 39(11):7399–7407.

Giannetti, B. F., Y. Ogura, S. H. Bonilla, and C. M. V. B. Almeida. 2011b. Energy assessment of a coffee farm in Brazilian Cerrado considering in a broad form the environmental services, negative externalities and fair price. *Agricultural Systems* 104(9):679–688.

Gierlinger, S., and F. Krausmann. 2012. The physical economy of United States of America. *Journal of Industrial Ecology* 16(3):365–377.

Gil, J., M. Siebold, and T. Berger. 2015. Adoption and development of integrated crop–Livestock–Forestry systems in Mato Grosso, Brazil. *Agriculture, Ecosystems and Environment* 199:394–406.

Glansdorff, P., and I. Prigogine. 1971. *Thermodynamic Theory of Structure, Stability and Fluctuations*. New York: Wiley Interscience.

GLASOD (The Global Assessment of Human Induced Soil Degradation). 1991. Geneva: Digital Database from UNEP/GRID.

Gliessman, S. R. 1998. *Agroecology. Ecological processes in sustainable agriculture*. Chelsea: Ann Arbor press.

Golluscio, R. 2009. Receptividad ganadera: Marco teórico y aplicaciones prácticas. *Ecología Austral* 19:215–232.

Gómez, J. A., and J. V. Giráldez. 2008. *Erosión y degradación de suelos*. Sevilla, Spain: Consejería de Agricultura y Pesca (Junta de Andalucía).

Gómez, J. A., M. G. Guzmán, J. V. Giráldez, and E. Fereres. 2009. The influence of cover crops and tillage on water and sediment yield, and on nutrient, and organic matter losses in an olive orchard on a sandy loam soil. *Soil and Tillage Research* 106:137–144.

González de Molina, M. 2010. A guide to studying the socio-ecological transition in European agriculture. Sociedad Española de Historia Agraria. *DT-SEHA*, 10. No.6.

González de Molina, M. 2002. Environmental constraints on agricultural growth in 19th century Granada (Southern Spain). *Ecological Economics* 41:257–270.

González de Molina, M. 2013. Agroecology and politics. How to get sustainability? About the necessity for a political agroecology. *Agroecology and Sustainable Food Systems* 37(1):45–59.

González de Molina, M., D. Soto, and J. Infante, et al. 2016. The evolution of the Spanish agriculture during the 20th century from the point of view of biophysical macro magnitudes. In *International Conference, Lisbon*, 27–30 January 2016. Old and New Worlds: The Global Challenges of Rural History. ISCTE-IUL.

González De Molina, M., D. Soto, J. Infante-Amate, and E. Aguilera. 2014. Crecimiento agrario en España y cambios en la oferta alimentaria (1900–1933). *Historia Social* 80:157–183.

González de Molina, M., D. Soto, J. Infante-Amate, and E. Aguilera. 2013. ¿Una o varias transiciones? Nuevos datos sobre el consumo alimentario en España (1900–2008). In *XIV Congreso de Historia Agraria* (Badajoz, noviembre 2013). Session B.1: La transición nutricional en perspectiva comparada: Mitos y realidades. Badajoz, November 7 and 8, 2013. (www.seha.info).

González de Molina, M., J. Infante-Amate, and G. I. Guzmán. 2014. Del manejo tradicional al manejo orgánico del olivar: Aplicaciones prácticas del conocimiento histórico. *Revista de Historia* 70:37–68.

González de Molina, M., and V. M. Toledo. 2011. *Metabolismos, Naturaleza e Historia*. Barcelona, Spain: Icaria editorial.

González de Molina, M., and V. M. Toledo. 2014. *Social Metabolisms: A Theory on Socio-Ecological Transformations*. New York, NY: Springer.

González de Molina, M., and G. I. Guzmán. 2006. *Tras los pasos de la insustentabilidad. Agricultura y medio ambiente en perspectiva histórica (siglos XVIII-XX)*. Barcelona, Spain: Icaria.

González González, G. 1993. El enfoque energético en la producción de hierba. *Revista Pastos* XXIII(1):3–44.

González Rebollar, J. L., A. B. Robles Cruz, M. C. Morales Torres, P. Fernández García, C. Passera Sassi, and J. Boza López. 1993. Evaluación de la capacidad sustentadora en pastos semiáridos del sureste ibérico. In Congreso Nuevas fuentes de alimentación para alimentación animal IV. *Congresos y Jornadas* 30:31–45.

González Vázquez, E. 1944. *Alimentación de la ganadería y los pastizales españoles*, 1st Edition Madrid, Spain: Ediciones Técnicas.

González, A. M., S. Bonachela, M. D. Fernández, and J. Gázquez. 2002. Use of three irrigation strategies in a greenhouse-grown green bean crop in the Almeria coast. Las Palmerillas: Publicaciones Cajamar. Available at http://www.publicacionescajamar.es /pdf/series-tematicas/centros-experimentales-las-palmerillas/use-of-three-irrigation -strategies.pdf (accessed on April 3, 2011).

Graboski, M. S. 2002. *Fossil Energy Use in the Manufacture of Corn Ethanol*. Chesterfield, MO: National Corn Growers Association.

Green, M. 1987. Energy in Pesticide Manufacture, Distribution, and Use. In *Energy in Plant. Nutrition and Pest Control.*, ed. Z. R. Helsel, pp. 165–196. New York, NY: Elsevier.

Griffin, T., M. Liebman, and J. Jr Jemison. 2000. Cover crops for sweet corn production in a short-season environment. *Agronomy Journal* 92:144–151.

Grigg, D. 2002. The worlds of tea and coffee: Patterns of consumption. *Geo Journal* 57(4):283–294.

Grisso, R. D., M. F. Kocher, and D. H. Vaughan. 2004. Predicting tractor fuel consumption. In *Biological Systems Engineering: Papers and Publications*. Paper 164. Available at http://digitalcommons.unl.edu/biosysengfacpub/164 (accessed on April 21, 2014).

Grisso, R. D. 2007. Nebraska Tractor Test Data. Available at http://filebox.vt.edu/users /rgrisso/Pres/Nebdata_07.xls (accessed on February 20, 2014).

Guerra, S. C. 2013. ¿Qué debemos tener en cuenta para incorporar la caña de azúcar en la dieta de nuestros animales? Ed: Instituto Nacional de Tecnología Agropecuaria (Argentina). Available at http://inta.gob.ar/documentos/que debemos tener en cuenta para incorporar la caña de azúcar en la dieta de nuestros animales (accessed on October 10, 2014).

Guerrero, A. 1987. *Cultivos herbáceos extensivos.* Madrid, Spain: Mundi-Prensa.

Guilford, M. C., C. A. S. Hall, P. O'Connor, and C. J. Cleveland. 2011. A new long-term assessment of energy return on investment (EROI) for U.S. oil and gas discovery and Production. *Sustainability* 3(10):1866–1887. doi:10.3390/su3101866.

Gupta, A. K., and C. A. S. Hall. 2011. A review of the past and current state of EROI data. *Sustainability* 3:1796–1809.

Guseo, R. 2011. Worldwide cheap and heavy oil productions: A long-term energy model. *Energy Policy* 39(9):5572–5577. doi:10.1016/j.enpol.2011.04.060.

Gutowski, T. G., S. Sahni, J. M. Allwood, M. F. Ashby, and E. Worrell. 2013. The energy required to produce materials: Constraints on energy-intensity improvements, parameters of demand. *Philosophical Transactions of the Royal Society A-Mathematical Physical and Engineering Sciences* 371 2012.0003. doi:10.1098/rsta.2012.0003.

Guzmán, G. I., and A. M. Alonso. 2008. A comparison of energy use in conventional and organic olive oil production in Spain. *Agricultural Systems* 98:167–176. doi:10.1016/j.agsy.2008.06.004.

Guzmán, G. I., and L. Foraster. 2011. El manejo del suelo y las cubiertas vegetales en el olivar ecológico. In *El Olivar Ecológico*, G. I. Guzmán (coord.), 51–94. Sevilla, Spain: Consejería de Agricultura y Pesca and Mundi-Prensa.

Guzmán, G. I., and M. González de Molina. 2007. Transición socio-ecológica y su reflejo en un agroecosistema del sureste español (1752–1997). *Revista Iberoamericana de Economía Ecológica* 7:55–70.

Guzmán, G. I., and M. González de Molina. 2008. Transformaciones agrarias y cambios en el paisaje. Un estudio de caso en el sur peninsular. In *El paisaje en perspectiva histórica. Formación y transformación del paisaje en el mundo mediterráneo*, Monografía de Historia Rural (Sociedad Española de Historia Agraria), eds. R. Garrabou and J. M. Naredo, pp. 199–233. Zaragoza: SEHA-PUZ.

Guzmán, G. I., and M. González de Molina. 2009. Preindustrial agriculture versus organic agriculture: The land cost of sustainability. *Land Use Policy* 26(2):502–510.

Guzmán, G. I., and M. González de Molina. 2015. Energy efficiency in agrarian systems from an agro-ecological perspective. *Agroecology and Sustainable Food Systems* 39(8):924–952.

Guzmán, G. I., E. Aguilera, D. Soto, et al. 2014. Methodology and conversion factors to estimate the net primary productivity of historical and contemporary agro-ecosystems (I). *Sociedad Española de Historia Agraria-Documentos de Trabajo 1406.* www.seha.info.

Guzmán, G. I., M. González de Molina, and A. M., Alonso. 2011. The land cost of agrarian sustainability. An assessment. *Land Use Policy* 28:825–835.

Guzmán, G. I., M. González de Molina, and E. S. Guzmán. 1999. *Introducción a la Agroecología como desarrollo rural sostenible.* Madrid, Spain: Mundi-Prensa.

Haas, G., F. Wetterich, and U. Köpke. 2001. Comparing intensive, extensified and organic grassland farming in southern Germany by process life cycle assessment. *Agriculture, Ecosystems and Environment* 83:43–53.

Haber, H., K. H. Erb, and F. Krausmann. 2014. human appropriation of net primary production: patterns, trends, and planetary boundaries. *Review of Environment and Resources* 39:363–391.

Haberl, H. 2001. The energetic metabolism of societies. I: Accounting concepts. *Journal of Industrial Ecology* 5:11–33.

Haberl, H. 2006. The global socioeconomic energetic metabolism as a sustainability problem. *Energy* 31:87–99.

Haberl, H., K. H. Erb, and F. Krausmann. 2014. Human appropriation of net primary production: Patterns, trends, and planetary boundaries. *Review of Environment and Resources* 39:363–391.

Haberl, H., K. H. Erb, F. Krausmann, et al. 2011. Global bioenergy potentials from agricultural land in 2050: Sensitivity to climate change, diets and yields. *Biomass and Bioenergy* 35(12):4753–4769.

Haberl, H., K. H. Erb, F. Krausmann, et al. 2007. Quantifying and mapping the human appropriation of net primary production in earth's terrestrial ecosystems. *Proceedings of the National Academy of Sciences of the USA* 104:12942–12947. doi:10.1073/pnas.0704243104.

Haberl, H., M. Fischer-Kowalski, F. Krausmann, H. Weisz, and V. Winiwarter. 2004. Progress towards sustainability? What the conceptual framework of material and energy flow accounting (MEFA) can offer. *Land Use Policy* 21:199–213.

Hall, C. A. S., R. Powers, and W. Schoenberg. 2008. Peak oil, EROI, investments and the economy in an uncertain future. In *Biofuels, Solar and Wind as Renewable Energy Systems*, ed. D. Pimentel, pp. 109–132. New York, NY: Springer.

Hall, C. A. S. 2011. Introduction to special issue on new studies in EROI (energy return on investment). *Sustainability* 3:1773–1777.

Hall, Charles. A. S., C. J. Cleveland, and R. Kaufmann. 1986. *Energy and Resource Quality: The Ecology of the Economic Process.* New York, NY: Wiley-Interscience.

Hall, C. A. S., J. G. Lambert, and S. B. Balogh. 2014. EROI of different fuels and the implications for society. *Energy Policy* 64:141–152. doi:10.1016/j.enpol.2013.05.049.

Hall, C. A. S., S. Balogh, and D. J. R. Murphy. 2009 What is the minimum EROI that a sustainable society must have? *Energies* 2(1):25–47. doi:10.3390/en20100025.

Hall, Carolyn. 1976. *El café y el desarrollo histórico-geográfico de Costa Rica.* San José: Editorial Costa Rica-Universidad Nacional.

Hanlan, T. G., R. A. Ball, and A. Vandenberg. 2006. Canopy growth and biomass partitioning to yield in short-season lentil. *Canadian Journal of Plant Science* 86(1):109–119.

Harmand, J. M., H. Avila, E. Dambrine, et al. 2007. Nitrogen dynamics and soil nitrate retention in a Coffea Arabica-Eucalyptus deglupta agroforestry system in Southern Costa Rica. *Biogeochemistry* 85:125–139.

Haugen-Kozyra, H., N. G. Juma, and M. Nyborg. 1993. Nitrogen partitioning and cycling in barley–soil systems under conventional and zero tillage in central Alberta. *Canadian Journal of Soil Science* 73:183–196 in-text citation .

Hay, R. K. M. 1995. Harvest index: A review of its use in plant breeding and crop physiology. *Annals of Applied Biology* 126(1):197–216.

Hayami, Y., and V. W. Ruttan. 1971. *Agricultural Development: An International Perspective.* Baltimore, MD: Johns Hopkins University Press.

Häyhää, T., and P. P. Franzese. 2014. Ecosystem services assessment: A review under an ecological-economic and systems perspective. *Ecological Modelling* 289:124–132.

Heichel, G. H. 1980. Assessing the fossil energy costs of propagating agricultural crops. In *Handbook of energy utilization in agriculture*, ed. D. Pimentel. Boca Raton, FL: CRC Press.

Heller, M. C., and G. A. Keoelian. 2003. Assesing the sustainability of the US food system: A life cycle perspective. *Agricultural Systems* 76:1007–1041.

Helsel, Z. R. 1992. Chapter 13—Energy and Alternatives for Fertilizer and Pesticide Use. In *Energy in Farm Production*, ed. Richard, C. F., pp. 177–201. Amsterdam: Elsevier.

Hernández Díaz-Ambrona, C. G. 1999. *Aplicación de modelos en los sistemas agrícolas de secano de la meseta central: Simulación de rotaciones y modelado de la arquitectura de la planta en leguminosas*. ETSIA Universidad Politécnica de Madrid, Spain: Tesis doctoral.

Hernández Díaz-Ambrona, C., A. Etienne, and J. M. Valderrama. 2008. Producciones potenciales de herbáceas, de bellota y carga ganadera en las dehesas de Extremadura. *Pastos* XXXVIII (2):243–258.

Herrero, M., B. Henderson, P. Havlík, et al. 2016. Greenhouse gas mitigation potentials in the livestock sector. *Nature Climate Change* 6:452–446. doi:10.1038/nclimate2925.

Herrero, M., P. Havlík, H. Valin, et al. 2013. Biomass use, production, feed efficiencies, and greenhouse gas emissions from global livestock systems. *Proceedings of the National Academy of Sciences of the USA* 110 (52):20888–20893. doi:10.1073/pnas.1308149110.

Hilbert, D. W., and J. Canadell. 1995. Biomass partioning and resource allocation of plants from Mediterranean-type ecosystems: Possible responses to elevated atmospheric CO_2. In *Global Change and Mediterranean-Type Ecosystems*, eds. J. M. Moreno and W. C. Oechel, 76–101, New York, NY: Springer-Verlagx.

Hilje L, L. E. Castillo, L. Thrupp, and C. Wesseling. 1987. *El Uso de los Plaguicidas en Costa Rica*. San José, Costa Rica: Ed. Heliconia/UNED.

Hilje, L., V. Cartín, and E. March. 1989. El combate de plagas agrícolas dentro del contexto histórico costarricense. *Manejo Integrado de Plagas* 14:68–86.

Hirst, E. 1973. Energy intensiveness of passenger and freight transport modes 1950–1970. Oak Ridge, TN, USA: Oak Ridge National Laboratory - National Science Foundation.

Ho, M. –W. 2013. Circular thermodynamics of organisms and sustainable systems. *Systems* 1:30–49.

Ho, M.-W., and R. Ulanowicz. 2005. Sustainable systems as organisms? *BioSystems* 82:39–51.

Ho, M. –W. 1998. Are sustainable economic systems like organisms? In *Sociobiology and Bioeconomics,* ed. P. Koslowski, pp. 237–258. New York, Berlin: Springer-Verlag.

Hole, D. G., A. J. Perkins, J. D. Wilson, I. H. Alexander, P. V. Grice, and A. D. Evans. 2005. Does organic farming benefit biodiversity? *Biological Conservation* 122:113–130.

House, G. J., B. R. Stinner, D. A. Crossley, and E. P. Odum. 1984. Nitrogen cycling in conventional and no-tillage agroecosystems—Analysis of processes. *Journal of Applied Ecology* 21:991–1012.

Hu, H., and P. Kavan. 2014. Energy consumption and carbon dioxide emissions of China's non-metallic mineral products industry: Present state, prospects and policy analysis. *Sustainability* 6(11):8012–8028. doi:10.3390/su6118012.

Huang, C. H., L. Zong, M., Buonanno, X. Xue, T. Wang, and A. Tedeschi. 2012. Impact of saline water irrigation on yield and quality of melon (Cucumis melo cv. Huanghemi) in northwest China. *European Journal of Agronomy* 43:68–76.

ICAFE. 2007. *Censo Cafetalero. 2003–2006. Principales resultados. Instituto del Café de Costa Rica*. San José: Instituto Nacional de Estadística y Censos.

ICAFE. 2010. Costa Rica renovará el 30 % del área cafetalera. Revista Infomativa ICAFÉ, II: 13. Available at http://www.icafe.cr/wp-content/uploads/revista_informativa/Revista -II-Sem-10.pdf (accessed on June 13, 2016).

ICAITI. 1967. *Informe sobre el mercado de fertilizantes en Centroamérica. Anexo 1*. Instituto Centroamericano de Investigación y Tecnología Industrial.

IEA. 2004. *Energy Statistics Manual*, ed. International Energy Agency. Paris: OECD Publishing.

IEA. 2007. Tracking industrial energy efficiency and CO_2 emissions. In *Energy Indicators*, ed. International Energy Agency. Paris: OECD Publishing.

IEA. 2012. *World Energy Outlook 2012*. Paris: International Energy Agency.

IEA. 2015. *Energy Statistics and Balances of Non-OECD Countries and Energy Statistics of OECD Countries, and United Nations, Energy Statistics Yearbook*. ed. International Energy Agency.

Infante-Amate, J., and L. Parcerisas. 2013. El carácter de la especialización agraria en el Mediterráneo español. El caso de la viña y el olivar en perspectiva comparada (1850-1935). In *XIV Congreso Internacional de Historia Agraria*, Badajoz, 7–9 de Noviembre, 2013.

Infante-Amate, J., and M. González de Molina. 2013. 'Sustainable degrowth' in agriculture and food: An agro-ecological perspective on Spain's agri-food system (year 2000). *Journal of Cleaner Production* 38:27–35.

Infante-Amate, J., and W. Picado. 2016. La transición socio-ecológica en el café costarricense. Flujos de energía, materiales y uso del tiempo (1935–2010). In *International Conference: Old and New Worlds: The Global Challenges of Rural History*, Lisbon, 27–30 January, 2016.

Infante-Amate, J., D. Soto Fernandez, E. Aguilera, et al. 2015. The Spanish transition to industrial metabolism. Long-term material flow analysis (1860–2010). *Journal of Industrial Ecology* 19(5):866–876. doi:10.1111/jiec.12261.

Infante-Amate, J., D. Soto, I. Iriarte Goñi, et al. 2014b. *La producción de leña en España y sus implicaciones en la transición energética: Una serie a escala provincial (1900–2000)*. DT-AEHE N°1416. Available at http://econpapers.repec.org/paper/ahedtaehe/1416.htm (accessed on April 2, 2015).

Infante-Amate, J., E. Aguilera, and M. González De Molina. 2014a. La gran transformación del sector agroalimentario español. Un análisis desde la perspectiva energética (1960–2010). *Sociedad Española de Historia Agraria. DT-SEHA* 1403. Available at https://ideas.repec.org/p/seh/wpaper/1403.html (accessed on April 2, 2015).

Infante-Amate, J. 2014. *¿Quién levantó los olivos? Historia de la especialización olivarera en el sur de España (ss. XVIII-XX)*. Madrid, Spain: Ministerio de Agricultura, Alimentación y Medio Ambiente.

Infoagro. 2013. El cultivo del avellano. http://www.infoagro.com/frutas/frutos_secos/avellana2.htm.

Ingelmo, F., García, J., Ibáñez, A. 1994. Efectos de una cubierta herbácea en las características físicas de un huerto de cítricos. In *I Congreso de la Sociedad Española de Agricultura Ecológica*, Toledo, September, 1994.

Instituto Nacional de Estadística (INE) 1999. *Censo agrario de 1999*. www.ines.es.

Instituto Nacional de Estadística (INE) 2009. *Censo agrario de 2009*. www.ines.es.

Instituto para la Diversificación y Ahorro de la Energía (IDAE) 2007. *Manuales de Energías Renovables 2*. Madrid, Spain: Energía de la biomasa. Ministerio de Industria, Turismo y Comercio.

Iodice, R. 2013. Estudio del metabolismo social y la salud del suelo en cinco producciones familiares tamberas en transición agroecológica de la cuenca del río Luján, Buenos Aires, Argentina. Master's thesis. Baeza, Spain: Universidad Internacional de Andalucía. Available at http://dspace.unia.es/bitstream/handle/10334/3427/0611_Iodice.pdf?sequence=1 (accessed on June 10, 2015).

Iriarte Goñi, I. & Infante Amate, J. 2014. *Primera aproximación al consumo de biomasa como combustible en España. Una visión de largo plazo* (1850–2000). Paper presented at the III Seminario Historia Agraria: Madrid, November.

Isherwood,K. F. 2003. Fertilizer use in Western Europe: Types and amounts. In *Agricultural Sciences. In Encyclopedia Of Life Support Systems (EOLSS).* ed. Rattan, L. Paris, France, [http://www.eolss.net] (accessed on November 20, 2012): Developed under the auspices of the UNESCO, Eolss publishers.

Jaška, P., P. Linhart, and R., Fuchs. 2015. Neighbour recognition in two sister songbird species with a simple and complex repertoire—A playback study. *Journal of Avian Biology* 46:151–158.

Jenssen, T. K., and G. Kongshaug. 2003. *Energy consumption and greenhouse gas emissions in fertiliser production.* London: International Fertiliser Society Meeting,

Jerez-Valle, C., P. A. García-López, M. Campos, and F. Pascual 2015. Methodological considerations in discriminating olive-orchard management type using olive-canopy arthropod fauna at the level of order. *Spanish Journal of Agricultural Research* 13(4):1–14.

Jiménez, E., and P. Martínez. 1979. Estudios ecológicos del agroecosistema cafetalero, 2: Producción de materia orgánica en diferentes tipos de estructura [*Coffea arabica*, Mexico]. *Biótica* 4(3):109–126.

Jones, M. R. 1989. Analysis of the use of energy in agriculture—Approaches and problems. *Agricultural Systems* 29:339–355.

Jørgensen, S. E., and B. D. Fath. 2004. Application of thermodynamic principles in ecology. *Ecological Complexity* 1:267–280.

Jørgensen, S. E., B. D. Fath, S. Bastianoni, et al. 2007. *A New Ecology: Systems Perspective.* the Netherlands: Elsevier.

Kader, A.A. 2013. Castaña: Recomendaciones para mantener la calidad postcosecha. Available at http://postharvest.ucdavis.edu/frutasymelones/Casta%C3%B1a/ (accessed on September 20, 2015).

Kahn, R. S., M. J. Figueroa, F. Creutzig, C. Dubeux, J. Hupe, and S. Kobayashi. 2012. Chapter 9—Energy end-use: Transport. In *Global Energy Assessment—Toward a Sustainable Future*, pp. 575–648. Cambridge and New York: Cambridge University Press and Laxenburg, Austria: The International Institute for Applied Systems Analysis.

Kaltsas, A. M., A. P. Mamolos, C. A. Tsatsarelis, G. D. Nanos, and K. L. Kalburtji. 2007. Energy budget in organic and conventional olive groves. *Agriculture, Ecosystems and Environment* 122 (2):243–251.

Kamakaté, F., and L. Schipper. 2009. Trends in truck freight energy use and carbon emissions in selected OECD countries from 1973 to 2005. *Energy Policy* 37 (10):3743–3751. doi:10.1016/j.enpol.2009.07.029.

Karras, G. 2010. Combustion emissions from refining lower quality oil: What is the global warming potential? *Environmental Science and Technology* 44(24):9584–9589. doi:10.1021/es1019965.

Kay, J. J., A. H. Regier, M. Boyle, and G. Francis. 1999. An ecosystem ap-proach for sustainability: Addressing the challenge for complexity. *Futures* 31:721–742.

Kellenberger, D., H., J. Althaus, N. Jungbluth, and T. Künniger. 2007. Life cycle inventories of building products. In *Final report ecoinvent data v2.0 No. 7.* Dübendorf, CH: Swiss Centre for Life Cycle Inventories.

Kemanian, A. R., C. O. Stöckle, D. R. Huggins, and L. M. Viega. 2007. A simple method to estimate harvest index in grain crops. *Field Crops Research* 103(3):208–216.

Khan, S. A., R. L. Mulvaney, and T. R. Ellsworth. 2014. The potassium paradox: Implications for soil fertility, crop production and human health. *Renewable Agriculture and Food Systems* 29(1):3–27. doi:10.1017/s1742170513000318.

Kim, S., B. E. Dale, and P. Keck. 2014. Energy requirements and greenhouse gas emissions of maize production in the USA. *Bioenergy Research* 7 (2):753–764. doi:10.1007 /s12155-013-9399-z.

Kisselle, K. W., C. J. Garrett, S. Fu, P. F. Hendrix, D. A. Crossley Jr., D. C. Coleman, and R. L. Potter. 2001. Budgets for root-derived C and litter-derived C: Comparison between conventional tillage and no tillage soils. *Soil Biology and Biochemistry* 33:1067–1075.

Koestler, A. 1967. *The Ghost in the Machine.* New York, NY: Macmillan Company.

Kool, A., M. Marinussen, and H. Blonk. 2012. *LCI Data for the Calculation Tool Feedprint for Greenhouse Gas Emissions of Feed Production and Utilization.* GHG Emissions of N, P and K fertilizer production. ed. Blonk Consultants.

Koppelaar, R. 2012. *World Energy Consumption—Beyond 500 Exajoules.* Available at http:// www.theoildrum.com/node/8936. Available at http://www.theoildrum.com/node/8936.

Koppelaar, R. H. E. M., and H. P. Weikard. 2013. Assessing phosphate rock depletion and phosphorus recycling options. *Global Environmental Change* 23:1454–1466.

Koutroubasa, S. D., D. K. Papakosta, and A. Doitsinis. 2004. Cultivar and seasonal effects on the contribution of pre-anthesis assimilates to safflower yield. *Field Crops Research* 90:263–274.

Kovanda, J., and T. Hak. 2011. Historical perspectives of material use in Czechoslovakia in 1855–2007. *Ecological Indicators* 11(5):1375–1384.

Krausmann, F. 2004. Milk, manure, and muscle power. Livestock and the transformation of preindustrial agriculture in Central Europe. *Human Ecology* 32(6):735–772. doi:10.1007/s10745-004-6834-y.

Krausmann, F., K.-H. Erb, S. Gingrich, C. Lauk, and H. Haberl. 2008a. Global patterns of socioeconomic biomass flows in the year 2000: A comprehensive assessment of supply, consumption and constraints. *Ecological Economics* 65(3):471–487.

Krausmann, F., and H. Haberl. 2002. The process of industrialization from the perspective of energetic metabolism: Socioeconomic energy flows in Austria 1830–1995. *Ecological Economics* 41(2):177–201.

Krausmann, F., H. Haberl, K.-H. Erb, M. Wiesinger, V. Gaube, and S. Gingrich. 2009. What determines geographical patterns of the global human appropriation of net primary production? *Journal of Land Use Science* 4:15–33.

Krausmann, F., H. Haberl, N. B. Schulz, K.-H. Erb, E. Darge, and V. Gaube. 2003. Land-use change and socioeconomic metabolism in Austria part I: Driving forces of land-use changes 1950–1995. *Land Use Policy* 20(1):1–20.

Krausmann, F., H. Schandl, and R. P. Sieferle. 2008b. Socioecological regime transition in Austria and United Kingdom. *Ecological Economics* 65:187–201.

Krausmann, F., K.-H. Erb, S. Gingrich, C. Lauk, and H. Haberl. 2008a. Global patterns of socioeconomic biomass flows in the year 2000: A comprehensive assessment of supply, consumption and constraints. *Ecological Economics* 65(3):471–487.

Krausmann, F., K. –H. Erb, S. Gingrich, et al. 2013. Global human appropriation on net primary production in the 20th century. *Proceedings of the National Academy of Sciences of the USA* 110(25):10324–10329.

Krausmann, F., S. Gingrich, and R. Nourbakhch-Sabet. 2011. The metabolic transition in Japan. *Journal of Industrial Ecology* 15(6):877–892.

Krausmann, F., S. Gingrich, N. Eisenmenger, K.-H. Erb, H. Haberl, and M. Fischer-Kowalski. 2009a. Growth in global materials use, GDP and population during the 20th century. *Ecological Economics* 68 (10):2696–2705. doi:10.1016/j. ecolecon.2009.05.007.

Kunelius, H. T., H. W. Johnston, and J. A. Macleod. 1992. Effect of undersowing barley with Italian ryegrass or red-clover on yield, crop composition and root biomass. *Agriculture, Ecosystems and Environment* 38:127–137.

Kuskova, P., S. Gingrich, and F. Krausmann. 2008. Long term changes in social metabolism and land use in Czechoslovakia, 1830–2000: An energy transition under changing political regimes. *Ecological Economics* 68(1):394–407.

Kyle, P., P. Luckhow, K. Calvin, W. Emanuel, M. Nathan, and Y. Zhou. 2011. GCAM 3.0 Agriculture and Land Use: Data Sources and Methods. US Department of Energy. PNNL-21025.

Lacasta, C., and R. Meco. 2011. La rotación en cultivos herbáceos de secano. In *Agricultura ecológica en secano*, eds. R. Meco Murillo, Lacasta Dutoit, C., and M. M. Moreno Valencia. Madrid, Spain: Ministerio de Medio Ambiente, Rural y Marino and Mundi-Prensa.

Lal, R., J. A. Delgado, P. M. Groffman, N. Millar, C. Dell, and A. Rotz. 2011. Management to mitigate and adapt to climate change. *Journal of Soil and Water Conservation* 66:276–285. doi:10.2489/jswc.66.4.276.

Lassaletta, L., G. Billen, B. Grizzetti, J., Garnier, A. M. Leach, and J. N. Galloway. 2013. Food and feed trade as a driver in the global nitrogen cycle: 50-year trends. *Biogeochemistry* 118:225–241.

Latawiec, A. E., B. B. N. Strassburg, J. F. Valentim, F. Ramos, and H. N. Alves-Pinto. 2014. Intensification of cattle ranching production systems: Socioeconomic and environmental synergies and risks in Brazil. *Animal* 8(8):1255–1263.

Lawson, B., and D. Rudder. 1996. *Building Materials, Energy and the Environment: Towards Ecologically Sustainable Development*: Royal Australian Institute of Architects.

Layke, C., E. Matthews, C. Amann, et al. 2000. *The Weight of Nations: Material Outflows From Industrial Economies*. Washington DC: World Resources Institute.

Lázaro, G. J. 2000. *El curado del tabaco Burley en España*. Spain: CETARSA.

Lazos, E. S. 1991. Composition and oil characteritics of apricot, peach and cherry kernel. *Grasas y Aceites* 42(2):127–131.

Le Houerou, H. N., and C. H. Hoste. 1977. Rangeland production and annual rainfall relations in the Mediterranean Basin and in the African Sahelo-Sudanian zone. *Journal of Range Management* 30:181–189.

Leach, G. 1976. *Energy and Food Production*. London: IPC Science and Technology.

Leip, A., F. Weiss, T. Wassenaar, et al. 2010. *Evaluation of the Livestock Sector's Contribution to the EU Greenhouse Gas Emissions (GGELS)–final report*. ed. Joint Research Centre: European Commission.

Lemckert, A., and J. J. Campos. 1981. *Producción y consumo de leña en las fincas pequeñas de Costa Rica*. Turrialba, Costa Rica: Centro Agronómico Tropical de Investigación y Enseñanza.

Lenzen, M. 2008. Life cycle energy and greenhouse gas emissions of nuclear energy: A review. *Energy Conversion and Management* 49(8):2178–2199. doi:10.1016/j.enconman.2008.01.033.

Leyshon, A. J. 1991. Effect of rate of nitrogen fertilizer on the aboverground and belowground biomass of irrigated bromegrass in Southwest Saskatchewan. *Canadian Journal of Plant Science* 71:1057–1067.

Liberato, J. R., F. X. R. Vale, and C. D. Cruz. 1999. Técnicas estatísticas de análise multivariada e a necessidade de o fitopatologista conhecê-las. *Fitopatologia Brasileira* 24:5–8.

Liebowitz, S. J., and S. E. Margolis. 1995. Path dependence, lock-in and history. *Journal of Law, Economics, and Organization* 11:205–226.

Lindborg, R., and O. Eriksson. 2004. Historical landscape connectivity affects present plant species diversity. *Ecology* 85(7):1840–1845.

López Bellido, L. 1991. *Cultivos herbáceos. Vol. I. Cereales.* Madrid, Spain: Mundi-Prensa.

López Pérez, D. 1998. Determinación de costes de cultivo en la Vega de Granada. Trabajo Profesional Fin de Carrera. Escuela Técnica Superior de Ingenieros Agrónomos y Montes. Universidad de Córdoba. Inédito.

López, M. L., and W. Picado 2012. Plantas, fertilizantes y transición energética en la caficultura contemporánea de Costa Rica. Bases para una discusión. *Revista de Historia* 65–66:17–51.

López-Díaz, M. L., V. Rolo, and G. Moreno. 2013. Matorralización de la dehesa: Implicaciones en la productividad total del sistema. In *Actas VI Congreso Forestal Español*, Ed: Sociedad Española de Ciencias Forestales, Vitoria-Gasteiz, Spain.

Lotka, A. 1956. *Elements of Mathematical Biology.* New York, NY: Dover Publication.

Lynch, J., P. Marschner, and Z. Rengel. 2012. Effect of internal and external factors on root growth and development. In *Mineral Nutrition of Higher Plants*, ed. P. Marchner 3rd Edition. 331–346. London: Academic Press-Elsevier.

Mäder, P., A. Fliebbach, D. Dubois, L. Gunst, P. Fried, and U. Niggli. 2002. Soil Fertility and biodiversity in organic farming. *Science* 296:1694–1697.

Madlool, N. A., R. Saidur, M. S. Hossain, and N. A. Rahim. 2011. A critical review on energy use and savings in the cement industries. *Renewable and Sustainable Energy Reviews* 15(4):2042–2060. doi:10.1016/j.rser.2011.01.005.

MAGRAMA (Ministerio de Agricultura Alimentación y Medio Ambiente). 2013. *Anuario de Estadística Agraria 2013.* Madrid, Spain: MAGRAMA.

MAGRAMA (Ministerio de Agricultura Alimentación y Medio Ambiente). 2015. *Encuesta sobre superficies y rendimientos de cultivo. Informe sobre regadíos en España.* Madrid, Spain: MAGRAMA.

MAGRAMA (Ministerio de Agricultura, Alimentación y Medio Ambiente) 2016. Anuario de Estadística, 2014. Available at http://www.magrama.gob.es/es/estadistica/temas/publicaciones/anuario-de-estadistica/default.aspx#para1 (accessed on February 4, 2016).

Malcolm, G. M., G. G. T. Camargo, V. A. Ishler, T. L. Richard, and H. D. Karsten. 2015. Energy and greenhouse gas analysis of northeast U.S. dairy cropping systems. *Agriculture, Ecosystems and Environment* 199:407–417.

Malhotra, N. K. 2008. *Investigación de mercados.* Mexico: Pearson Educación.

Maltby, C. 1980. *Use of Pesticides in Latin America*, Rep. UNIDO/IOD.353. Vienna: United Nations Industrial Development Organization.

Marcelis, L. F. M., E. Brajeul, A. Elings, A. Garate, E. Heuvelink, and P. H. B. de Visser. 2005. Modelling nutrient uptake of sweet pepper. In *Proceedings of the International Conference on Sustainable Greenhouse Systems*, eds. Straten, G. V., G. P. A. Bot, W. T. M. Van Meurs, L. M. F. Marcelis, Vols 1 and 2: pp. 285–292.

Margalef, R. 1979. *Perspectivas de la teoría ecológica.* Barcelona, Spain: Editorial Blume.

Margalef, R. 1980. *La biosfera. Entre la termodinámica y el juego.* Barcelona, Spain: Kairós.

Margalef, R. 1993. *Teoría de los Sistemas Ecológicos.* Barcelona, Spain: Universitat de Barcelona.

Markussen, M. V., and H. Østergård. 2013. Energy analysis of the Danish food production system: Food-EROI and fossil fuel dependency. *Energies* 6(8):4170–4186.

Marozzi, M., G. Bellavista, and I. Varela. 2004. Análisis comparativo de dos fincas productoras de café orgánico utilizando los métodos del balance energético y agroeconómico. *Economía y Sociedad* 9(24):97–118.

Martín Bellido, M., M. E. Díaz, J. P. Gonzalo, and T. L. Carrión. 1986. *Metodología para la determinación de la carga ganadera de pastos extensivos.* Madrid, Spain: INIA.

Martin, G., and M. A. Magne. 2015. Agricultural diversity to increase adaptive capacity and reduce vulnerability of livestock systems against weather variability—A farm-scale simulation study. *Agriculture, Ecosystems and Environment* 199:301–311.

Martin, G., and M. Willaume. 2016. A diachronic study of greenhouse gas emissions of French dairy farms according to adaptation pathways. *Agriculture, Ecosystems and Environment* 221:50–59.

Martínez, F., O. Merino, A. Martín, D. G. Martín, and J. Merino. 1998. Belowground structure and production in a Mediterranean sand dune shrub community. *Plant and Soil* 201:209–216.

Martinez-Alier, J. 2011. The EROI of agriculture and its use by the Via Campesina. *Journal of Peasant Studies* 38(1):145–160.

Martínez-Alier, J., G. Munda, and J. O'Neill. 1998. Weak comparability of values as a foundation for ecological economics. *Ecological Economics* 26:277–286.

Martinez-Alier, J., and J. Roca-Jusmet. 2000. *Economía Ecológica y Política Ambiental.* Mexico D.F. (Mexico): Fondo de Cultura Económica.

Marull, J., E. Tello, E., N. Fullana, et al. 2015. Long-term bio-cultural heritage: Exploring the intermediate disturbance hypothesis in agroecological landscapes (Mallorca, c. 1850–2012). *Biodiversity and Conservation* 24(13):3217–3251.

Maseda, F., F. Díaz, and C. Álvarez. 2004. Family dairy farms in Galicia (NW Spain): Classification by some family and land factors relevant to quality of life. *Biosystems Engineering* 87(4):509–521.

Mataix, V. J., and M. Mañas Almendros. (eds.). 1998. *Tabla de composición de alimentos españoles.* Granada: Universidad de Granada.

Maturana, H. R., and F. J. Varela. 1980. *Autopoiesis and Cognition. The Realization of the Living.* Boston, MA: Boston Studies in the Philosophy and History of Science.

Maynard, L. A., J. K. Loosli, H. F. Hintz, and R. G. Warner. 1979. *Animal Nutrition.* 7th Edition. New York, NY: McGraw-Hill Book Company.

McGarigal, K., E. Landguth, and S. Stafford. 2000. *Multivariate Statistics for Wildlife and Ecology Research.* New York, NY: Springer-Verlag.

McNeill, J. R. 2000. *Something New Under the Sun: An Environmental History of the Twentieth Century World.* New York, NY: W. W. Norton and Company.

McNeill, J. R., and W. H. McNeill. 2004. *Las Redes Humanas: Una historia global del mundo.* Barcelona, Spain: Editorial Crítica.

MEA (Millenium Ecosystem Assessment) 2005. Synthesis Report. Washington, DC: Island Press. Available at www.milleniumassesssment.org/ (accessed on October 18, 2013).

Meadows, D. H., D. L. Meadows, J. Randers, and W. W. Behrens III. 1972. *The Limits to Growth: A Report for the Club of Rome's Project on the Predicament of Mankind.* Ed. The Club of Rome: Universe Books.

Meco, R., M. M. Moreno, and C. Lacasta 2010. Productividad de sistemas de secano semiárido en manejo ecológico. In *La reposición de la fertilidad en los sistemas agrarios tradicionales,* eds. R. Garrabou and M. González de Molina, pp. 85–108. Barcelona, Spain: Icaria.

Meineri, G., and P. G. Peiretti. 2005. Determination of gross energy of silages. *Italian Journal of Animal Science* 4(2):147–149.

Melby, P. 1995. *Simplified Irrigation Design,* 2nd Edition. New York, NY: John Wiley & Sons Inc.

Méndez, V. E., C. M. Bacon, R. Cohen, and S. R. Gliessman (eds.) 2015. *Agroecology: A Transdisciplinary, Participatory and Action-Oriented Approach*. Boca Raton, FL: CRC Press.

Merlo, C. M. 2007. Comportamiento productivo del café (*Coffea arabica L.* variedad caturra), el poró (*Erythrinapo eppigiana*) el amarillon (*Terminalia amazónica*) y el Cashá (*Choroleuco neurycydum*) en sistemas agroforetales bajo manejo convencional y organicos en Turrialba, Costa Rica: Tesis de Maestría. CATIE.

Merrill, A. L., and B. K. Watt. 1973. Energy values of food. Basis and derivation. Agriculture Handbook n° 74. *Human Nutrition Research Branch*. Washington DC: Agricultural research service. USDA.

Milchunas, D. G., and W. K. Lauenroth 1993. Quantitative effects of grazing on vegetation and soils over a global range of environments. *Ecological Monographs,* 63:327–366.

MINETUR (Ministerio de Energía Industria y Turismo). 2015. *Balances de energía final (1990–2013)*. Madrid, Spain: MINETUR.

Ministerio de Agricultura (MA). 1959, 1960a, 1961, 1962, 1963. *Anuario Estadístico de la producción agrícola campaña 1958–59, 1959–60, 1960–61, 1961–62, 1962–63*. Madrid, Spain: Ministerios de Agricultura.

Ministerio de Agricultura (MA). 1960b. *Censo de la ganadería española 1960*. Madrid, Spain: Ministerios de Agricultura.

Ministerio de Agricultura, Alimentación y Medio Ambiente (MAGRAMA). 2006, 2007, 2008, 2009, 2010, 2011. *Anuario de Estadística*. Available at http://www.magrama.gob.es/es/estadistica/temas/publicaciones/anuario-de-estadistica/#para4 (accessed on May 12, 2014).

Ministerio de Agricultura, Alimentación y Medio Ambiente (MAGRAMA). 2014. *Diagnóstico del Sector Forestal Español. Análisis y Prospectiva-Serie Agrinfo/ Medioambiente n° 8*. Available at http://www.magrama.gob.es/es/ministerio/servicios/ (accessed on October 20, 2015).

Ministerio de Agricultura, Alimentación y Medio Ambiente (MAGRAMA). 2015. *Encuesta sobre superficie y rendimiento de cultivos. Informe sobre regadíos en España. 2014*. Available at http://www.magrama.gob.es/es/estadistica/temas/estadisticas-agrarias /Regadios2014_tcm7-359782.pdf (accesed on September 12, 2015).

Ministerio de Agricultura, Alimentación y Medio Ambiente (MAGRAMA). 2008. *Encuesta sobre superficie y rendimiento de cultivos. Resultados 2008*. Available at http://www .magrama.gob.es/es/estadistica/temas/estadisticas-agrarias/boletin2008_tcm7-14342 .pdf (accessed on April 9, 2015).

Ministerio de Agricultura, Alimentación y Medioambiente (MAGRAMA). 2012. Capítulo 10. Agricultura. Inventario Nacional de Emisiones a la Atmósfera 1990–2011. Volumen 2: Análisis por actividades SNAP. p. 147. Available at http://www.magrama.gob.es /es/calidad-y-evaluacion-ambiental/temas/sistema-espanol-de-inventario-sei-/10 _Agricultura_tcm7-219790.pdf (accessed on December 7, 2013).

Ministerio de Agricultura, Industria, Comercio y Obras Públicas (MAICOP). (undated). *Noticias estadísticas sobre la producción agrícola española por la Junta Consultiva Agronómica, 1902*. Madrid, Spain: MAICOP.

Ministerio de Agricultura, Industria, Comercio y Obras Públicas (MAICOP). 1905. *Prados y pastos. Resumen hecho por la Junta Consultiva Agronómica de las memorias sobre dicho tema remitidas por los ingenieros jefes del servicio agronómico provincial*. Madrid, Spain: MAICOP.

Ministerio de Agricultura, Pesca y Alimentación (MAPA). 1995. Informe sobre la situación de los recursos fitogenéticos en España. International Conference and Programme for Plant Genetic Resources, Madrid, Spain: MAPA.

Ministerio de Economía y Competitividad. 2015. DATACOMEX- Estadísticas del comercio exterior español. Madrid, Spain: Ministerio de Economía y Competitividad. Available at http://datacomex.comercio.es/principal_comex_es.aspx (accessed on April 2, 2015).

Ministerio de Fomento. 1892. *La ganadería en España. Avance sobre la riqueza pecuaria en 1891, formada por la Junta Consultiva Agronómica, conforme a las memorias reglamentarias que en el citado año han redactado los ingenieros del Servicio Agronómico.* Madrid, Spain: Ministerio de Fomento.

Ministerio de Fomento. 1912. *Memoria relativa a los servicios de la Dirección General de Agricultura, Minas y Montes.* Madrid, Spain: Ministerio de Fomento.

Ministerio de Fomento. 1913. *Avance estadístico de la riqueza que en España representa la producción media anual de árboles y arbustos frutales, tubérculos, raíces y bulbos. Resumen hecho por la Junta Consultiva Agronómica de las memorias de 1910, remitidas por los ingenieros del Servicio Agronómico Provincial.* Madrid, Spain: Ministerio de Fomento.

Ministerio de Fomento. 1914a. *Avance estadístico de la riqueza que en España representa la producción media anual de las plantas hortícolas y plantas industriales. Resumen hecho por la Junta Consultiva Agronómica de las memorias de 1911, remitidas por los ingenieros del Servicio Agronómico Provincial.* Madrid, Spain: Ministerio de Fomento.

Ministerio de Fomento. 1914b. *Avance estadístico de la riqueza que en España representa la producción media anual de pastos, prados, y algunos aprovechamientos y pequeñas industrias zoógenas anexas. Resumen hecho por la Junta Consultiva Agronómica de las memorias de 1912, remitidas por los ingenieros del Servicio Agronómico Provincial.* Madrid, Spain: Ministerio de Fomento.

Ministerio de Fomento. 1915. *Avance estadístico de la riqueza que en España representa la producción media anual en el decenio de 1903 a 1912 de cereales y leguminosas, vid y olivo y aprovechamientos diversos derivados de estos cultivos. Resumen hecho por la Junta Consultiva Agronómica de las memorias de 1913, remitidas por los ingenieros del Servicio Agronómico Provincial.* Madrid, Spain: Ministerio de Fomento.

MITYC (Ministerio de Industria, Turismo y Comercio). 2009. *Estadística de la industria de energía eléctrica 2008.* Madrid, Spain: Secretaría General Técnica.

Mokani, K., R. J. Raison, and A. S. Prokushkin. 2006. Critical analysis of root: Shoot ratios in terrestrial biomes. *Global Change Biology* 12:84–96. doi:10.1111/j.1365-2486. 2005.001043.x.

Mondelaers, K., J. Aertsens, and G. Van Huylenbroeck. 2009. A meta-analysis of the differences in environmental impacts between organic and conventional farming. *British Food Journal* 111(10):1098–1119.

Mondino, M. H., and O. A. Peterlin. 2003. Respuesta del cultivo de algodón (*Gossypium hirsutum* L.) sembrado en surcos ultraestrechos a la aplicación de fertilizantes nitrogenados. 1. Rendimiento y sus componentes. In Trabalhos do IV Congreso Brasileiro do Algodão. 15 a 18 de Setembro no Goiânia. Available at http://www.cnpa.embrapa.br /produtos/algodao/publicacoes/trabalhos_cba4/336.pdf (accessed on April 13, 2013).

Monreal, C. R. 2012. Caracterización nutricional de variedades locales de tomate en producción ecológica y relación con el consumo. *Tesis de Máster Agricultura, Ganadería y Silvicultura Ecológica.* Baeza, España: Universidad Internacional de Andalucía.

Montenegro, G. E. J. 2005. *Efecto de la dinámica de la materia de nutrientes de la biomasa de tres tipos de árboles de sombra en sistemas de manejo de café orgánico y convencional.* Dissertation, Turrialba, CR: CATIE. p. 67.

Moore, S. R. 2010. Energy efficiency in small-scale biointensive organic onion production in Pennsylvania, USA. *Renewable Agriculture and Food Systems* 25(3):181–188.

Mora Delgado, J., C. Ramírez, and O. Quirós. (2006) Análisis beneficio-costo y cuantificación de la energía invertida en sistemas de caficultura campesina en Puriscal, Costa Rica. *Agronomía Costarricense* 30(2):71–82.

Morales, M., M. Luis, A. R. González, G. C. García, and M. O. Ramuco. 2012. Inventario 2011 del cultivo del aguacate y evaluación del impacto ambiental forestal en el estado de Michoacán. Morelia, Mexico: Centro de Investigaciones en Geografía Ambiental, UNAM–COFUPRO.

Moreiras, O., Á. Carbajal, L. Cabrera, and C. Cuadrado. 2011. *Tablas de composición de alimentos.* Madrid, Spain: Pirámide.

Mosquera-Losada, M. R., A. Rigueiro-Rodríguez, and J. McAdam, (eds.). 2004. *Proceedings of an International Congress on Silvopastoralism and sustainable management.* Lugo, Spain, April. Walingford: CABI Publishing.

Mulder, K., and N. J. Hagens. 2008. Energy return on investment: Toward a consistent framework. *AMBIO* 37(2):74–79.

Muner, L. H. D., O. Masera, M. J. Fornazier, C. V. D. Souza, and M. D. D. de Loreto. 2015. Energetic sustainability of three arabica coffee growing systems used by family farming units in Espírito Santo state. *Engenharia Agrícola* 35(3):397–405.

Muñoz-Romero, V., L. L. Bellido, and J. R. López-Bellido. 2011. Faba bean root growth in a Vertisol: Tillage effects. *Field Crops Research* 120 (3):338–344.

Murgueitio, E., R. Barahona, J. D. Chará, M. X. Flores, R. M. Mauricio, and J. J. Molina. 2015. The intensive silvopastoral systems in Latin America sustainable alternative to face climatic change in animal husbandry. *Cuban Journal of Agricultural Science* 49:541–554.

Murphy, D. J., and C. A. S. Hall.x 2010. Year in review—EROI or energy return on (energy) invested. *Annals of the New York: Academy of Sciences* 1185:102–118. doi:10.1111/j.1749-6632.2009.05282.x.

Murphy, D. J., and C. A. S. Hall. 2011. Energy return on investment, peak oil, and the end of economic growth. In *Ecological Economics Reviews*, ed. R. Costanza, K. Limburg, and I. Kubiszewski, 52–72.

Murphy, D. J., C. A. S. Hall, M. Dale, and C. Cleveland. 2011a. Order from chaos: A preliminary protocol for determining the EROI of fuels. *Sustainabilty* 3:1888–1907.

Murphy, H. T., and J. Lovett-Doust. 2004. Context and connectivity in plant metapopulations and landscape mosaics: Does the matrix matter? *Oikos* 105(1):3–14.

Murray, J., and D. King. 2012. Climate policy: Oil's tipping point has passed. *Nature* 481(7382):433–435.

Muschler, R. 1999. *Árboles en cafetales, Módulo de Enseñanza Agroforestal*, 5. Costa Rica: CATIE.

Naredo, J. M. 1999. El enfoque eco-integrador y su sistema de razonamiento. In *Desarrollo Económico y Deterioro Ecológico*, eds. J. M. Naredo and A. Valero, pp. 47–56. Madrid, Spain: Visor y Fundación Argentaria.

Naredo, J. M. 2000. El metabolismo de la sociedad industrial y su incidencia planetaria. *In Economía, Ecología y Sostenibilidad en la Sociedad Actual*, eds. J. M. Naredo and F. Parra, 913–229. Madrid, Spain: Siglo XXI Editores.

Naredo, J. M., and P. Campos. 1980. Los balances energéticos de la agricultura española. *Agricultura y Sociedad* 15:163–255.

National Research Council (NRC). 1998. *Nutrient Requirements of Swine.* 10th Edition. Washington, DC: National Academy Press.

National Research Council (NRC). 2001. *Energy. Nutrient Requirements of Dairy Cattle.* 7th Revised Edition. Washington, DC: National Academy Press.

Nehring K., Haenlein G.F.W., 1973. Feed evaluation and ration calculation based on net energy fat. *Journal of Animal Science* 36:949–964.

Nemecek, T., A. Heil, O. Huguenin, et al. 2007. Life Cycle Inventories of Agricultural Production Systems. In *Ecoinvent Reports*, ed. Agroscope FAL Reckenholz and FAT Taenikon. Dübendorf, CH: Swiss Centre for Life Cycle Inventories.

Norton, L., P. Johnson, A. Joys, et al. 2009. Consequences of organic and non-organic farming practices for field, farm and landscape complexity. *Agriculture, Ecosystems and Environment* 129:221–227.

OAS 1970. *La situación de los fertilizantes en Costa Rica.* Washington, DC: Organization of American States.

Olaeta, J. A., M. Schwartz, P. Undurraga, and S. Contreras. 2007 Utilización de la semilla de palta (*Persea americana* Mill.) CV. Hass como producto agroindustrial. In Proceedings VI World Avocado Congress (Actas VI Congreso Mundial del Aguacate) 2007. Viña Del Mar, Chile. 12 – 16 Nov. 2007.

Ordóñez, J. A. B., B. H. J. De Jong, F. García-Oliva, et al. 2008. Carbon content in vegetation, litter, and soil under 10 different land-use and land-cover classes in the Central Highlands of Michoacan, Mexico. *Forest Ecology and Management* 255(7):2074–2084.

Ozdogan, M. 2011. Exploring the potential contribution of irrigation to global agricultural primary productivity. *Global Biogeochemical Cycles* 25(3): **GB3016. doi:10.1029/2009GB003720.

Pagiola, S., and G. Platais. 2002. *Payments for Environmental Services, Environment Strategies, No. 3.* Washington, DC: The World Bank.

Pagiola, S., K. V. Ritter, and J. Bishop. 2004. Assessing the economic value of ecosystem conservation, Environment Department, Paper No. 101. Washington, DC: The World Bank.

Pajarón, M., M. Soriano, and L. Hurtado. 1996. El manejo de cubiertas vegetales en el olivar ecológico. In *Actas II Congreso de la Sociedad Española de Agricultura Ecológica, Navarra*, septiembre de 1996.

Pardo, G., R. Moral, E. Aguilera, and A. Del Prado. 2015. Gaseous emissions from management of solid waste: A systematic review. *Global Change Biology* 21 (3):1313–1327.

Parra, P. 2013. Nuez del nogal (*Juglans regia* L.) Análisis de la cadena alimentaria. Dirección Nacional de Alimentos. Available at http://www.alimentosargentinos.gov.ar/contenido/revista/ediciones/37/cadenas/Frutas_secas_nuez.htm (accessed on April 22, 2014).

Parras-Alcántara, L., L. Díaz-Jaimes, and B. Lozano-García. 2015. Organic farming affects C and N in soils under olive groves in Mediterranean areas. *Land Degradation and Development* 26(8):800–806.

Passera Sassi, C. B., J. L. G. Rebollar, A. B. R. Cruz, and L. I. Allegretti. 2001. Determinación de la capacidad sustentadora de pastos de zonas áridas y semiáridas del sureste ibérico, a partir de algoritmos. In *Biodiversidad en pastos*. XLI Reunión Científica de la SEEP. Alicante, 23–27 de Abril de 2001. CIBIO. pp. 611–617.

Patón, D., J. Cabezas, T. Buyolo, L. Fernández-Pozo, F. M. Venegas, and C. Crisóstomo. 2005. Calidad nutritiva del pastizal mediterráneo de ecosistemas de la reserva de la biosfera de Monfragüe. In Actas XLV Reunión Científica de la SEEP (Sesión: Ecología y Botánica de Pastos). pp. 869–874.

Pattee, H. H. 1995. Evolving self-reference: Matter, symbols, and semantic closure. *Communication and Cognition-Artificial Intelligence* 12:9–28.

Patzek, T. W. 2004. Thermodynamics of the corn-ethanol biofuel cycle. *Critical Reviews in Plant Sciences* 23(6):51–567. doi:10.1080/07352680490886905.

Paustian, K., J. Lehmann, S. Ogle, D. Reay, G. P. Robertson, and P. Smith. 2016. Climate-smart soils. *Nature* 532:49–57. doi:10.1038/nature17174

Peiretti, P.G.,Valente, M.E., Canale, A., Ciotti, A., 1994. Previsione dell'energia lorda dell'insilato di Trifoglio violetto in funzione dei prodotti di fermentazione e dell'azoto totale. *Atti Soc. Ital. Scienze Vet.* 48:1489–1493.

Pelletier, N., M. Ibarburu, and X. Hongwei. 2014. Comparison of the environmental footprint of the egg industry in the United States in 1960 and 2010. *Poultry Science* 93(2):241–255. doi:10.3382/ps.2013-03390.

Pendergrast, M. 1999. *Uncommon Grounds: The History of Coffee and How it Transformed Our World*. New York, NY: Basic Books.

Pérez, V. M. 1977. Veinticinco años de investigación sistemática del cultivo del café en Costa Rica: 1950–1975. *Revista Agronomía Costarricense* 1(2):169–185.

Pérez-Soba, M., B. Elbersen, M. Kempen, et al. 2015. *Agricultural Biomass as Provisioning Ecosystem Service: Quantification of Energy Flows*. European Commission: Technical Report Joint Research Centre. EUR27538 EN. doi:10.2788/679096.

Perfecto, I., and J. Vandermeer. 2010. The agroecological matrix as alternative to the land-sparing/agriculture intensification model. *Proceedings of the National Academy of Sciences of the USA* 107:5786–5791.

Perfecto, I., R. A. Rice, R. Greenberg, and M. E. Van der Voort. 1996. Shade coffee: A disappearing refuge for biodiversity. *BioScience* 46(8):598–608.

Pervanchon, F., C. Bockstaller, and P. Girardin. 2002. Assessment of energy use in arable farming systems by means of an agro-ecological indicator: The energy indicator. *Agricultural Systems* 72:149–172.

Petr, J., J. Lipavský, and D. Hradecká. 2002. Production process in old and modern Spring Barley varieties. *Die Bodenkultur* 53(1):19–27.

Phalan, B., M. Onial, A. Balmford, and R. E. Green. 2011. Reconciling food production and biodiversity conservation: Land sharing and land sparing compared. *Science* 333:1289–1291. doi:10.1126/science.1208742.

Phelps, J., Roman Carrasco, L., Webb, E. L., Pin Koh, L, Pascual, U. 2013. Agricultural intensification escalates future conservation costs. *PNAS*, 110 (19):7601–7606.

Philpott, S. M., W. J. Arendt, I. Armbrecht, et al. 2008. Biodiversity loss in Latin American coffee landscapes: Review of the evidence on ants, birds, and trees. *Conservation Biology* 22:1093–1105.

Piat, D. M. C. 1989. Materias primas alternativas vegetales en la fabricación de piensos compuestos en España. In *Nuevas fuentes de alimentos para la producción animal III,* Colección: Congresos y Jornadas, n° 12-1989, coords. Cabrera, A. G., E. M. Alcaide, and A. G. Varo, 71–175. Sevilla, Spain: DGIEA, Consejería de Agricultura y Pesca. Junta de Andalucía.

Picado, W. 2000. *La expansión del café y el cambio tecnológico desigual en la agricultura del cantón de Tarrazú, Costa Rica*. Universidad Nacional, Costa Rica: Master's thesis in Applied History.

Pimentel, D. 2003. Ethanol fuels: Energy balance, economics, and environmental impacts are negative. *Natural Resources Research* 12(2):127–134. doi:10.1023/a:1024214812527.

Pimentel, D. 2004. Livestock production and energy use. In *Encyclopedia of Energy*, ed. R. Matsumura. San Diego, CA: Elsevier.

Pimentel, D. 2006. *Impacts of organic farming on the efficiency of energy use in agriculture.* Ed: The Organic Center. www.organicvalley.coop/.../pdf/ENERGY_SSR.pdf

Pimentel, D., and M. Burgess. 1980. Energy inputs in corn production. In *Handbook of Energy Utilization in Agriculture*, ed. D. Pimentel. Boca Raton, FL: CRC Press.

Pimentel, D., and M. Pimentel. 2003. Sustainability of meat-based and plant-based diets and the environment. *American Journal of Clinical Nutrition* 78(suppl):660S–663S.

Pimentel, D., G. Berardi, and S. Fast. 1983. Energy efficiency of farming systems: Organic and conventional agriculture. *Agriculture, Ecosystems and Environment* 9(4):359–372.

Pimentel, D., P. Hepperly, J. Hanson, D. Douds, and R. Seidel. 2005. Environmental, energetic, and economic comparisons of organic and conventional farming systems. *Bioscience* 55 (7):573–592.

Pimentel, D., W. Dazhong, and M. Giampietro. 1990. Technological changes in energy use in U.S. agricultural production. In *Agroecology*, ed. S. R. Gliessman, New York, NY: Springer-Verlag.

Pimentel, D., and M. Pimentel. 1979. *Food, Energy and Society.* London: Edward Arnold.

Pinilla, V. 2001. El comercio exterior en el desarrollo agrario de la España contemporánea: Un balance. *Historia Agraria* 23:13–37.

Pinilla, V., and D. Gallego. 1996. Del librecambio matizado al proteccionismo selectivo: el comercio exterior de productos agrarios y alimentos en España entre 1849 y 1935. *Revista de Historia Económica* 14:371–420.

Piratla, K., S. Ariaratnam, and A. Cohen. 2012. Estimation of CO_2 emissions from the life cycle of a potable water pipeline project. *Journal of Management in Engineering* 28(1):22–30. doi:10.1061/(ASCE)ME.1943-5479.0000069.

Podobnik, B. 2006. *Global Energy Shifts: Fostering Sustainability in a Turbulent Age.* Filadelfia: Temple University Press.

Poiani, K. A., B. D. Richter, M. G., Anderson, and H. E. Richter. 2000. Biodiversity conservation at multiple scales: Functional sites, landscapes, and networks. *BioScience* 50 (2):133–146.

Ponisio, L. C., L. K. M'Gonigle, K. C. Mace, J. Palomino, P. de Valpine, and C. Kremen. 2015. Diversification practices reduce organic to conventional yield gap. *Proceedings of the Real Society B* 282(1799) doi:10.1098/rspb.2014.1396.

Ponte, S. 2002. Thelatte revolution'? Regulation, markets and consumption in the global coffee chain. *World Development* 30(7):1099–1122.

Pou, A., J. Gulías, M. Moreno, M. Tomás, H. Medrano, and J. Cifre. 2011. Cover cropping in vitis vinifera l. Cv. Manto negro vineyards under mediterranean conditions: Effects on plant vigour, yield and grape quality. *Journal International des Science de la Vigne et du Vin* 45(4):1–12.

Poudel, D. D., W. R. Horwath, W. T. Lanini, S. R. Temple, and A. H. C. van Bruggen. 2002. Comparison of soil N availability and leaching potential, crop yields and weeds in organic, low-input and conventional farming systems in northern California. *Agriculture, Ecosystems and Environment* 90:125–137.

Pracha, A. S., and T. A. Volk. 2011. An edible energy return on investment (EEROI). Analysis of wheat and rice in Pakistan. *Sustainability* 3(12):2358–2391. doi:10.3390/su3122358.

Prados, E. L. 2003. El progreso económico de España (1850–2000). Madrid, Spain: Fundación BBVA.

Prieto, P. A., and C. Hall. 2013. *Spain's Photovoltaic revolution. The energy Return on Investment. SpringerBriefs in Energy.* New York, NY: Springer.

Prigogine, I. 1955. Thermodynamics of irreversible processes and fluctuations. *Temperature* 2:215–232.

Prigogine, I. 1978. Time structure and fluctuations. *Science* 201:777–785.

Prigogine, I. (1947). *Etude Thermodynamique des Phenomenes Irreversibles*. Paris, France: Liège.

Prigogine, I. 1962. *Non-Equilibrium Statistical Mechanics*. New York, NY: Interscience.

Prigogine, I. 1983. *¿Tan sólo una ilusión? Una exploración del caos al orden*. Barcelona, Spain: Tusquets.

Prince, S. D., J. Haskett, M. Steininger, H. Strand, and R. Wright. 2001. Net primary production of the U.S. Midwest croplands from agricultural harvest yield data. *Ecological Applications* 11(4):1194–1205.

Pujol, J., M. González de Molina, L. Fernández Prieto, D. Gallego, and R. Garrabou. 2001. *El pozo de todos los males. Sobre el atraso en la agricultura española contemporánea*. Barcelona, Spain: Crítica.

Pulido, G. F., and M. E. Sánchez. 1994. Análisis de los recursos de pastoreo aportados por el medio en dos dehesas características del suroeste de Badajoz (España). *Archivos de Zootecnia* 43(163):239–249.

Pulido, F., R. Sanz, and D. Abel, et al. 2007. *Los bosques de Extremadura. Evolución, ecología y conservación*. Mérida, Spain: Junta de Extremadura.

Rafiqul, I., C. Weber, B. Lehmann, and A. Voss. 2005. Energy efficiency improvements in ammonia production—Perspectives and uncertainties. *Energy* 30(13):2487–2504. doi:10.1016/j.energy.2004.12.004.

Rahn, C. R., and R. D. Lillywhite. 2002. A study of the quality factors affecting the short-term decomposition of field vegetable residues. *Journal of the Science of Food and Agriculture* 82(1):19–26.

Ramankutty, N., and J. Rhemtulla. 2012. Can intensive farming save nature? *Frontiers in Ecology and the Environment* 10:455.

Ramírez, C. A., and E. Worrell. 2006. Feeding fossil fuels to the soil: An analysis of energy embedded and technological learning in the fertilizer industry. *Resources, Conservation and Recycling* 46(1):75–93. doi:10.1016/j.resconrec.2005.06.004.

Ramírez, F. 2011. *Importación de plaguicidas en Costa Rica. Periodo 2007–2009*. Instituto Regional de Estudios en Sustancias Tóxicas. Available at http://cep.unep.org/repcar/informacion-de-paises/costa-rica/Impoortaciones_07-09_REPCar.pdf (accessed on February 10, 2014).

Ramos-Martin, J. 2003. Empiricism in ecological economics: A perspective from complex systems theory. *Ecological Economics* 46:387–398.

Rappaport, R. A. 1971. The flow of energy in an agricultural society. *Scientific American* 225 (3):117–22 passim.

Raupp, J., C. Pekrun, M. Oltmanns, and U. Köpke (eds). 2006. Long-term Field Experiments in Organic Farming. ISOFAR. *Scientific Series*. No 1. Berlin, Germany: Verlag Dr. Köster.

Redclift, M., and G. Woodgate (eds). 1997. *The International Handbook of Environmental Sociology*. Cheltenham: Edward Elgar.

REE (Red Eléctrica de España). 2015. *Balances de energía eléctrica 1990–2014*. ed. Red Eléctrica Española. Madrid, Spain.

Renjifo, A. 1992. El café en Costa Rica. *Economía Cafetalera* 7:29–77.

Repullo, M. A., R. Carbonell, J. Hidalgo, A. Rodríguez-Lizana, and R. Ordóñez. 2012. Using olive pruning residues to cover soil and improve fertility. *Soil and Tillage Research* 124:36–46.

Rios, A., and A. I. Carriquiry. 2007. Control de Lolium multiflorum y Avena fatua en trigo. La malherbología en los nuevos sistemas de producción agraria. In *Actas del XI Congreso de la Sociedad Española de Malherbología*. Albacete, 7–9 de noviembre de 2007. 299–304.

Risku-Norja, H. 1999. The total material requirement-concept applied to agriculture: A case study from Finland. *Agricultural and Food Science in Finland* 8:393–410.

Risku-Norja, H., and I. Mäenpää. 2007. MFA model to assess economic and environmental consequences of food production and consumption. *Ecological Economics* 60:700–711. doi:10.1016/j.ecolecon.2006.05.001.

Robledo, A., A. Martínez, M. D. Megías, A. B. Robles, M. Erena, P. García, S. Ríos, and E. Correal. 2007. Productividad y valor nutritivo de los pastos. Tipificación, cartografía y evaluación de los recursos pastables de la región de Murcia. Serie Informes 18, pp. 63–88. Murcia: Consejería de Agricultura y Agua.

Robles, A. B. 2008. En el conjunto de las Sierras Béticas: Pastos, producción, diversidad y cambio global. In *Pastos, clave en la gestión de los territorios: Integrando disciplinas*, pp. 31–51. Sevilla, Spain: Junta de Andalucía.

Robles, A. B., J. L. González Rebollar, C. B. Passera, and J. Boza López. 2001. Pastos de zonas áridas y semiáridas del sureste ibérico. *Archivos de zootecnia* 50(192):501–515.

Robles, A. B., J. Ruiz-Mirazo, M. E. Ramos, and J. L. González-Rebollar. 2008. Role of live-stock grazing in sustainable use, naturalness promotion in naturalization of marginal ecosystems of Southeastern Spain (Andalusia). *Advances in Agroforestry* 6:211–231.

Roccuzzo, G., D. Zanotellib, M. Allegra, et al. 2012. Assessing nutrient uptake by field-grown orange tres. *European Journal of Agronomy* 41:73–80.

Rodale Institute. 2011. *The Farming Systems Trials. Celebrating 30 Years.* Available at http://rodaleinstitute.org/assets/FSTbooklet.pdf (accessed on February 11, 2016).

Rodríguez Martín, A. Juan, M. L. Arias, and J. M. G. Corbi. 2009. Metales pesados, materia orgánica y otros parámetros de los suelos agrícolas y pastos de España. Madrid, Spain: INIA-MAGRAMA, MCI.

Rodríguez, N., and D. A. Zambrano. 2010. Los subproductos del café: Fuente de energía renovable. *Cenicafé* 393:1–8.

Rojas, A. 1979. *Demanda de fertilizantes: Un estudio econométrico.* Tesis de Grado para optar al título de Ingeniero Agrónomo, Universidad de Costa Rica, Costa Rica.

Romero, S. 2006. Aporte de biomasa y reciclaje de nutrientes en seis sistemas agroforestales de café (Coffea arabica var. Caturra), con tres niveles de manejo. Tesis de Maestría: CATIE.

Romijn, M., and E. Wilderink. 1981. *Fuelwood Yield from Coffee Prunings in the Turrialba Valley.* Turrialba: CATIE.

Rosecrance, R., and C. J. Lovatt. 2003. Seasonal Patterns of Nutrient Uptake and Partitioning as a Function of Crop Load of the 'Hass' Avocado, p. 9. Available at https://www.cdfa.ca.gov/is/docs/Rosecrance00[1].pdf (accessed on February 2, 2016).

Rosen, R. 1985. *Anticipatory Systems: Philosophical, Mathematical and Methodological Foundations.* New York, NY: Pergamon Press.

Rosen, R. 2000. *Essays on Life Itself.* New York, NY: Columbia University Press.

Rosset, P. M., and M. A. Altieri. 1997. Agroecology versus inputs substitution: A fundamental contradiction of sustainable agriculture. *Society and Natural Resources* 10:283–295.

Rotz, C. A. 1987. A standard model for repair costs of agricultural machinery. *Applied Engineering in Agriculture* 3(1):2–9.

Ruiz González., D. A., and J. A. Vega Hidalgo. 2007. Modelos de predicción de la hume-dad de los combustibles muertos: Fundamentos y aplicación. Madrid, Spain: Instituto Nacional de Investigación y Tecnología Agraria y Alimentaria (Ministerio de Educación y Ciencia).

Rutherford, P. M., and N. G. Juma. 1989. Shoot, root, soil and microbial nitrogen dynamics in two contrasting soils cropped to barley (*Hordeum vulgare* L.). *Biology and Fertility of Soils* 8:134–143.

Ruzzenenti, F., and R. Basosi. 2009. Evaluation of the energy efficiency evolution in the European road freight transport sector. *Energy Policy* 37(10):4079–4085. doi:10.1016 /j.enpol.2009.04.050.

Sahlins, D. M., and E. R. Service (eds) 1960. *Evolution and Culture*. Ann Arbor, MI: University of Michigan Press.

Salazar, A., and C. Palm. 1987. Screening of leguminous tres for alley cropping on acid soils of the humid tropics. In *Gliricidia sepium (Jacq.) Walp.: Management and Improvement*. NFTA Special Publication 87-01, eds. Withington, D., N. Glover, and J. L. Brewbaker, pp. 61–67. Hawaii, USA: Nitrogen Fixing Tree Association.

Salo, T. 1999. Effects of band placement and nitrogen rate on dry matter accumulation, yield and nitrogen uptake of cabbage, carrot and onion. *Agricultural and Food Science in Finland* 2:157–232.

Samper, M., C. Naranjo, and P. Sfez. 2001. Entre la tradición y el cambio. *Evolución tecnológica de la caficultura costarricense*, pp. 25. Universidad Nacional: Instituto Panamericano de Geografía e Historia.

San Miguel Ayanz, A. (coord.) 2009. *Los pastos de la comunidad de Madrid. Tipología, Cartografía y Evaluación*. Madrid, Spain: Consejería de Medio Ambiente, Vivienda y Ordenación del Territorio.

Sánchez Vallduví, G. E., and S. J. Sarandón. 2011. Effects of Changes in Flax (*Linum usitatissimum* L.) Density and Interseeding with Red Clover (*Trifolium pratense* L.) on the Competitive Ability of Flax against Brassica Weeds. *Journal of Sustainable Agriculture* 35(8):914–926.

Sánchez-García, M., C. Royo, N. Aparicio, J. A. Martín-Sánchez, and F. Álvaro. 2013. Genetic improvement of bread wheat yield and associated traits in Spain during the 20th century. *Journal of Agricultural Science* 151:105–118.

Santos, J. J., L. Pratt, and J. M. Pérez. 1997. Uso de plaguicidas en la agroindustria de Costa Rica. Centro Latinoamericano para la Competitividad y el Desarrollo Sostenible. CEN 708.

Schader, C., A. Muller, N. E.-H. Scialabba, et al. 2015. Impacts of feeding less food-competing feedstuffs to livestock on global food system sustainability. *Journal of the Royal Society Interface* 12(113):20150891.

Schahczenski, J. J. 1984. Energetics and traditional agricultural systems: A review. *Agricultural Systems* 14:31–43.

Schandl, H., and N. B. Schulz. 2002. Changes in United Kingdom's natural relations in terms of society's metabolism and land use from 1850 to the present day. *Ecological Economics* 41(2):203–221.

Schandl, H., C. M. Grünbühel, H. Haberl, and H. Weisz. 2002. Handbook of Physical Accounting. Measuring bio-physical dimensions of socio-economic activities MFA–EFA–HANPP. Social Ecology Working Paper 73. Vienna: Institute for Interdisciplinary Studies of Austrian Universities (IFF).

Scheidel, A., and A. H. Sorman. 2012. Energy transitions and the global land rush: Ultimate drivers and persistent consequences. *Global Environmental Change* 22(3):559–794.

Schmidhuber, J. 2006. The EU Diet – Evolution, Evaluation and Impacts of the CAP. Documentos de FAO, Available at http://www.fao.org/fileadmin/templates/esa/Global _persepctives/Presentations/Montreal-JS.pdf (accessed on April 10, 2012).

Schmidt, A. 1971. *The Concept of Nature in Marx*. London: New Left Books.

Schramski, J. R., K. L. Jacobsen, T. W. Smith, M. A. Williams, and T. M. Thompson. 2013. Energy as a potential systems-level indicator of sustainability in organic agriculture: Case study model of a diversified, organic vegetable production system. *Ecological Modelling* 267:102–114.

Schrödinger, E. 1984. *¿Qué es la vida?* Barcelona, Spain: Tusquets

Schroll, H. 1994. Energy-flow and ecological sustainability in Danish agriculture. *Agriculture, Ecosystems and Environment* 51(3):301–310.

Schröter, M., D. N. Barton, R. P. Remme, and L. Hein. 2014. Accounting for capacity and flow of ecosystem services: A conceptual model and a case study for Telemark, Norway. *Ecological Indicators* 36:539–551.

Schuch, S., J. Bock, B. Krause, K. Wesche, and M. Schaefer. 2012. Long-term population trends in three grassland insect groups: A comparative analysis of 1951 and 2009. *Journal of Applied Entomology* 136(5):321–331.

SE. 2012. *Monografía del sector del aguacate en México: Situación actual y oportunidades de mercado.* Mexico: Secretaría de Economía. Available at http://www.economia.gob .mx/files/Monografia Aguacate.pdf.

Secretaria de Agricultura, G., D. Rural, Pesca y Alimentación (SAGARPA) 2013. Producción a partir de caña de azúcar. Available at http://www.bioenergeticos.gob.mx/index.php /bioetanol/prouccion-a-partir-de-cana-de-azucar.html (accessed on September 21, 2014).

Shepherd, K. D., E. Ohlsson, J. R. Okalebo, and J. K. Ndufa. 1996. Potential impact of agroforestry on soil nutrient balances at the farm scale in the East African Highlands. *Fertilizer Research* 44:87–99.

Siddique, K. H. M., E. J. M. Kirby, and M. W. Perry. 1989. Ear: Stem ratio in old and modern wheat varieties: Relationship with improvement in number of grains per Ear and Yield. *Field Crops Research* 21:59–78.

Siddique, K. H. M., R. K. Belford, and D. Tennant. 1990. Root: Shoot ratios of old and modern, tall and semi-dwarf wheats in a mediterranean environment. *Plant and Soil* 121:89–98.

Sieferle, R. P. 2011. Cultural evolution and social metabolism. *Geografiska Annaler Series B, Human Geography* 93:315–324.

Sieferle, R. P. 2001. Qué es la Historia Ecológica. In *Naturaleza Transformada. Estudios de Historia Ambiental en España,* eds. M. González de Molina, and Alier, J. M.Barcelona, Spain: Icaria.

Sieferle, R. P. 2001a. *The Subterranean Forest: Energy Systems and the Industrial Revolution.* Cambridge, MA: The White Horse Press.

Siles, P., J. –M. Harmand, and P. Vaast. 2010. Effects of Inga densiflora on the microclimate of coffee (*Coffea arabica* L.) and overall biomass under optimal growing conditions in Costa Rica. *Agroforestry Systems* 78:269–286.

Simpson, J. 1997. *La agricultura española (1765–1965): la larga siesta.* Madrid, Spain: Alianza.

Singh, S. J., F. Krausmann, S. Gingrich, H. Haberl, K. –H. Erb, and P. Lanz. 2012. India's biophysical economy, 1961–2008. Sustainability in a national and global context. *Ecological Economics* 76:60–69. doi:10.1016/j.ecolecon.2012.01.022.

Singh, S. J., H. Haberl, M. Chertow, M. Mirtl, and M. Schmid (eds). 2012a. *Long-Term Socio-Ecological Research: Studies in Society-Nature Interactions Across Spatial and Temporal Scales* (Vol. 2). New York, NY: Springer Science and Business Media.

Sisson, A. B 1979. *Survey of Fuelwood Use in Selected Sites of Central America.* Guatemala: ROCAP.

Skinner, C., Gattinger, A., Muller, A., Mäder, P., Fließbach, A., Stolze, M., Ruser, R. and U. Niggli. 2014. Greenhouse gas fluxes from agricultural soils under organic and non-organic management—A global meta-analysis. *Science of the Total Environment* 468–469:553–563.

Smil, V. 1994. *Energy in World History.* Boulder, CO: Westview Press.

Smil, V. 1999. *Energies: An Illustrated Guide to the Biosphere and Civilization*. Cambridge, MA: The MIT Press.

Smil, V. 2001a. *Energías. Una guía ilustrada de la biosfera y la civilización*. Barcelona, Spain: Editorial Crítica.

Smil, V. 2001b. *Enriching the Earth: Fritz Haber, Carl Bosch, and the Transformation of World Food Production*. Cambridge, MA: The MIT Press.

Smil, V. 2010. *Energy transitions: History, Requirements, Prospects*. Santa Barbara, CA: Praeger.

Smil, V. 2013a. *Harvesting the Biosphere: What We Have Taken from Nature*. Cambridge, MA: The MIT Press.

Smil, V. 2013b. *Making the Modern World: Materials and Dematerialization*. Hoboken, NJ: John Wiley & Sons.

Smith, G. S., J. G. Buwalda, and C. J. Clark. 1988. Nutrients dynamics of a kiwifruit ecosystem. *Scientia Horticulturae* 37:87–109.

Smith, L. G., A. G. Williams, and B. D. Pearce. 2015. The energy efficiency of organic agriculture: A review. *Renewable Agriculture and Food Systems* 30(3):280–301.

Smith, P. 2004. Carbon sequestration in croplands: The potential in Europe and the global context. *European Journal of Agronomy* 20:229–236.

Soon, Y. K. 1988. Root distribution of and water uptake by field-grown barley in a Black Solod. *Canadian Journal of Soil Science* 68:425–432.

Soroa, J. M. 1953. *Prontuario del agricultor y del ganadero*. Madrid, Spain: Dossat.

Soto, D., A. Infante-Amate, G. I. Guzmán, et al. 2016. The Social Metabolism of Biomass in Spain, 1900–2008: From food to feed-oriented changes in the Agro-ecosystems. *Ecological Economic* 128:130–138. Available at 10.1016/j.ecolecon.2016.04.017.

Spielmann, M., and R. W. Scholz. 2005. Life cycle inventories of transport services. *International Journal of Life Cycle Assessment* 10(1):85–94. doi:10.1065/lca.10.181.10.

Spielmann, M., R. Dones, and C. Bauer. 2007. *Life Cycle Inventories of Transport Services. Final Report Ecoinvent v2.0 No. 14*. Dübendorf, CH: Swiss Centre for Life Cycle Inventories.

StatSoft. 2011. STATISTICA (data analysis software system), version 10. Available at www.statsoft.com (accessed on September 13, 2014).

Steffan-Dewenter, I., M. Kessler, J. Barkmann, et al. 2007. Tradeoffs between income, biodiversity, and ecosystem functioning during tropical rainforest conversion and agroforestry intensification. *Proceedings of the National Academy of Sciences of the USA* 104 (12):4973–4978.

Steinfeld, H., P. Gerber, T. Wassenaar, V. Castel, M. Rosales, and C. de Haan. 2006. *Livestock's Long Shadow. Environmental Issues and Options*. Rome: FAO.

Stopford, M. 2009. *Maritime Economics,* 3rd Edition. New York, NY: Routledge.

Stout, B. A., and M. McKiernan. 1992. Chapter 11: New technology—Energy implications. In *Energy in Farm Production*, ed. Richard, C. F., pp. 131–170. Amsterdam: Elsevier.

Suh, S., M. Lenzen, G. J. Treloar, et al. 2004. System boundary selection in life-cycle inventories using hybrid approaches. *Environmental Science and Technology* 38(3):657–664. doi:10.1021/es0263745.

Swannack, T. M., and W. E. Grant. 2008. Systems ecology. *Encyclopedia of Ecology* 3477–3481. doi:10.1016/B978-008045405-4.00698-4.

Swanson, G. A., K. D. Bailey, and J. G. Miller. 1997. Entropy, social entropy and money: A living systems theory perspective. *Systems Research and Behavioral Science* 14(1):45–65.

Tacharntke, T., Y. Clough, L. Jackson, et al. 2012. Global food security, biodiversity conservation and the future of agricultural intensification. *Biological Conservation* 151:53–59.

Tan, P., M. Steinbach, and V. Kumar. 2006. *Introduction to Data Mining. Chapter 8. Cluster
Analysis: Basic Concepts and Algorithms.* Reading, MA: Pearson Addison-Wesley.

Tay Oroxom, J. M. 2007. Evolución Tecnológica de la Fabricación de Equipos Domésticos
para combustión de leña como consecuencia del tipo de materiales utilizados. Ensayo de
eficiencia, Estudio especial de graduación, Escuela de estudios de postgrado, Facultad
de Ingeniería, Universidad de San Carlos de Guatemala, Guatemala. Available at http://
es.scribd.com/doc/23882418/Estufas-ahorradoras-de-lena-Guatemala (accessed on
December 6 de 2013).

Taylor, E. L., G. Holley, and M. Kirk. 2007. Pesticide development. A brief look at the history.
South Regional Extension Forestry 10:1–7.

Tello, E., E. Galán, G. Cunfer, et al. 2015. A proposal for a workable analysis of Energy
Return On Investment (EROI) in agroecosystems. Part I: Analytical approach. *Social
Ecology Working Paper 156. IFF-Social Ecology.* Available at https://www.uni-klu
.ac.at/socec/inhalt/1818.htm (accessed on January 15, 2015).

Tello, E., E. Galán, V. Sacristán, et al. 2016. Opening the black box of energy throughputs in
farm systems: A decomposition analysis between the energy returns to external inputs,
internal biomass reuses and total inputs consumed (the Vallès County, Catalonia, c. 1860
and 1999). *Ecological Economics* 121:160–174. doi:10.1016/j.ecolecon.2015.11.012.

Tello, E., R. Garrabou, X. Cussó, and J. R. Olarieta. 2008. Una interpretación de los cambios
de uso del suelo desde el punto de vista del metabolismo social agrario. La comarca
catalana del Vallès, 1853–2004. *Revista Iberoamericana de Economía Ecológica*
7:97–115.

Tello, E. 2005. *La Historia cuenta. Del crecimiento económico al desarrollo humano sostenible.* Barcelona, Spain: El Viejo Topo.

Theurl, M. C. 2008. CO_2-Bilanz der Tomatenproduktion: Analyse acht verschiedener
Produktionssysteme in Österreich, Spanien und Italien. In *Social Ecology Working
Paper 110.* Vienna, Austria: Institute of Social Ecology.

Theurl, M. C., H. Haberl, K. –H. Erb, and T. Lindenthal. 2013. Contrasted greenhouse gas
emissions from local versus long-range tomato production. *Agronomy for Sustainable
Development* 34(3):593–602. doi:10.1007/s13593-013-0171-8.

Thomassen, M. A., K. J. van Calker, M. C. J. Smits, G. L. Iepema, and I. J. M. de Boer. 2008.
Life cycle assessment of conventional and organic milk production in the Netherlands.
Agricultural Systems 96:95–107.

Thornton, P. K., and M. Herrero. 2010. The inter-linkages between rapid growth in livestock
production, climate change, and the impacts on water resources, land use, and deforestation. *Policy Research Working Paper 5178.* The World Bank Group.

Tittonell, P. 2014. Livelihood strategies, resilience and transformability in African Agroecosystems. *Agricultural Systems* 126:3–14.

Tittonell, P., E. Scopel, N., Andrieu, et al. 2012. Agroecology-based aggradation-
conservation agriculture (ABACO): Targeting innovations to combat soil degradation
and food insecurity in semi-arid Africa. *Field Crops Research* 132:168–174.

Toledo, V. M. 1990. The ecological rationality of peasant production. In *Agroecology and
Small Farm Development*, eds. Altieri, M. A., and S. B. Hecht, pp. 53–60. Boca Raton,
FL: CRC Press.

Toledo, V. M. 1993. La racionalidad ecológica de la producción campesina. In *Ecología,
campesinado e Historia,* eds. Sevilla, E., and M. González de Molina, pp. 197–218.
Madrid, Spain: La Piqueta.

Toledo, V. T., and M. González de Molina. 2007. El metabolismo social: las relaciones entre la sociedad y la naturaleza. In *El paradigma ecológico en las ciencias sociales*, eds. Garrido, F., M. González de Molina, J. L. Serrano, and J. L. Solana, pp. 85–112. Barcelona, Spain: Icaria.

Toledo, V. M., J. Carabias, C. Mapes, and C. Toledo. 1985. *Ecología y autosuficiencia alimentaria: hacia una opción basada en la diversidad biológica, ecológica y cultural de México*. México: Siglo XXI.

Torrellas, M., A. Anton, J. C. Lopez, et al. 2012. LCA of a tomato crop in a multi-tunnel greenhouse in Almeria. *International Journal of Life Cycle Assessment* 17 (7):863–875. doi:10.1007/s11367-012-0409-8.

Tran, T. S., and M. Giroux. 1998. Fate of 15N-labelled fertilizer applied to corn grown on different soil types. *Canadian Journal of Soil Science* 78(4):597–605.

Tscharntke, T., Y. Clough, S. A. Bhagwat, et al. 2011. Multifunctional shade-tree management in tropical agroforestry landscapes—A review. *Journal of Applied Ecology* 48(3):619–629.

Tuomisto, H. L., I. D. Hodge, P. Riordan, and D. W. Macdonald. 2012. Comparing energy balances, greenhouse gas balances and biodiversity impacts of contrasting farming systems with alternative land uses. *Agricultural Systems* 108:42–49.

Turco, P. H. N., M. S. T. Esperancini, and O. C. Bueno. 2012. Eficiência energética da produção de café orgânico na região sul de Minas Gerais. *Revista Energia na Agricultura* 27:86–95.

Tyrtania, L. 2008. La indeterminación entrópica Notas sobre disipación de energía, evolución y complejidad. *Desacatos* 28:41–68.

Ugalde, L. 1982. *Algunos aspectos sobre el consumo y producción de leña en el cantón de Turrialba. Seminario de Desarrollo Agroindustrial de Turrialba*. pp. 25–27 de noviembre de 1982. Turrialba (Costa Rica): CATIE.

Ulanowicz, R. E. 2004. On the nature of ecodynamics. *Ecological Complexity* 1:341–354.

Ulanowicz, R. E. 1983. Identifying the structure of cycling in ecosystems. *Mathematical Biosciences* 65:210–237.

Ulanowicz, R. E. 1986. *Growth and Development: Ecosystem Phenomenology*. New York, NY: Springer-Verlag.

UN. 1952. World energy supplies in selected years, 1929–1950. In *Statistical Papers Series*. New York, NY: Statistical Office of the United Nations http://unstats.un.org/unsd/energy/yearbook/Series_J_No_1.World_Energy_Supplies-1929-1950.pdf (accessed on September 20, 2016).

Unkovich, M., J. Baldock, and M. Forbes. 2010. Chapter 5–Variability in Harvest Index of grain crops and potential significance for carbon accounting: Examples from Australian Agriculture. *Advances in Agronomy* 105:173–219.

Urbano, P. 1992. *Tratado de Fitotecnia General*. Madrid, Spain: Mundi-Prensa.

Valente, M.E., Canale, A., Ciotti, A., Peiretti, P.G. 1991. Previsione dell'energia grezza dell'insilato di erba medica in funzione dei prodotti di fermentazione e dell'azoto totale. *Atti Soc. Ital. Scienze Vet.* 45:1659–1663.

Van der Woude, J. H. A. 2013. Is 50% Energy Efficiency Improvement in Glass Product/Production Chain Feasible? Needs of the NL Glass Industry for the future. GlassTrend—ICG seminar and workshop "Innovation in glass production," Eindhoven, the Netherlands.

Vandermeer, J. H. 1990. Intercropping. In *Agroecology*, eds. Caroll, C. R., J. H. Vandermeer, and P. M. Rosset, pp. 481–516. New York, NY: McGraw Hill.

Vanwalleghem, T., J. Infante-Amate, M. González de Molina, D. Soto Fernández, and J. A. Gómez. 2011. Quantifying the effect of historical soil management on soil erosion rates in Mediterranean olive orchards. *Agriculture, Ecosystems and Environment* 142:341–351.

Vásquez Panizza, R. A. 2013. Sobre el nogal. Available at http://www.agronomia.uchile.cl
/webcursos/cmd/22005/rvasquez/nogal.htm#generalidades.

Vecina, A., and G. I. Guzmán. 1997. Determinación del período crítico de competencia de la
flora arvense en dos cultivos hortícolas. In *Actas IV Congreso de la Sociedad Española
de Agricultura Ecológica (SEAE)*. Córdoba, Spain: septiembre de 1997.

Veermäe, I., J. Frorip, E. Kokin, et al. 2012. Energy Consumption in Animal Production.
European Union: Enpos Interreg IV Programme. Available at http://enpos.weebly.com
/uploads/3/6/7/2/3672459/energy_consumption_in_animal_production.pdf   (accessed
on June 20, 2014).

Viales, R., and A. Montero. 2010. *La construcción sociohistórica de la calidad del café y
del banano de Costa Rica*. Un análisis comparado 1890–1950. San José, CA: Editorial
Alma Máter.

Villamil, E. L., U. Y. Merlín, M. E. Gavito, M. Astier, and M. Devoto. 2016 *Relationship Between
Indicators of Management Practices and Diversity of Flower Visitors and Herbaceous
Plants in Conventional and Organic Avocado Orchards of Michoacán*, México (in press.)

Vitousek, P. M., P. R. Ehrlich, A. H. Ehrlich, and P. A. Matson. 1986. Human appropriation
of the products of photosynthesis. *Bioscience* 36(6):363–373.

Voivontas, D., D. Assimacopoulos, and E. G. Koukios. 2001. Assessment of biomass potential
for power production: A GIS based method. *Biomass and Bioenergy* 20:101–112.

Vos, W., and H. Meekes. 1999. Trends in European cultural landscape development:
Perspectives for a sustainable future. *Landscape and Urban Planning* 46:3–14.

Wallerstein, I. 1974. *The Modern World System. Capitalist Agriculture and the Origins of the
European World-economy in the Sixteenth Century*. London: Academic Press.

Wang, A. 2008. Estimation of Energy Efficiencies of U.S. Petroleum Refineries. *CTR/ANL*
March.

Weber, C. L., and H. S. Matthews. 2008. Food-miles and the relative climate impacts of food
choices in the united states. *Environmental Science & Technology* 42 (10):3508–3513.
doi:10.1021/es702969f.

Wehrden, H. von, D. J. Abson, M. Beckmann, A. F. Cord, S. Klotz, and R. Seppelt. 2014.
Realigning the land-sharing/land-sparing debate to match conservation needs:
Considering diversity scales and land use history. *Landscape Ecology* 29:941–948.

Weiske, A., A. Vabitsch, J. E. Olesen, et al. 2006. Mitigation of greenhouse gas emissions
in European conventional and organic dairy farming. *Agriculture, Ecosystems and
Environment* 112:221–232.

Weisz, H. 2007. Combining social metabolism and input-output analyses to account for ecologically unequal trade. In *Rethinking Environmental History. World-System History
and Global Environmental Change*, eds. Hornborg, A., J. R. McNeill, and J. Martínez-
Alier, pp. 289–306. Lanhan: Altamira Press.

Wendel, C. H. 1985. *Nebraska Tractor Tests Since 1920*. Sarasota, CA: Crestline Publishing Co.

Wendel, C. H. 2005. *Standard Catalog of Farm Tractors 1890–1980,* 2nd Edition. Lola, WI:
Krause Publications.

White, W. 2008. Economic history of tractors in the United States. In *EH.Net Encyclopedia*,
ed. R. Whaples. Online publication.

Willer, H., and J. Lernoud (eds.) 2016. *The World The World of Organic Agriculture. Statistics
and Emerging Trends 2016*. FiBL, Frick and IFOAM – Organics International, Bonn.
Available at http://www.organic-world.net/yearbook/yearbook-2016.html (accessed on
February 29, 2016).

Williams, J. D., D. K. McCool, C. L. Reardon, C. L. Douglas Jr., S. L. Albrecht, and R. W. Rickman. 2013. Root: Shoot ratios and belowground biomass distribution for Pacific Northwest dryland crops. *Journal of Soil and Water Conservation* 68:349–360.

Wirsenius, S. 2003. The biomass metabolism of the food system. A model-based survey of the global and regional turnover of food biomass. *Journal of Industrial Ecology* 7(1):47–80.

Witzke, H., and S. Noleppa. 2010. EU Agricultural Production and Trade: Can More Efficiency Prevent Increasing "Land Grabbing" Outside of Europe? OPERA Re-search Center. Available at http://www.appgagscience.org.uk/linkedfiles/Final_Report_Opera.pdf (accessed on April 5, 2013).

Wolman, A. 1965. The metabolism of cities. *Scientific American* 213(3):178–193.

Wood, R., M. Lenzen, C. Dey, and S. Lundie. 2006. A comparative study of some environmental impacts of conventional and organic farming in Australia. *Agricultural Systems* 89:324–348.

Worrell, E., and C. Galitsky. 2008. *Energy Efficiency Improvement and Cost Saving Opportunities for Cement Making.* Berkeley: Ernest Orlando Lawrence Berkeley National Laboratory.

Wright, D. H. 1990. Human impacts on energy flow through natural ecosystems, and implications for species endangerment. *Ambio* 19(4):189–194.

Wrigley, E. A. 1993. *Continuidad, cambio y azar. Carácter de la revolución industrial en Inglaterra.* Barcelona, Spain: Crítica.

Wullschleger, S. D., T. M. Yin, S. P. DiFazio, et al. 2005. Phenotypic variation in growth and biomass distribution for two advanced-generation pedigrees of hybrid poplar. *Canadian Journal of Forest Research* 35:1779–1789.

Wurtenberger, L., T. Koellner, and C. R. Binder. 2006. Virtual land use and agricultural trade: Estimating environmental and socio-economic impacts. *Ecological Economics* 57:679–697.

Xu, J. G., and N. G. Juma. 1992. Aboveground and belowground net primary production of 4 barley (*Hordeum vulgare* L.) cultivars in Western Canada. *Canadian Journal of Plant Science* 72:1131–1140.

Xu, J. G., and N. G. Juma. 1993. Aboveground and belowground transformation of photosynthetically fixed carbon by 2 barley (*Hordeum vulgare* L.) cultivars in a Typic Cryoboroll. *Soil Biology and Biochemistry* 25:1263–1272.

Zan, C. S., J. W. Fyles, P. Girouard, and R. A. Samson. 2001. Carbon sequestration in perennial bioenergy, annual corn and uncultivated systems in southern Quebec. *Agriculture, Ecosystems and Environment* 86:135–144.

Zerbini, E., and B. Shapiro. 1997. *Feeding Draught Milking Cows in Integrated Farming Systems in the Tropics—Ethiopian Highlands Case Study.* Second FAO Electronic Conference on Tropical Feeds—Livestock feed resources within integrated farming systems.

Zhu, Z., S. Piao, R. B. Myneni, et al. 2016. Greening of the earth and its drivers. *Nature Climate Change.* doi:10.1038/nclimate3004.

Zoiopoulos, P., and I. Hadjigeorgiou. 2013. Critical overview on organic legislation for animal production: Towards conventionalization of the system? *Sustainability* 5:3077–3094.

Zotarelli, L., J. M. Scholberg, M. D. Dukes, R. Muñoz-Carpena, and J. Icerman. 2009. Tomato yield, biomass accumulation, root distribution and irrigation water use efficiency on a sandy soil, as affected by nitrogen rate and irrigation scheduling. *Agricultural Water Management* 96:23–34.

Index

Printed and bound by CPI Group (UK) Ltd, Croydon, CR0 4YY

23/10/2024

01778242-0019